Indigenous Empowerment through Human-Machine Interactions

Indigenous Empowerment through Human-Machine Interactions: The Challenges and Strategies from Business Lenses

EDITED BY

INDRAJIT GHOSAL
Brainware University, India

SWATI GUPTA
Universal AI University, India

SUDHIR RANA
Gulf Medical University, UAE

AND

DIPAK SAHA
Institute of Engineering and Management, Kolkata, India

emerald
PUBLISHING

United Kingdom – North America – Japan – India – Malaysia – China

Emerald Publishing Limited
Emerald Publishing, Floor 5, Northspring, 21-23 Wellington Street, Leeds LS1 4DL

First edition 2025

British Library Cataloguing in Publication Data
A catalogue record for this book is available from the British Library

ISBN: 978-1-83608-069-5 (Print)
ISBN: 978-1-83608-068-8 (Online)
ISBN: 978-1-83608-070-1 (Epub)

Printed and bound by CPI Group (UK) Ltd, Croydon, CR0 4YY

To my parents, who taught me the value of dreams and hard work. My God who is my power and my blessings.
To my wife, for standing by me through every page, every stage. To my children their love; and my scholar, students their wishes and love. And to all the readers, may these words bring you as much joy in reading as they did in writing. (IG)

To my family, for their unwavering support, encouragement, and love. To my God with their blessing. My friends provided laughter and comfort during challenging times. To my mentors, whose guidance has been invaluable throughout this journey. (SG)

Dedicated to the dreamers, the doers, and those who never give up. Dedicated to scholars, family members, and students. (SR)

To my family, for their unwavering support, encouragement, and love. To my friends, who provided laughter and comfort during challenging times. (DS)

Contents

About the Editors

Dr. Indrajit Ghosal currently serves as an Associate Professor and a Research Coordinator in Brainware University, Kolkata (India) in the Department of Management. Dr. Indrajit has been awarded PhD in Online Marketing (Management) in 2020. He was awarded FRPM in 2017. He has done his Master's on MCA and undergraduate with Bsc IT. He has more than 16+ years' experience in teaching. He has published 73 research papers in many reputed journals listed in ABDC, ABS, UGC care and indexed in Scopus, WOS, and many more and published 6 patents. He has presented papers in 37 international, 7 national, and 4 case studies in many conferences. He is the author of 6 books and serving as an editor in 11 books such as (CRC-Taylor & Francis, Emerald, Wiley, Bentham Science, Springer, and Nova Science) and published 12 book chapters. He assists three PhD scholars, four MPhil scholars already awarded under his supervision. He serves as a reviewer in many reputed journals indexed in Scopus, WOS with listed ABDC, ABS. As a Keynote Speaker/Resource person/Chairperson, he has been invited to 31 conferences/workshop/seminar. He is a member of ACM (Meerut Chapter), SSRN, MTTF (Bhatinda, Punjab), life member, and executive member of Education Research and Development Association (ERDA), member of IARA (Guwahati), IARA (Tiruchirappalli) and World Leadership Academy. His area of research is digital marketing, consumer behavior, online marketing, entrepreneurship, and ecommerce.

Swati Gupta is a distinguished academician and mentor, with over 19 years of experience in higher education and industry. As an Associate Professor at Universal AI University, she excels in teaching entrepreneurial finance, leadership, and corporate finance. Specializing in cutting-edge financial technologies such as blockchain, AI, and decentralized finance, she has published 11 peer-reviewed papers and authored/coauthored 4 books. Her patents on blockchain and AI-powered financial systems highlight her contributions to innovative research. As the founder of Finaccopedia, an EdTech platform, Dr. Gupta merges research and practice to empower learners with advanced finance skills. Her upcoming book exemplifies her dedication to advancing financial scholarship and fostering global academic and professional excellence.

Sudhir Rana is the Program Director and an Associate Professor at the College of Healthcare Management and Economics, Gulf Medical University, UAE. He has established FIIB Business Review (Sage Publishing) which is a fast growing multidisciplinary journal as Editor-in-Chief. He is also the co-Editor-in-Chief of

South Asian Journal of Marketing (Emerald Publishing Limited) and Senior Editor of *International Journal of Emerging Markets* (Emerald Publishing Limited). He is the editor of two series: *Review of Management Literature* (Emerald Publishing Limited) and *Advances in Emerging Markets and Business Operations* (Taylor and Francis). His areas of research are consumer behavior, emerging markets, and business education. He published in *Journal of Business Research*, *Journal of Knowledge Management*, *Journal of Consumer Behaviour*, and *Journal of Strategic Marketing*, among others.

Dipak Saha is currently associated as an Experienced Professor of Marketing and Supply Chain Management with a demonstrated history of leadership in the respective discipline with the Department of Management, Institute of Engineering and Management, Kolkata, India, and has over 20 years of corporate and academic experience. He has published several articles in reputed and renowned scholarly journals. He is the co-editor of the book *Perspectives in Marketing, Innovation and Strategy* with Father of Marketing, Professor Philip Kotler. He is a consultant and trainer in the field of strategic decision-making. He is the recipient of the AMP-Academic Excellence Award 2020, Research Excellence Award 2023. Dr. Dipak Saha is the Series Editor of the Series Contemporary Management Practices (T & F, Routledge Publication).

About the Contributors

Nouran Ajabnoor is a dedicated academic as well as Dy CEO at the Applied College, Jazan University, Saudi Arabia. Her passion for teaching and research is evident in her contributions to the university and her engagement with students. Dr. Ajabnoor's work reflects a commitment to educational excellence and the advancement of knowledge in her field.

Gina B. Alcoriza is a Faculty Member of the Tourism Management program at the University of the Philippines, Asian Institute of Tourism. Previously, she served as the Campus Coordinator for International Affairs and Program Director of the BS Hospitality Management program at Partido State University. With 14 years as the Chairperson of the Tourism Management Program at the University of Santo Tomas-Legazpi, she also has four years of industry experience in travel and tour agencies. Dr. Alcoriza holds a degree in Tourism from the Polytechnic University of the Philippines and a Doctor of Business Administration from the University of Santo Tomas-Legazpi. A Certified Guest Service Professional, she is the National Vice President of the Union of the Filipino Tourism Educators and a Board member of the Philippine Association of Researchers in Tourism and Hospitality. She has published research on community-based tourism, ecotourism, and is a peer reviewer for several international journals.

Andi Asrifan, SPd, MPd, is the Head of the Institute for Cooperation and International Affairs at Universitas Muhammadiyah Sidenreng Rappang, and has an extensive academic background, including a doctorate in English Language Education. His articles and research address various education, language, and technological themes. The COVID-19 epidemic has affected educational practices, EFL classroom methods, and technological integration in teaching and learning. Dr. Asrifan's engagement in international conferences, webinars, and seminars in India, the Philippines, and Pakistan shows his dedication to education. His branding, public speaking, and literacy teaching courses and certifications demonstrate his commitment to professional development. Beyond academics, Dr. Asrifan has led organizational and educational development as an external examiner for PhD dissertations and a scientific journal reviewer. His awards, professional memberships, and international association ambassadorships demonstrate his influence and dedication to educational achievement and sustainable development.

Erisa Avdimetaj completed her Bachelor's degree at the Haxhi Zeka Faculty in Accounting and Auditing. Currently, she is a second-year Master's student in the same field and faculty. Erisa works as an Insurance Officer at Raiffeisen Bank, where she initially joined as an intern and then transitioned to a full-time position. During her tenure at the bank, she has demonstrated excellent analytical skills and a strong commitment to her duties, combining academic knowledge with practical experience to deliver exceptional services. This blend of education and professional experience makes Erisa a valuable asset in the financial services industry, showcasing a strong dedication to growth and excellence.

Gül Erkol Bayram is currently an Associate Professor and the Vice-Dean in Tourism Faculty, Department of Tour Guiding, University of Sinop, Sinop, Turkey. Dr. Erkol Bayram has worked as an internal trainer and teacher in the tour guiding arena. Her doctorate is in Tourism Management from the Sakarya University, Turkey, and she completed her dissertation research on Tour Guiding in Turkey. Her core subjects are tourism, tour guiding, tourism policy and planning, and women studies. Erkol Bayram has also worked as a professional tour guide in the tourism sector. She has many book chapters in the international arena and published her books as an editor and author in the national arena. She has been invited for many talks/lectures/panel discussions by different universities and countries. She is a member of eminent journals in advisory board editor, and she published articles in tourism and management areas.

Pradipta Banerjee is an accomplished academic with a strong publication record, having worked as a Professor of Department of Commerce and Dean of Faculty Council of Sidho Kanho Birsha University, West Bengal, India. He is also an active member of IAA and review boards for various indexed journals. In addition to his academic work, Dr. Pradipta Banerjee has completed training programs at leading universities and has participated in various projects. His contributions to the field have earned him recognition and awards, both locally and globally, as a distinguished student and professor in global education.

Divya Bansal, Associate Professor at Amity University Online, Noida, has more than 19 years of experience in teaching Business Management students at UG and PG levels. She is a committed and competent educator who shares a passion for each student's academic success. Her research areas have focused on big data analytics, trends in supply chain management, blockchain technology, cultural marketing, etc. She is a critical player in business, strategy, and the environment, and her recent publications explore the phenomenon of artificial intelligence in various domains.

Kamaladevi Baskaran is a distinguished professional with over 19 years of international experience in industry, applied research, and academia. She is the Head of MBA and EMBA program at Amity Business School, Amity University Dubai, UAE. She is a Fellow of the Chartered Institute of Marketing and a prolific author, having published three management books. Notably, she holds a patent for an innovative Online Security Verification Model aimed at enhancing e-commerce payment security. Her research spans e-commerce, business analytics, entrepreneurship, and business research methods, with over 50 peer-reviewed articles and conference presentations. Dr. Kamaladevi has actively collaborated on projects involving digital innovation and predictive analytics, focusing on real-world problems in online retail. She has served as a keynote speaker at various international conferences across multiple countries, showcasing her expertise. Additionally, she is deeply involved in professional and community outreach initiatives, reflecting her commitment to advancing knowledge and practice in her fields of interest.

Indira Bhati has a teaching experience of 14 years. Her areas of interest are entrepreneurship development, start-up policies, project management, and consumer behavior. She has published research papers and articles in various journals of repute. She is a life member of "The Indian Commerce Association." She has been associated as an expert with various entrepreneurship development organizations.

Naboshree Bhattacharya is an Assistant Professor at Amity University Jharkhand with more than 12 years of experience in teaching Economics at UG level. Her specialization is in healthcare economics with a broader research focus on development economics. Her research output includes articles, book chapters, and books on behavior economics, healthcare economics, microeconomics, and Indian economy.

Indranil Bose, a Fellow of the Higher Education Academy (UK), has held senior academic roles in India and abroad. Dr. Bose has also advised PhD programs at Indian universities such as Symbiosis International University and VIT University, making him a significant figure in international higher education.

Luis Cardoso, PhD in Modern Languages and Literature, in the speciality of Comparative Literature from the University of Coimbra, Portugal. Adjunct Professor at the Department of Languages and Communication Sciences. Researcher at CARE – Research Center on Health and Social Sciences, Portalegre, and at the Center for Comparative Studies of the University of Lisbon, Portugal.

Abhijit Chandratreya is the Deputy Director in the Department of Post Graduate Research Center at SCES's Indira Institute of Management, Pune, for the past 16 years. He graduated in Production Engineering from Birla Vishwakarma Mahavidyalaya (Sardar Patel University), Anand, Gujarat, India. He secured a Master of Business Administration in Human Resource Development from IGNOU. He has been awarded a PhD in the field of Marketing Management from Savitribai Phule Pune University, India. He has presented several papers in national and international journals, conferences, and symposiums and has published 15 research papers both in referred and Scopus/ABDC-listed journals. He has published three patents. He is a certified trainer of CSCP, CPTP, CLDM, MTF, and CID certified by MiddleEarth HR. He has also published two chapters in books published by IGI Global. His main areas of interest include marketing, HRM, SCM, statistics, business analytics, and research methodologies.

Mrinal Kanti Das is the Head of the Department of Commerce at Kanchrapara College, Kanchrapara, West Bengal. Before that, he was at the Department of Business Administration, JIS College of Engineering, Kalyani, West Bengal, as a faculty member of marketing. He has over 18 years of academic experience. One scholar has already been awarded a PhD in the marketing domain under his direct supervision. He has authored five books and has contributed research articles in *various* journals indexed in Scopus, ABDC, and UGC-CARE, and he has also indexed edited volumes and proceedings in Scopus to his credit. He is an editor of *Perspectives in Human Resources* (Routledge Publication, Taylor & Francis Group, USA), *Perspectives in Agile Sustainable Practices and Business Value* (Routledge Publication, Taylor & Francis Group, USA), and *Futuristic Trends in Management* (*Iterative International* Publishers). He is a Series Editor of Sustainable Management Practices, *Bentham Science*, UAE.

Blerta Dragusha is a Lecturer at the University of Shkodra "Luigj Gurakuqi," Faculty of Economy, Department of Finance and Accounting since 1998 and the Dean of the Faculty of Economics since 2016. Prof. Dragusha has a BA degree in Business Administration, University of Shkodra "Luigj Gurakuqi," an MBA degree from Southern Illinois University in Carbondale, USA, and a PhD from the Department of Finance, Faculty of Economics, University of Tirana. As of October 2023, she holds the title "Professor." She also has teaching experience at some European and American universities as well. Prof. Dragusha has published several articles in national and international scientific journals and has participated in many national and international conferences. Her field of study is finance and investment.

Syed Mohammad Faisal is an illustrious faculty member at the Applied College, Jazan University, Saudi Arabia. With a robust background in his field, Dr. Faisal is dedicated to advancing academic excellence and research innovation. His contributions to the university have been instrumental in shaping the educational landscape, inspiring students, and fostering a culture of scholarly pursuit.

Debarshi Ghosh is an Assistant Professor at Meghnad Saha Institute of Technology with over 13 years of academic experience and 8 years in industry.

Vijay Prakash Gupta, an experienced educator and researcher with over 17 years of expertise, currently serves as an Associate Professor and PhD Coordinator at GLA University, Mathura. He has done PhD in Management, MBA, MA Economics, Qualified UGC NET, and many certification courses from IITs and IIMs. He is the editor of various books and journals and has published approx. 20 research papers in various journals indexed in ABDC, Scopus, Web of Science, and UGC CARE. His accomplishments include two patents and engagements as a keynote speaker at seminars, conferences, and as an expert evaluator for PhD thesis and Viva-Voce in prestigious institutes. Dr. Gupta has authored 5 textbooks, 5 edited books, and contributed chapters to approx. 10 books.

Isnaini Harahap is a Doctor of Islamic Economics and a Professor (Assistant) at State Islamic University of Sumatera Utara, Medan Indonesia. Isnaini Harahap currently works at the Faculty of Islamic Economics and Business, State Islamic University of Sumatera Utara, Medan Indonesia. Isnaini does research in Islamic Economics, Islamic Banking and Finance, and Business Ethics. Their most recent publication is "Conference Paper Students" Comprehension of Arabic and English Terminology in Islamic Finance: A Comparative Study."

Mohammad Irfan is currently working as an Associate Professor at NSB Academy Business School, Bangalore. He was associated with the School of Business, AURO University, Surat, Gujarat, for five years. He holds an MBA (Finance) and an MCom (Account and Law). Dr. Irfan has done this PhD from the Central University of Haryana.

Rajiv Iyer is an accomplished academic and industry expert with over 14 years of experience. He holds a BE, ME, and PhD and is an Associate Professor of Computer Science and Engineering at Amity University, Mumbai. Dr. Iyer is AWS certified in cloud, machine learning, and data analytics, as well as certified in process mining (Celonis) and a Google Machine Learning Facilitator. He is also an ISO internal auditor, Six Sigma Yellow Belt, and SCRUM certified. With 22 research papers published in national and international journals, Dr. Iyer has authored several books in electronics and telecommunications. He is a registered patent agent, hackathon mentor, and reviewer for international journals. Additionally, he has taken on leadership roles in conferences and workshops, receiving numerous accolades for his contributions to academia and industry.

Richa Jain is currently working as Assistant Professor at Prestige Institute of Management and Research, Bhopal, Madhya Pradesh, India. She has more than 15 years of rich experience in the education field. She holds a PhD in Management and an MBA with a dual specialization in Finance and Retail Management. Her areas of specialization are finance and sustainability. She has published several papers in conferences and seminars. She is the author of a book named *Financial Accounting*. She has attended many workshops and faculty development programs at prestigious institutes like IIMs.

Rishabh Jain has a teaching experience of more than 19 years. His areas of research are Fintech, entrepreneurship, social entrepreneurship and corporate finance. He has published papers in Scopus and ABDC journals of repute. He has also worked as a reviewer for Q1 SCOPUS journals.

Ansh Jindal is pursuing MBA (Finance) at Chitkara Business School, Chitkara University, Punjab, India. He has successfully obtained a Bachelor's degree in Business Administration with a specialization in Financial Technology. He has participated in various competitions and won cash prizes. He has been granted more than 12 design and utility patents. He is mostly interested in the fields of banking, finance, and marketing. He has completed his summer internship program from 4 July to 19 July 2022 at Nottingham University, Malaysia, while pursuing his BBA (Fintech) degree. During his MBA studies, he participated in a remarkable Global Immersion Program in the United Arab Emirates (Dubai and Abu Dhabi).

Priya Jindal is currently working as an Associate Professor and Assistant Dean of Commerce at Chitkara Business School, Chitkara University, Punjab, India. She holds Master's degrees in Commerce and Economics. Her areas of interest include banking, finance, insurance, and general management. She had achieved distinction in her MPhil in 2009. She earned her doctorate in Management in 2014 from Maharishi Markandeshwar University, Mullana, Haryana, India. She has contributed approximately 17 years to teaching. She has supervised five PhD scholars and two MPhil candidates. Numerous research papers, book chapters, and conference papers (including IEEE) have been indexed in Scopus. She has filed over 14 design patents and 12 utility patents. Two copyrights have been granted in her name. Dr. Jindal is an editor of five edited books under IGI Global, Nova Science, and CRC publications, all of which have been indexed in Scopus.

Amit Joshi has teaching and industry experience of more than 18 years. His specialization lies in the areas of security analysis and portfolio management, business statistics, project management, and managerial economics. He has presented and published more than 50 research papers in various seminars and journals of repute. He has been a part of an editorial board of an international journal in applied financial management and is an editor of a refereed management journal. He holds the life membership of "The Society of Management Education, Kurukshetra."

Logasakthi Kandasamy is an academician with rich experience in the area of human resources and organizational behavior. He has published research papers on artificial intelligence versus emotional intelligence, occupational intelligence, occupational stress, and pedagogy development. Majorly, his research work addresses the contemporary business and social issues. He has published research articles in reputed journals which are indexed in Scopus, SCI, and ABDC.

Ahmad Khalid Khan is a prominent educator and researcher at the Applied College, Jazan University, Saudi Arabia. Renowned for his expertise and commitment to academic growth, Dr. Khan has significantly impacted both his students and colleagues. His work not only enhances the academic environment at Jazan University but also contributes to the broader academic community.

Early Ridho Kismawadi is a Lecturer at the Department of Islamic Banking, Faculty of Islamic Economics and Business, IAIN Langsa, Aceh, Indonesia. In 2018, he finished his doctoral studies at the State Islamic University of North Sumatra, with a focus on Sharia economics. At the Langsa State Islamic Institute (IAIN Langsa), Aceh, Indonesia, he was named head of the study programs for Islamic economics law (2023), Islamic banking (2020), and Islamic financial management (2019). His areas of interest in study are Islamic economics, banking, finance, financial economics, and applied econometrics. He has published articles in national and international journals. He is also a reviewer of several reputable international journals such as *Finance Research Letters*, *Financial Innovation*, *Journal of Islamic Marketing*, *Cogent Economics & Finance*, *Cogent Business & Management*, *Journal of Islamic Accounting and Business*

Research, and *Heliyon*. He has also presented his papers at various local and international seminars.

Rajneesh Kler, currently an Associate Professor of Economics at Amity University, Tashkent, and a Postdoctoral Research Fellow at Infrastructure University, Malaysia, holds a PhD from NUEPA and a degree from the University of Glasgow.

Ravishankar Krishnan is an accomplished academic with a rich background in logistics, supply chain management, and human resources. He holds a PhD from Anna University, Chennai, and has extensive experience in teaching and research. His commitment to research is evident from his involvement in significant projects and publications in Scopus, ABDC, and Web of Science. Additionally, he has completed several value-added courses and certifications, enhancing his expertise and contribution to the field.

Ruanda Kukalaj earned her Bachelor's degree in Accounting and Auditing from the Haxhi Zeka Faculty in Peja. She is now a second-year Master's student in the same field at the same institution. Currently, she works as a credit analyst at Rural Credit of Kosovo. Ruanda has extensive practical experience in commercial banks and various businesses in Kosovo, working as a financial analyst and accountant. Her hands-on experience has significantly enriched her knowledge, evident in her scientific projects.

Enkeleda Lulaj holds a PhD in Economic Sciences with a focus on Finance and serves as an Assistant Professor at the Faculty of Business, Haxhi Zeka University in Kosovo. With extensive international experience, she has been a Visiting Professor at universities worldwide and a featured speaker at various academic events. Dr. Lulaj is an accomplished academic with a strong publication record, having worked as an editor, reviewer, and author for prestigious academic journals. She is also an active member of editorial and review boards for journals indexed in Scopus and Web of Science. In addition to her academic work, Dr. Lulaj has completed training programs at leading universities and has participated in various projects. Her contributions to the field have earned her recognition and awards, both locally and globally, as a distinguished student and professor in global education.

Rajesh Manchanda has more than 18 years of teaching experience in undergraduate and postgraduate programs in leading business schools and five years of experience in reputed multinational companies in the field of sales and marketing. He is also involved in research and has more than 10 publications to his credit in national and international journals and has attended various A.I.C.T.E. sponsored workshops. His areas of interest include marketing research, consumer behavior, research methodology, retail management, and international marketing management.

Vedprakash C. Maralapalle boasts a strong academic background with a PhD in Civil Engineering from NMIMS University, Mumbai. He brings over 10 years of experience in the teaching field, currently serving as an Assistant Professor in the

Department of Civil Engineering at ASET, Amity University, Mumbai. Dr. V. Maralapalle is a licensed Structural Engineer with the Mumbai Municipal Corporation, demonstrating his practical expertise in the field. His innovative spirit is evident through his eight patent filings, with six already granted and others under review. He actively contributes to scholarly discourse, having published five papers in internationally recognized, peer-reviewed journals (indexed in SCI/Scopus) and six papers in national and international conferences. Dr. Maralapalle has also authored two published books, further solidifying his expertise within the field. His dedication extends beyond the classroom, as he serves as a reviewer for six international journals. Additionally, Dr. Maralapalle has actively engaged students through workshops and seminars.

Soumya Mukherjee is an Associate Professor in Business Administration at Techno India (Hooghly Campus), Chinsurah, West Bengal, India. He is an accomplished educator with demonstrated ability and experience to teach, motivate, and direct students while maintaining high interest and achievement. He has more than 15 years of academic experience. He has authored three books and has published several research articles in peer-reviewed, Scopus-indexed, ABDC-listed journals. He has also contributed many articles in Scopus-indexed edited volumes and conference proceedings. He is also one of the editors of *Perspectives in Agile Sustainable Practices and Business Value* (Routledge Publication, Taylor & Francis Group, USA). He is associated with different management institutions and universities as a visiting faculty member, paper setter, examiner, and moderator.

Muthmainnah is an Assistant Professor at Universitas Al Asyariah Mandar west Sulawesi Indonesia. She is a lecturer, global speaker and international leader. Now, she is the Deputy Director of language and Character development institute at her university. In the current year, she is active as international member, international coordinator conference, and international trainer. She is also the author of 36 national and international books Springer, Wiley, Emerald, Taylor and Francis https://sinta.kemdikbud.go.id/profile/books. She is also as International Board for Professors and Expert of Scientific Innovation Research Group, EGYPT (www.sirg.club), International board of TEFL KUWAIT (Member of Large) representing ASIAN, she is also a Country Head, ambassador and country director for many international organization representing Indonesia.

Amina Omrane is an Associate Professor, authorized to supervise researchers in Management Science at the University of Sfax, and researcher at ECSTRA research center, based in IHEC-Carthage, Tunisia. She teaches postgraduate and Master's courses related to research methods, strategic management, business plans, entrepreneurship, as well as human/personal development. She has published more than 45 research papers in international conferences/congresses' proceedings and numerous ranked and indexed journals, such as IJBG, IJESM, IJBE, JAB, TCR, La RIPME, IJMP, FIIB Business Review, and La Revue des Sciences de Gestion. She has also served as a reviewer for Emerald, T&F, and Inderscience journals as a researcher at ECSTRA center, in IHEC-Carthage.

Author of five books on management science, entrepreneurship, and human development, eight book chapters, she has also coedited many collective books on sustainable entrepreneurship, innovation and renewable energy, and digitalization, higher education, and new business models.

Durdana Ovais holds a PhD in Management and an MBA with a specialization in Human Resource Management, with over 13 years of teaching experience in esteemed institutions such as BSSS Institute of Advanced Studies and The Bhopal School of Social Sciences. Dr. Ovais has established herself as a seasoned academician. She has been recognized for her scholarly contributions with 45 research papers published in international journals, including 10 in Scopus-indexed publications, and 8 papers in national journals. Additionally, she has authored four books and contributed chapters to 09 publications. Dr. Ovais is a recipient of four prestigious Best Paper Awards, including two international accolades. She has patented three inventions in the fields of biomedical engineering and mechanical engineering. With a keen focus on sustainability, corporate social responsibility, and human resource management, her work particularly delves into the concept of ESG.

Niyasha Patra is an Assistant Professor in Computational Sciences at Brainware University. She has previously worked at iLEAD, Adamas University.

Hardyal Kour Rainu is an MBA graduate specializing in Business Analytics from Amity University Dubai, UAE. Her academic research focuses on the social media marketing strategies on indigenous consumer behavior in the UAE. Through her expertise in business analytics, she applies quantitative and qualitative methodologies to examine how data-driven marketing strategies can optimize customer engagement, brand loyalty, and overall business performance. Hardyal's proficiency in analytics has enabled her to rigorously analyze market trends, consumer behavior patterns, and the effectiveness of social media campaigns. Her work contributes to a deeper understanding of the role digital platforms play in shaping consumer decisions and business outcomes. In addition to her academic research, she has practical experience in marketing and customer service, which has further strengthened her critical thinking and problem-solving abilities. Her research provides actionable insights for businesses seeking to enhance their digital marketing efforts in diverse and rapidly evolving markets.

Ruchika Rastogi is currently working as an Associate Professor in the Department of Business Administration, Pranveer Singh Institute of Technology, Kanpur. She has been teaching Managerial Economics, Strategic Management, and Operations Management at the postgraduate level since 2009. She has published several research articles in Scopus and UGC Care indexed journals. She has published one book titled *International Logistics Management*. Her areas of research are behavioral economic and marketing. She has conducted research development programs at different institutes in India.

Nilanjan Ray is an Associate Professor in Management Studies at JIS University, Kolkata, with over 15 years of teaching and 7 years of research experience. Dr. Ray is a life member of various professional bodies and has organized several academic conferences.

Anju Rohilla is an Assistant Professor at the Faculty of Management and Commerce, SRM University. She has 5 years of teaching experience. She has done a PhD (Management) from Bhagat Phool Singh Mahila Vishwavidyala, Khanpur Kalan, Sonipat. She has done MBA from Guru Jambeshwar University of Science and Technology in Hisar. Her area of interest is banking fraud. She has delivered 15 research paper presentations at the national conference. She has published several research papers in national and international journals. She has delivered workshops on case study analysis.

Chinhita Sanyal serves as an Assistant Professor at the Department of Management, Brainware University, Kolkata, where she is dedicated to nurturing the next generation of business leaders. She is a member of Indian Accounting Association (IAA). Specializing in finance, she leverages her extensive academic background and professional expertise to teach a variety of finance-related subjects, ranging from corporate finance to financial markets and investment strategies. In addition to her teaching responsibilities, Chinhita is deeply involved in academic research, focusing on areas like financial management, investment behavior, and risk analysis. As an ongoing research scholar, she is currently pursuing her PhD, contributing new insights to the field of finance through her scholarly work. She has one Scopus and four UGC care publications and a few other peer-reviewed publications. Her role also involves mentoring students, guiding them in their research projects, and preparing them for successful careers in the world of business and finance.

Suraj Sharma is an Assistant Professor at Adamas University's School of Business. He has previously worked at Brainware University and iLEAD.

Shebin Sharief is a dynamic academician and industry expert renowned for his expertise in management, particularly in the domain of Artificial Intelligence–enabled Supply Chain Management. With a PhD in Management and specializations in Marketing and Supply Chain and Logistics Management, Dr. Shebin brings over 4.5 years of impactful teaching experience to his role. Recognized for his exceptional ability to simplify intricate concepts, Dr. Shebin Sharief's teaching style is marked by clarity and precision. Leveraging nearly 3 years of industrial experience, he seamlessly integrates real-world practices into academia, enriching the learning experience.

Meeta Siddhu is a Design Facilitator at Symbiosis Institute of Design, Pune, and has completed her PhD in documentation, product development, and training of handicraft artisans from The Maharaja Sayajirao University of Baroda, Vadodara. Her areas of interest include design education, traditional textiles, sustainability, and artisan training. Since 2018, she has also been involved with curriculum planning and textbook writing under the "Apparel, Made Ups and

Home Furnishing Sector" for the vocational education division of NCERT, Govt. of India. A keen researcher herself, she has also published research papers and book chapters on sustainability, traditional textiles, handicrafts, and artisan issues. Having worked pan-India, she has also conducted training programs, workshops, webinars, and presentations for design facilitators, students, craft enthusiasts and has participated in national and international conferences too. She also holds a professional certification for university teaching.

Madhuri Singh is a PhD scholar at the Symbiosis Institute of Design, Symbiosis International (Deemed University). Her areas of interest include Indian heritage crafts, traditional Indian embroideries, and artisan training and craft product development. She has been working at grassroots level as part of government projects and schemes such as ODOP, design development workshop, skill education, and other schemes. She aims to contribute to the preservation and revival of traditional crafts while ensuring the sustainable livelihoods for the artisan communities too.

Vishal Kumar Tyagi has accomplished his graduation (BCA) from CCS University, Meerut, MBA from Dr. A.P.J. Abdul Kalam Technical University Uttar Pradesh with specialization in Marketing Management and Information Technology, and LLB from CCS University, Meerut. He is currently pursuing a PhD in Management from Amity University. Additionally, he is certified in Marketing Research and Analysis and has completed AICTE training in various subjects.

Foreword

The intersection of technology and Indigenous empowerment is a fascinating and critical area of exploration. As someone deeply interested in the convergence of technology and cultural preservation, I am excited to introduce this edited volume, *Indigenous Empowerment through Human–Machine Interactions: The Challenges and Strategies from Business Lenses*.

In compiling this book, I have had the privilege of working with a team of distinguished scholars and practitioners who share a commitment to bridging traditional Indigenous practices with modern technological advancements. Our aim with this volume is to demonstrate how human–machine interactions (HMI) can be leveraged to empower Indigenous communities, addressing both socio-economic challenges and cultural preservation needs.

The chapters within this book explore a range of HMI applications tailored to Indigenous contexts. From preserving cultural heritage and revitalizing endangered languages to enhancing economic opportunities through digital platforms, each contribution offers valuable insights into the practical and theoretical aspects of integrating technology with Indigenous traditions.

As I reviewed the diverse and impactful work presented here, it became clear that creating culturally sensitive and inclusive technologies is not just a goal but a necessity. The authors provide compelling evidence of how HMI solutions can be designed to respect and support Indigenous values while fostering sustainable development.

I am confident that this book will be an invaluable resource for researchers, policymakers, and technologists who are dedicated to advancing Indigenous empowerment through technology. It is my hope that the insights and strategies shared in these pages will inspire further exploration and collaboration, leading to a more equitable and respectful integration of technology in Indigenous communities.

Prof P. K. Biswas
Professor Emeritus and Advisor to Pro Chancellor Office, JLU
Former Vice Chancellor, Jagran Lakecity University
Former Vice Chancellor, Skyline University Nigeria
Former Director, LBSIM, Delhi and IFMR, Chennai
Former Dean, IMT Ghaziabad and Former Professor of IIFM

Preface

In the ever-evolving landscape of technology and business, the convergence of HMI and indigenous empowerment represents a transformative shift. This convergence, explored through the business lens in *Indigenous Empowerment through Human–Machine Interactions: The Challenges and Strategies from Business Lenses*, is reshaping traditional approaches to indigenous development and creating new pathways for sustainable growth. As organizations and communities increasingly embrace the transformative potential of technology, this book emerges not just as a scholarly work but as a guiding framework through the complexities of integrating indigenous wisdom with cutting-edge technological advancements.

In recent years, we have witnessed significant advancements in AI, robotics, and other technologies, which are increasingly being applied to various sectors, including those involving indigenous communities. Simultaneously, there is a growing recognition of the importance of preserving indigenous knowledge, culture, and practices. The synergy between these two forces offers both challenges and opportunities – challenges that require careful navigation and opportunities that, if harnessed correctly, can lead to meaningful empowerment. This new paradigm demands exploration, understanding, and a strategic approach to unlocking the potential of HMI in the context of indigenous empowerment.

Comprehensive Exploration

Indigenous Empowerment through Human–Machine Interactions offers a comprehensive exploration of how HMI can be leveraged to support indigenous communities. From the application of AI in preserving indigenous languages to the integration of traditional knowledge with modern business practices, the book delves into the multifaceted opportunities and challenges of this transformation.

Practical Insights

The book provides practical insights for professionals and policymakers seeking to leverage technology for the benefit of indigenous communities. Real-world examples and case studies illustrate how HMI are not merely theoretical concepts but tangible forces that can drive economic development, cultural preservation, and social inclusion.

Navigational Guide

By addressing ethical considerations, business strategies, and the socioeconomic impacts of technology, this book serves as a navigational guide for readers. It empowers them to approach the complexities of HMI with a balanced perspective, ensuring that indigenous empowerment remains at the forefront of technological integration.

Future Perspectives

Indigenous Empowerment through Human–Machine Interactions looks not only at current trends but also at the future. It contributes to the ongoing dialogue about how technology can be aligned with the values, needs, and aspirations of indigenous communities. The book offers insights into potential future developments, challenges, and opportunities that lie on the horizon.

The primary goal of this book is to demystify the intricate relationship between HMI and indigenous empowerment. By providing clear and accessible explanations, it aims to make the convergence of these two areas understandable to a diverse audience. Readers, regardless of their background, will gain a foundational understanding of how technology can be thoughtfully integrated into indigenous contexts. This objective seeks to empower individuals and organizations to engage with and contribute to discussions on the role of technology in fostering indigenous empowerment.

Another key objective is to offer practical, actionable insights for business leaders, technologists, and community advocates. By presenting real-world case studies, industry best practices, and examples of successful initiatives, the book serves as a guide for applying strategic approaches to HMI in indigenous settings. Professionals and policymakers will gain actionable knowledge, enabling them to develop strategies that support both technological advancement and the preservation of indigenous heritage. This objective bridges the gap between theoretical understanding and practical application, fostering a holistic approach to indigenous empowerment.

The book seeks to empower its readers with the knowledge required for informed decision-making in this rapidly evolving field. By analyzing the ethical, socioeconomic, and business implications of HMI, the objective is to provide a comprehensive view that aids in strategic planning for the future. Readers will be equipped with insights to address the ethical considerations, navigate the complex socioeconomic landscape, and stay ahead of emerging trends. This objective aims to shape a cohort of informed decision-makers who can actively contribute to the future direction of HMI in the context of indigenous empowerment.

Organization of the Book

Chapter 1: Examining the Influence of Social Media Marketing Strategies on Indigenous Consumer Behavior in the UAE
Hardyal Kour Rainu, Kamaladevi Baskaran

This research investigates the impact of social media marketing on indigenous consumer behavior and business performance in the UAE, employing mixed methods to analyze platform preferences, cultural influences, and business outcomes.

Chapter 2: Empowering Women: Challenges, Progress
Dr. Nilanjan Ray, Niyasha Patra, Suraj Sharma, Debarshi Ghosh, Dr. Indranil Bose, Dr. Rajneesh Kler

This study examines women's empowerment in India, exploring historical constraints, current policies, and challenges like gender-based violence. It evaluates government initiatives focused on education, skill development, and economic empowerment.

Chapter 3: Bridging Theory to Practice: Empowering Indigenous Businesses Through Innovative Analysis of Audited Financial Statements
Erisa Avdimetaj, Enkeleda Lulaj, Gina B. Alcoriza

This chapter analyzes the impact of audited financial statements on indigenous businesses in Kosovo, revealing significant variability in financial performance and emphasizing the importance of transparency, reliability, and innovative strategies in financial reporting.

Chapter 4: Exploring the Dynamics of Ethnoconsumerism: The Interplay Between Indigenous and Mainstream Consumer Behavior in Cross-cultural Contexts
Naboshree Bhattacharya, Dr. Divya Bansal

This chapter explores ethno-consumerism by examining how cultural identity shapes consumer behavior in Indigenous and mainstream contexts, highlighting the interplay between cultural values and economic interactions to inform inclusive economic strategies and sustainable consumption.

Chapter 5: How ChatGPT Shapes Knowledge Acquisition and Career Trajectories in Higher Education: Decoding Students' Perceptions to Achieve Quality Education
Ravishankar Krishnan, Logasakthi Kandasamy

This research explores ChatGPT's impact on learners' knowledge and career trajectories using the extended Technology Acceptance Model. Findings emphasize ChatGPT's ease of use, perceived usefulness, and trust as key factors shaping learner attitudes and influencing academic and career outcomes. The study suggests that technology developers and educators should integrate these insights to enhance curriculum design and classroom delivery.

Chapter 6: Cross-Cultural Leadership in Tech-Driven World
Dr. Richa Jain, Dr. Durdana Ovais

This study examines cross-cultural leadership in a technology-driven world through a systematic literature review. It highlights the significance of cultural intelligence, ethical behavior, and adaptability in navigating diverse teams and leadership challenges in a globalized business environment.

Chapter 7: Overcoming Challenges and Embracing Opportunities in the Integration of Modern Financial Tools Preserving: Cultural Heritage in Finance
Dr. Syed Mohammad Faisal, Dr. Ahmad Khalid Khan, Dr. Nouran Ajabnoor

This study explores the challenges and opportunities of modernizing financial systems within traditional cultures, highlighting the need for empathic innovation that respects cultural values. Using a multidisciplinary approach, the research identifies potential issues such as identity loss and marginalization while also presenting possibilities like using modern financial tools to preserve cultural heritage. The findings underscore the importance of culturally sensitive integration of financial innovations, offering insights for policymakers, financial institutions, and cultural organizations to balance progress with cultural preservation.

Chapter 8: Evaluating Business Performance Through Audited Financial Statements: Considerations for Future Trends in Indigenous Finance
Ruanda Kukalaj, Enkeleda Lulaj, Blerta Dragusha

This chapter analyzes the influence of audited financial statements on business performance, using a seven-year dataset and correlation analysis. Findings reveal significant positive correlations between financial variables, highlighting complex interactions that affect financial performance. The study emphasizes the role of indigenous finance in shaping these outcomes and suggests further research on external factors influencing financial performance in indigenous contexts.

Chapter 9: Revolutionizing the Indigenous Financial System With the Utilization of Artificial Intelligence
Anju Rohilla, Ansh Jindal, Priya Jindal

This chapter examines AI's transformative role in the indigenous financial sector, enhancing personalized services, fraud prevention, and operational efficiency, while enabling competitiveness through innovative tools and digitalization.

Chapter 10: The Role of Education in Indigenous Empowerment
Dr. Ruchika Rastogi, Dr. Swati Gupta

This paper investigates how education empowers Indigenous communities by preserving knowledge, fostering self-determination, and bridging socioeconomic gaps. It advocates for decolonized, culturally relevant approaches and collaborative efforts to overcome challenges.

Chapter 11: Leveraging the Internet of Things (IoT) for Financial Empowerment in Indigenous Communities
Chinhita Sanyal, Dr. Pradipta Banerjee, Dr. Indrajit Ghosal

This research explores IoT's potential to boost financial empowerment in Indigenous communities by improving financial access. It addresses challenges and offers policy recommendations for inclusive and culturally sensitive solutions.

Chapter 12: Influence of Indigenous Marketing Practices on Consumer Behavior and Brand Loyalty in Retail
Dr. Vijay Prakash Gupta, Dr. Shebin Sharief, Vishal Kumar Tyagi, Dr. Rajesh Manchanda

This chapter explores how indigenous marketing practices, integrating cultural elements and emotional appeal, influence consumer behavior and brand loyalty in retail. It highlights the role of cultural sensitivity and modern technology.

Chapter 13: Techno-Entrepreneurship: Catalyst for Profitability and Efficiency in the Handicraft Industry
Dr. Mrinal Kanti Das, Dr. Soumya Mukherjee, Dr. Amina Omrane

This study highlights how techno-entrepreneurship enhances the handicraft industry by reducing costs, boosting profits, and improving efficiency. It offers new insights into leveraging technology for sustainable growth and organizational success.

Chapter 14: Integrating AI and ML Ttechnologies Into Indigenous Financial Systems: Applications of AI and ML in Indigenous Financial Management
Rajiv Iyer, Vedprakash C. Maralapalle

This chapter explores how AI and ML can enhance indigenous financial systems by improving decision-making, risk management, and customer relations. It addresses challenges such as data quality and cultural resistance and provides global examples and future directions for integration and financial inclusion.

Chapter 15: Empowering Indigenous Communities Through Education
Meeta Siddhu, Madhuri Singh

This study evaluates digitalization's impact on artisan communities in Chitrakoot, Uttar Pradesh, focusing on the ODOP Scheme. Employing a descriptive design with case studies and thematic analysis, it reveals digitalization's effects on market access, product efficiency, and socioeconomic factors, addressing policy effectiveness and gaps in existing literature.

Chapter 16: Integrating Artificial Intelligence in Islamic Financial Management: Opportunities and Challenges in Maintaining Shariah Compliance
Early Ridho Kismawadi, Dr. Mohammad Irfan, Isnaini Harahap

This study explores integrating AI technology into Islamic financial management, focusing on sharia compliance. Using literature analysis, it identifies AI's potential to enhance efficacy, transparency, and precision while highlighting sharia compliance and ethical challenges. The study calls for a regulatory framework aligned with sharia principles and collaboration between financial experts and scholars. Recommendations include expanding training for practitioners and future research on specific AI strategies' impact on sharia adherence and operational efficacy.

Chapter 17: Bridging Cultures: A Framework for Inclusive Indigenous Technologies
Dr. Abhijit Chandratreya

This chapter explores integrating technology design with Indigenous cultural processes to create inclusive technologies. Utilizing a qualitative approach, it analyzes case studies and interviews with Indigenous leaders and technologists. The findings highlight conventional technology's tendency to marginalize Indigenous perspectives, proposing a framework for co-design that incorporates Indigenous knowledge and culture. The research emphasizes the need for culturally sensitive technology and offers a unique framework to bridge cultural divides, enhancing digital inclusion and socioeconomic benefits for Indigenous communities.

Chapter 18: Cultural Perspectives on Wealth and Prosperity in Indigenous Communities
Dr. Indira Bhati, Dr. Amit Joshi, Mr. Rishabh Jain

HMI technology bridges traditional Indigenous practices and contemporary needs by preserving cultural knowledge, fostering economic activity, and enhancing community well-being. HMI addresses challenges from colonization, land loss, and forced assimilation, offering solutions like archiving oral histories, revitalizing endangered languages, and facilitating global craft sales. Effective HMI implementation can sustain traditional knowledge, build sustainable economies, and improve well-being, provided Indigenous communities have governance control and robust data security measures.

Chapter 19: The Effect of Family Financial Support on Facilitating Students' Academic Success: A Mixed Methods Analysis During University Closure and Reopening Toward Indigenous Empowerment
Enkeleda Lulaj, Luis Cardoso, Muthmainnah, Gül Erkol Bayram, Andi Asrifan

This research examines the impact of family financial support (FAM) on student academic success across Kosovo, Portugal, and Indonesia, with a focus on how self-financing (SF) and various independent variables influence performance during and after COVID-19. Using descriptive analysis, one-way ANOVA (OWA), and General Linear Model (GLM) on data from 216 students, the study reveals significant differences in family financial support and its effect on student performance. The findings highlight the importance of adapting financial support to new learning environments, particularly in VARFAM5 and VARFSFPC1.

In Conclusion

As editors of *Indigenous Empowerment through Human–Machine Interactions: Challenges and Strategies from Business Lenses*, we are honored to present this pivotal volume that bridges technological innovation with Indigenous cultural empowerment. This collection provides a comprehensive examination of how HMI can be harnessed to address the unique challenges faced by Indigenous communities while fostering their socioeconomic development.

The chapters within this book offer a rich array of perspectives on the integration of HMI technologies with Indigenous cultural practices. Each contribution explores the intersection of technology and business strategies, highlighting both the potential and limitations of these interactions. The authors have meticulously analyzed case studies, theoretical frameworks, and practical applications to illuminate how HMI can support Indigenous empowerment, from preserving cultural heritage to enhancing economic opportunities.

Our goal in compiling this volume was to offer a valuable resource that not only examines current advancements but also provides actionable strategies for overcoming obstacles in the deployment of HMI technologies. We aim to encourage dialogue between technologists, policymakers, and Indigenous leaders to foster solutions that are both innovative and respectful of cultural traditions.

We hope this book serves as a catalyst for continued research and collaboration in the field. By advancing our understanding of how HMI can be used to empower Indigenous communities, we aspire to contribute to a more inclusive and equitable technological landscape. Thank you for engaging with this important work; we trust that the insights shared will inspire meaningful progress and innovation.

Editors

Dr. Indrajit Ghosal, Brainware University, India
Dr. Swati Gupta, Universal AI University, Maharashtra, India
Dr. Sudhir Rana, Gulf Medical University, UAE
Dr. Dipak Saha, Institute of Engineering and Management, Kolkata, India

Acknowledgments

I would like to acknowledge all my fellow editors for their continuous hard work, support, and motivation to complete this book, especially Dr. Swati Gupta. I would also like to show my gratitude and love to my family members, my loving wife (Arpita), and my two sons (Amitrajit and Adhrit). I would also like to show my gratitude to all my research colleagues and all my scholars. I also express my gratitude to my PhD Supervisor Professor (Dr.) Pranay Kumar Roy Barman.

I want to express my gratitude to our HOD (Dept. of Management), Dean Academics Brainware University, Kolkata, and all the colleagues of Management Department.

Finally, I would like to thank my all-loving students for their constant support and wishes.

Indrajit Ghosal

Dr. Gupta as one of the editors of the Bock entitled *Indigenous Empowerment through Human–Machine Interactions: The Challenges and Strategies from Business Lenses*. I extend my heartfelt thanks to my fellow editors, Indrajit Ghosal, Sudhir Rana, and Dipak Saha, for their collaboration and insight. I am also grateful to the publishing team at Emerald. I would also like to thank my family members for their support and motivation.

Swati Gupta

Dr. Rana as one of the editors of the book entitled *Indigenous Empowerment through Human–Machine Interactions: The Challenges and Strategies from Business Lenses*. I would like to thank my fellow editors, Indrajit Ghosal and Dipak Saha, for their collaboration and insight. I am also grateful to the publishing team at Emerald. I would also like to thank my family members, my research colleagues, and scholars for their support and motivation. I would also like to extend my gratitude to our university colleagues, Gulf Medical University, UAE.

Sudhir Rana

Introducing *Indigenous Empowerment through Human–Machine Interactions: The Challenges and Strategies from Business Lenses*, I am happy to express my sincere thanks to all contributors for their contributions in making the book visible. I would also like to thank Prof. (Dr.) Satyajit Chakrabarti, Hon'ble Director, Institute of Engineering and Management, Kolkata, and Pro Vice-Chancellor of the University of Engineering and Management, Kolkata, India, for his constant encouragement and motivation.

Definitely, continuous team involvement and team hard work deserve appreciation. Thanks also goes to my friend, colleague, and family members for their outstanding support throughout this project.

Dipak Saha

Chapter 1

Examining the Influence of Social Media Marketing Strategies on Indigenous Consumer Behavior in the UAE

Hardyal Kour Rainu and Kamaladevi Baskaran

Amity University Dubai, UAE

Abstract

Social media has emerged as a crucial tool in the current digital era, fundamentally transforming consumer behavior and corporate performance. This research explores the interplay between social media marketing (SMM) strategies, consumer behavior, and business performance, focusing on indigenous consumers in the United Arab Emirates (UAE). In a market characterized by rapid digital transformation and diverse consumer demographics, the study aims to analyze the influence of various SMM approaches on indigenous consumer attitudes, perceptions, and purchasing behaviors. Employing a mixed-methods design, the research integrates both quantitative and qualitative methodologies. A structured questionnaire is utilized to collect quantitative data from indigenous consumers regarding their buying behaviors and perceptions of social media marketing. The study employs a stratified random sampling technique to ensure a comprehensive representation of demographic factors, including age, gender, income, and geography. Qualitative insights are gathered through interviews with marketing professionals, business owners, and social media influencers, providing a deeper understanding of their strategic experiences in this domain. The findings reveal that effective SMM strategies significantly enhance key business performance indicators such as market share, sales revenue, customer acquisition, and brand loyalty. Additionally, the research considers contextual factors such as cultural norms and technological infrastructure that may affect marketing effectiveness. This study offers actionable recommendations for businesses aiming to optimize their SMM efforts and engage effectively with indigenous consumers, thereby

Indigenous Empowerment through Human-Machine Interactions, 1–18

Published under exclusive licence by Emerald Publishing Limited
doi:10.1108/978-1-83608-068-820251001

contributing to sustainable business success in the digital age and enriching the academic discourse on marketing practices in the UAE.

Keywords: Social media marketing strategies; indigenous consumer behavior; online shopping; digital transformation; customer engagement; UAE

1. Introduction

Social media marketing (SMM) is essential for businesses aiming to engage indigenous consumers in the digital landscape, particularly in the United Arab Emirates (UAE), where social media usage is pervasive. This research investigates the effects of SMM on indigenous consumer behavior and business performance within the UAE context. Prominent platforms such as Instagram, Facebook, Twitter, and LinkedIn enable businesses to effectively reach their target audiences (Kumawat & Baskaran, 2023). As the retail industry in the UAE increasingly adopts digital marketing strategies, understanding these dynamics is crucial for influencing consumer perceptions and purchasing decisions.

This study specifically addresses the underexplored demographic of indigenous consumers in the UAE. It examines various facets of SMM, including influencer partnerships, targeted advertising, and customer engagement, to determine how these elements shape consumer preferences in the retail sector. Furthermore, this research aims to elucidate how SMM can drive business growth and success in the UAE market.

1.1 Social Media Growth in UAE

Social media platforms facilitate the creation and exchange of content, allowing users to connect and share interests. Over the past decade, social media has gained immense popularity, with millions of users leveraging these platforms to maintain relationships and share experiences. In the UAE, social media has become integral to daily life, particularly among younger demographics, driven by high smartphone penetration and internet access. Businesses in the UAE have recognized the importance of establishing a robust social media presence to engage consumers and promote their offerings.

As of recent data, approximately 99% of the UAE population uses the internet, with over 112% engaging on social media platforms, spending an average of nearly three hours daily on these sites (Global Media Insight, 2023). The growth in social media usage, reported to be 9% in 2020, highlights the shift in consumer engagement, compelling businesses to adapt their marketing strategies to effectively reach their audiences (Tahseen Consulting, 2023). Social media also enables companies to receive instant feedback from consumers, allowing them to promptly address issues and improve their products and services (Moncey & Baskaran, 2020).

1.2 Social Media Marketing

SMM involves the creation and distribution of content on various platforms to engage target audiences, drive website traffic, and enhance sales (Baskaran, 2022). The global surge in social media use has significantly boosted the popularity of this marketing approach. With 3.8 billion users worldwide, social media serves as an effective tool for brands to connect with consumers. Common strategies include social media advertising campaigns, the creation of engaging content – such as blogs and videos – and real-time interactions with customers through comments and messages. Success can be measured using metrics like engagement, reach, conversion rates, and return on investment (ROI).

1.3 Social Media Marketing Action Plan

To maximize effectiveness, SMM strategies should focus on indigenous customers. An effective action plan involves several key steps: defining corporate and marketing objectives, understanding customer demographics (age, location, interests, income), analyzing competitors, and assessing current marketing strengths and weaknesses. Creating a content release schedule, producing high-quality content, and monitoring results are also critical components of this strategy. SMM involves engaging with customers through meaningful discussions and timely responses, which is essential for building and maintaining a loyal follower base (Baskaran, 2019b). Analytics plays a crucial role in helping businesses refine their social media strategies by identifying content that resonates with audiences. By fostering relationships and monitoring performance, businesses can achieve their marketing goals through engaging social media practices (Baskaran, 2019a).

1.4 Social Media Marketing in the UAE's Retail Sector

SMM has grown to be a crucial component of the retail sector, giving companies a more efficient means of connecting with the target market and growing the indigenous customer base. The usage of social media platforms for marketing has significantly increased in the United Arab Emirates, a hub of many cultures and nations. SMM has made it possible for companies in the UAE to communicate with the customer base and gain insight into the requirements, wants, and preferences. SMM has become even more essential for businesses to stay competitive and boost performance with the growth of e-commerce and the move toward online purchasing (Baskaran, 2023a; 2023b).

2. Literature Review

The empirical research aims to investigate how SMM strategies affect the behavior of indigenous consumers and the performance of businesses in the UAE. It provides a solid foundation for understanding the current knowledge on this topic, highlighting areas that require further investigation in the UAE market by building upon existing research. This review helps establish a framework for the research paper's objectives and hypotheses, laying the groundwork for examining how SMM strategies influence indigenous consumer behavior and business performance.

2.1 Social Media Marketing in Small and Medium Enterprises (SMEs)

The research identified three clusters of SMEs based on their SMM strategies. The first cluster, calculative pragmatists, includes retail businesses that use social media to effectively target their audience, while minimizing advertising costs. The second cluster, cautious watchers, comprises business-to-business companies that cautiously experiment with new social media strategies and closely monitor their brand's online presence. Finally, the third cluster, proactive strategists, consists of SMEs that actively utilize social media to build brand awareness, engage with customers, and gather feedback through various platforms and tools. Each of these clusters demonstrates unique characteristics and tailored approaches to SMM, aligning with their respective business goals and target markets (Pentina & Koh, 2014). The study reveals the profound influence of social media on SMEs in the Arab region. These studies identify four primary impacts: (a) providing a crucial marketing platform suited for the Information Age, (b) serving to enhance sales, (c) offering innovative opportunities for cost reduction and increased profitability, and (d) establishing a standard for direct client communication in modern public relations (PR). These impacts underscore the significant role of social media for Arab SMEs, aligning with broader trends observed globally (Basri, 2016). The key findings of the paper include the importance of SMEs in developing and developed countries due to their significant contributions to employment and gross domestic product (GDP). The study highlights the positive impact of digital media on SMEs, such as improving business performance, increasing customer base, and enhancing communication (Hassan, 2018). It highlights the potential benefits of digital media for SMEs in the UAE, like reaching global customers, facilitating communication, and providing convenient online transactions. Additionally, the paper discusses the challenges SMEs face in adopting digital media and the need for strategies to address disadvantages like increased competition. The research shows that social media strategies affect consumer attitudes, driving higher sales, word-of-mouth (WOM) referrals, and profits. It stresses the importance of emotive brand posts, strategic positioning, and consumer engagement. Opinion leaders are key for spreading information, and marketing should target them. Social media sentiment impacts stock performance, and WOM messages adapt based on the receiver (Alves, 2016).

2.2 Social Media Influencer Marketing and Indigenous Consumer Behavior

The research highlights the positive impact of influencers in the UAE on the youth, who perceive influencers as trustworthy and appealing by assisting in product evaluation (Radwan et al., 2021). The key findings suggest the need to explore influencers' strategies in shaping market trends, examine their societal role and influence on behavior, and recognize influencer marketing as a significant communication tool in the economic sphere (Padmaja, 2021). The study aims to assess the effects of digital influencers on economic recessions, develop strategies to regulate purchasing behavior through social media influencers, and enhance sustainable economic growth. Survey results indicate a substantial number of participants are swayed by Instagram influencers, follow them for various reasons, and alter their online buying behavior based on influencer endorsements. The research underscores the substantial purchasing trends and preferences, particularly during economic downturns (Al Hosani & Snoussi, 2023). The Consumer Behavior Cycle describes the sequential stages individuals go through when making purchasing decisions. It begins with the recognition of a need or desire, prompting the consumer to seek information about potential solutions through internal recollection or external sources like reviews and advertisements. Subsequently, consumers evaluate alternatives based on criteria such as quality, price, and brand reputation before making a purchase decision. Following the transaction, consumers assess their satisfaction with the product or service, influencing their likelihood of repeat purchases and WOM recommendations. This cycle highlights the importance for businesses to align their marketing efforts to meet consumer needs effectively at each stage, fostering customer loyalty and optimizing business outcomes (Baskaran, 2019a).

2.3 Social Media Advertising and Brand Equity

The research discovered that factors such as entertainment, informativeness, relevance, and interactivity have a positive impact on brand experience and equity. It confirms the crucial role of social media advertising in shaping brand experience and equity in the retail sector (Hanaysha, 2022). The study emphasizes the significant effects of attitude, brand loyalty, brand image, and brand awareness on consumer purchase intention, especially among UAE consumers of digital lifestyle products. It highlights the need for policymakers in the industry to focus on restructuring and strategizing relevant policies and practices to align with advanced technologies. The integration of technological platforms to enhance consumer engagement and improve service quality is emphasized as a way to enhance firm performance and competitive advantage in the industry.

2.4 Digital Marketing Capabilities and Business Performance

The study found that digital marketing applications impact the performance of SMEs in the UAE, enabling firms to reach customers remotely. Adoption of digital marketing applications is driven by external competitive pressure, but environmental factors like competitive pressure and industry competitiveness don't directly affect business performance (Badghis & Aljandali, 2018). Despite this, the study suggests that businesses should leverage the digital advancements in marketing to effectively approach customers and thrive in a competitive environment. The study utilized quantitative data from 350 respondents and found that digital marketing (social media, online advertisement, content marketing) significantly impacts business performance enhancement and customer relationship management (CRM), with CRM mediating between digital marketing and business performance enhancement (Baskaran & Rajavelu, 2020). The research model successfully met reliability and validity criteria, with all constructs showing values above the required thresholds for internal consistency, reliability, and convergent validity (Pathak & Kushwaha, 2023). The study concluded that digital marketing emphasizes the need for continuous updates of marketing strategies to maintain a competitive relationship with customers (Nuseir & Refaea, 2022).

2.5 Social Media Adoption and Firm Performance

The study found that social media adoption did not impact on SME performance. This insight can guide SME managers and decision-makers to stay updated on social media trends and prepare to leverage social commerce as it becomes increasingly common. The main outcomes of the study involve examining the direct connections between SME performance and external factors such as national culture, Facebook, WhatsApp, and Email, which collectively account for approximately 81% of the variance in SME performance. Furthermore, the research identifies notable relationships between national culture and these external factors, explaining around 56% of the variance in national culture. The research also identified an indirect relationship between exogenous constructs and SME Performance mediated by national culture, with significant indirect effects for Facebook, WhatsApp, and Email on Performance. The final structural model achieved model acceptance with satisfactory fitness indexes after re-specification (Ghanem & Hamid, 2020).

2.6 Cultural Influence on Social Media Marketing

The study underscores the need to address the gap in analyzing cultural aspects and media distribution, highlighting the importance of understanding how visual images are perceived by consumers from diverse cultures. Facebook and WhatsApp dominate the region, offering insights into cultural norms regarding technology and gender, which influence social media usage and corporate marketing. Social media users are seen as emerging influential groups, significantly impacting media democratization and the transformation of PR (Baskaran & Vanithamani, 2014). Features like participation and openness on social media platforms play a vital role in bridging the gap between users and media, facilitating feedback and content sharing. The research investigates the relationship between social media features, culture, and media preferences, emphasizing the influence of cultural factors on social media usage in Arab countries. Social media is revolutionizing communication by providing interactive platforms that simplify interactions, collaborations, and content sharing (Milla & Mataruna, 2019).

2.7 Consumer Experience and Interaction

Brand awareness is crucial for developing trust among customers, as knowing about an organization's features fosters trust (Hashim, 2017). Social media interaction, e-WOM, and PR are vital for brand awareness and trust development, aligning with previous research findings. The study used a questionnaire based on past literature, achieving a valid response rate of 69.72% from 512 customers in the UAE retail sector. The structural equation modeling technique was employed, revealing that all proposed hypotheses were accepted, providing valuable insights for further research and policy development (Aljumah et al., 2022).

2.8 Strategic Social Media Marketing

The key findings of the paper "Importance of Strategic Social Media Marketing" include the necessity for organizations to allocate resources for creating viral content, monitoring user-generated content, and engaging with influencers to succeed on social media platforms (Choudhary & Baskaran, 2022; Sarkar & Ghosal, 2018). Influencer marketing is highlighted as integral for promoting products and services through individuals with a large follower base, impacting behavioral results like purchases or intent to buy. Social media platforms play a crucial role in enhancing customer relationships, influencing consumer behavior, and providing valuable marketing intelligence. Additionally, the study emphasizes the importance of social media in developing brand relationships, engaging consumers, and stimulating sales through interactive marketing strategies.

2.9 Research Framework

The research framework focuses on exploring how SMM strategies influence indigenous consumer behavior and business performance in the UAE. It begins with a clear articulation of the research problem, emphasizing the need to understand the specific dynamics of social media engagement among indigenous populations in a rapidly digitizing market. A comprehensive literature review grounds the study in existing theories of SMM and consumer behavior, highlighting gaps in understanding related to indigenous contexts. The research objectives are structured to investigate various aspects, including consumer attitudes, preferences, and the impact of marketing strategies on business metrics like sales and brand loyalty.

3. Research Methodology

3.1 Purpose of the Study

This study examines the effects of SMM strategies on consumer behavior and business performance in the UAE. Given the transformative nature of social media on business operations, the research aims to understand how these strategies can affect both indigenous consumer behavior and overall business performance in the UAE. Furthermore, the study intends to investigate the correlation between social media usage by businesses in the UAE and its consequences on financial, consumer, and ethical dimensions. Additionally, the purpose of research is as follows:

- Investigate the relationship between the usage of social media by businesses in the UAE and their impact on financial aspects such as limited-time offers, coupon codes, buy-one-get-one-free deals (BOGO), and discounts for referring friends.
- Examine the relationship between the age of indigenous consumers in the UAE and their impact on indigenous consumer aspects, including purchase intentions, brand perception, engagement and interaction, and consumer loyalty.
- Assess the relationship between the impact of annual household income in the UAE and its impact on online shopping such as ease of browsing, product discovery, checkout processes, and consumer support.
- Study the relationship between the usage of social media in the UAE and its impact on ethical aspects such as transparency, privacy, authenticity, and social responsibility.

The outcomes of this study could prove advantageous for businesses operating in the UAE, policymakers, and scholars. Grasping the ramifications of SMM strategies on consumer behavior and business performance can assist UAE-based businesses in formulating effective strategies tailored to their objectives, fostering customer attraction, engagement, and retention.

3.2 Research Design

This research adopts both quantitative and qualitative methods. This approach enables the collection of both numerical and subjective data with a sample size of 101, which is crucial for the generalizability of the results. Ensuring a sufficient sample size helps represent the population accurately. Data collection was carried out using a standardized questionnaire sent to the target audience, enabling data gathering from a large number of respondents in a structured and consistent manner. The standardized questionnaire helps ensure the reliability of the collected data. Data analysis was performed using regression models in Statistical Package for the Social Sciences (SPSS) to test the hypothesis, providing a systematic and organized method for analyzing the data and making statistical inferences, which enhances the validity and reliability of the findings.

4. Data Analysis

4.1 Measurement Model Assessment

The reliability of the measurement instrument used in the study was thoroughly assessed to ensure the internal consistency of the scale items. The evaluation of internal consistency, a crucial step in validating the measurement tool's credibility, was conducted using Cronbach's Alpha. For the given set of items, the Cronbach's Alpha value was recorded at 0.955, mirroring the value obtained when items were standardized. This exceptionally high Cronbach's Alpha value significantly surpasses the commonly accepted threshold of 0.7, indicating an exemplary level of internal consistency among the items.

According to Fig. 1.1, it can be observed that the most notable proportion of indigenous customers fall within the age bracket of 18–24 years and 25–34 years and represents 29.7% and 25.74% of the total customer base. The customers of age group between 35 to 44 years and under 18 years represent 15.84% and 14.85%. The age group between 45 to 54 years and 55 or above years makes up about 7.92% and 5.94% of the customer base.

This demographic information is significant as it suggests that businesses in the UAE could benefit from targeting younger audiences through SMM strategies. The data also reveal that customers who are 35 years old or above constitute a significant portion of the customer base.

According to Fig. 1.2, it can be observed that the major proportion of customers are female and male which are 46.53% and 33.66% of the total customer base. The remaining 19.8% comprises individuals who identified as "other" or chose "prefer not to say," collectively representing a smaller but significant portion of the customer demographic.

It suggests that both genders are important target markets for SMM strategies. Companies could develop social media campaigns that appeal to both male and female customers.

According to Fig. 1.3, it can be observed that the major proportion of indigenous customer's income is between Arab Emirates Dirham (AED) 50,000 and AED 100,000 and less than AED 50,000 which is 37.62% and 36.63% of the total sample size. This information is significant for businesses in the UAE as it provides insight into the income distribution of their indigenous customer base. It suggests that businesses should consider developing SMM strategies that cater to customers with lower income levels, as they constitute the majority of the customer base.

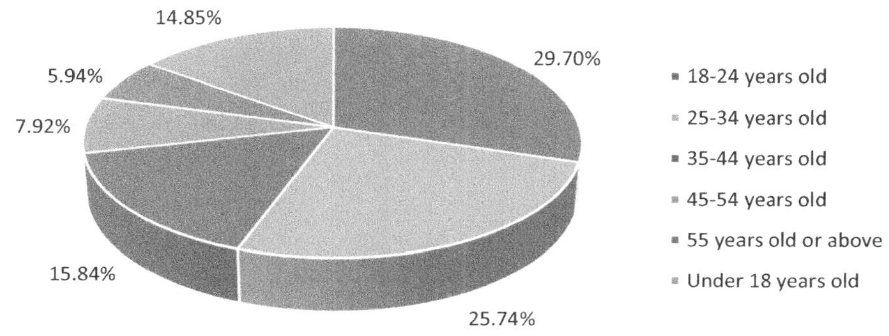

Fig. 1.1. Age Distribution.

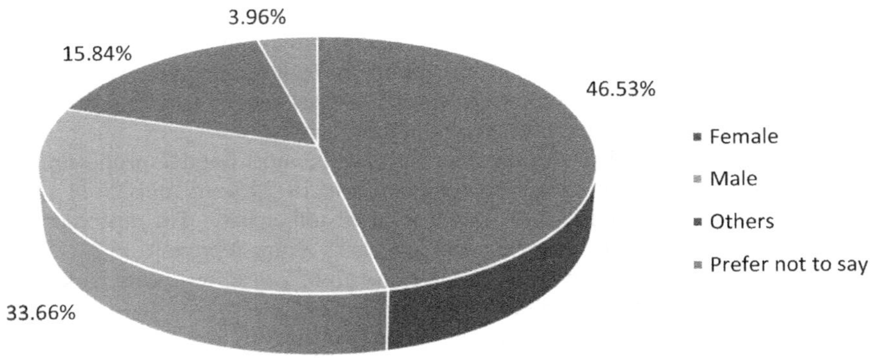

Fig. 1.2. Gender Distribution.

According to Fig. 1.4, it can be observed that the major proportion of sample uses social media for 30 minutes to 1 hour which are 36.63% of the total sample size. Then comes 1–2 hours and 2–4 hours which represent 27.72% and 15.84%. Lastly, less than 30 minutes and more than 4 hours include 10.89% and 8.91% of the sample size.

4.2 Hypothesis Testing

Null Hypothesis (H_{01}): There is no significant relationship between the usage of social media and financial promotions in the UAE.

Alternative Hypothesis (H_{a1}): There is a significant relationship between the usage of social media and financial promotions in the UAE.

The study seeks to empirically evaluate these hypotheses to offer insights into how social media usage impacts financial promotions in the UAE market. The

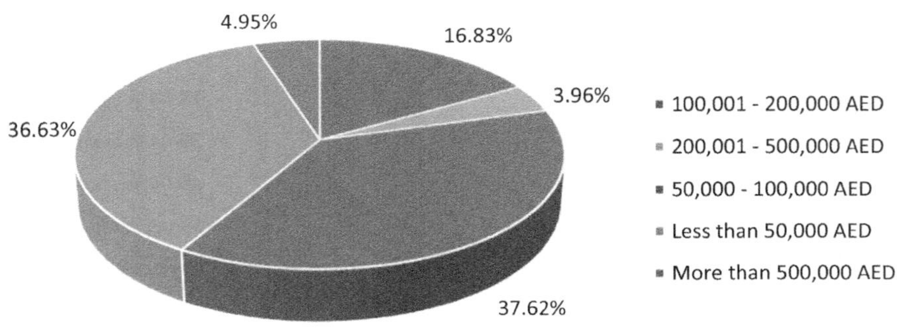

Fig. 1.3. Annual Income (Dirhams).

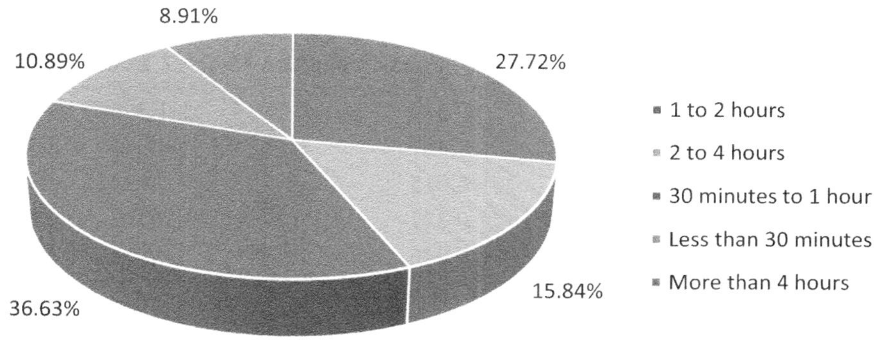

Fig. 1.4. Daily Usage of Social Media.

calculated *R* Square value is 0.455, which falls between 0.3 and 0.5. This suggests that the model is moderately effective in illustrating the relationship between the dependent variable (social media usage) and the independent variables (limited-time offers, coupon codes, BOGO, and discounts for referring friends). However, there is room for improvement to better capture this relationship. Therefore, the model in its current form may not be the best fit for the given data set, and further analysis and refinement may be required to improve its performance.

To establish statistical significance, the *p*-value or significance level should typically be less than 0.05. In Table 1.1, the *p*-value is <0.001, indicating a highly significant result. An *F*-ratio greater than 1 suggests an efficient model. In this case, the *F*-ratio value of 20.022, as shown in the table above, indicates a good fit for the model.

In Table 1.2, the hypothesis testing examines how various marketing strategies, such as limited-time offers, coupon codes, BOGO deals, and discounts for referring friends, affect the dependent variable. The *p*-value was utilized to determine whether to reject the null hypothesis, which assumes no significant difference or relationship between the dependent and independent variables.

The findings reveal that limited-time offers and BOGO have a notable impact on the dependent variable, as their *p*-values were less than 0.05, the predetermined

Table 1.1. ANOVA.

	DF	SS	MS	*F*	Significance *F*
Regression	4	57.677	14.419	20.022	<0.001
Residual	96	69.135	0.72		
Total	100	126.812			

Table 1.2. Computation of *p*-Value.

	Coefficients	Standard Error	*T* Stat	*p*-Value
Constant	0.397	0.283	1.399	0.165
Limited-time offers	0.407	0.101	4.044	<0.01
Coupon codes	0.015	0.119	0.122	0.903
Buy-one-get-one-free (BOGO)	0.321	0.118	2.714	0.008
Discounts for referring friends	0.046	0.099	0.458	0.648

significance level. Thus, the null hypothesis is rejected, indicating that these marketing strategies positively influence the dependent variable. In contrast, the *p*-values for coupon codes and discounts for referring friends were greater than 0.05, suggesting that these strategies do not significantly affect the dependent variable.

Null Hypothesis (H_{02}): There is no significant relationship between the age of indigenous consumers and consumer aspects in the UAE.

Alternative Hypothesis (H_{a2}): There is a significant relationship between the age of indigenous consumers and consumer aspects in the UAE.

The *R*-squared value of 0.099 from the regression analysis indicates a weak or low effect size. This suggests that the age of consumers explains only a small portion of the variation in consumer aspects in the UAE. As a result, these findings may not provide enough support for the alternative hypothesis, and further investigation may be necessary to establish a significant relationship between age and these consumer aspects.

In Table 1.3, the *p*-value is 0.073, indicating a nonsignificant result. The *F*-ratio measures how well the model predicts the variable, considering any inaccuracies in the model. An *F*-ratio greater than 1 suggests a more efficient model. In this table, the *F*-ratio value is 2.091, which is considered good.

In Table 1.4, the *p*-values for purchase intentions, brand perception, engagement, interaction, and customer loyalty were found to be 0.011, 0.117, 0.750,

Table 1.3. ANOVA.

	DF	SS	MS	*F*	Significance *F*
Regression	5	18.937	3.787	2.091	0.073
Residual	95	172.072	1.811		
Total	100	191.01			

Table 1.4. Computation of p-Value.

	Coefficients	Standard Error	T Stat	p-Value
Constant	1.395	0.541	2.577	0.011
Purchase intentions	0.265	0.168	1.581	0.117
Brand perception	0.058	0.182	0.320	0.750
Engagement	−0.096	0.208	−0.462	0.645
Interaction	0.037	0.206	0.181	0.857
Customer loyalty	0.244	0.175	1.4	0.165

0.645, 0.857, and 0.165, respectively. These findings indicate that these variables do not exert a statistically significant influence on the dependent variable. These results can be useful for businesses seeking to understand the factors that influence their customers' behavior and make informed decisions about their marketing and customer engagement strategies.

Null Hypothesis (H_{03}): There is no significant relationship between annual household income and online shopping including ease of browsing, product discovery, checkout processes, and consumer support in the UAE.

Alternative Hypothesis (H_{a3}): There is a significant relationship between annual household income and online shopping including ease of browsing, product discovery, checkout processes, and consumer support in the UAE.

The analysis showed a weak effect size with an R-squared value of 0.06, indicating that income explains only a small part of the variation in online shopping behavior.

According to Table 1.5, the p-value or significance level should ideally be less than 0.05 for statistical significance. The F-ratio indicates how well the model predicts the variable, accounting for any inaccuracies. An F-ratio greater than 1 suggests a more effective model. In Table 1.5, the F-ratio value is 1.538, which is considered favorable.

In Table 1.6, the p-values for these variables were found to be 0.457, 0.742, 0.968, and 0.297, respectively. In this study, all the p-values were greater than

Table 1.5. ANOVA.

	DF	SS	MS	F	Significance F
Regression	4	6.922	1.731	1.538	0.197
Residual	96	107.989	1.125		
Total	100	114.911			

Table 1.6. Computation of *p*-Value.

	Coefficients	Standard Error	*T* Stat	*p*-Value
Constant	1.121	0.391	2.867	0.005
Ease of browsing	0.1	0.134	0.747	0.457
Product discovery	0.049	0.148	0.33	0.742
Checkout processes	0.005	0.136	0.04	0.968
Consumer support	0.136	0.13	1.049	0.297

0.05, indicating that the null hypothesis cannot be rejected for any of the variables. This suggests that these variables have no significant impact on consumer behavior and business performance in the UAE. Therefore, the results suggest that Ease of Browsing, Product Discovery, Checkout Processes, and Consumer Support do not significantly impact consumer behavior and business performance in UAE. It is crucial to note that the interpretation of these findings is dependent on the *p*-values and does not incorporate other important factors such as statistical power, effect size, and sample size.

Null Hypothesis (H_{04}): There is no significant relationship between the usage of social media and ethical aspects in the UAE.

Alternative Hypothesis (H_{a4}): There is a significant relationship between the usage of social media and ethical aspects in the UAE.

In the regression analysis, the *R*-squared value of 0.209 indicates a moderate effect size. This suggests that the usage of social media, as an independent variable, moderately explains the variation in the dependent variable, ethical aspects in the UAE. These results support the alternative hypothesis (H_{a4}), suggesting a relationship between the variables. Further investigation is needed to understand the explicit nature of this relationship and how it impacts ethical considerations related to social media usage in the UAE.

According to Table 1.7, the *p*-value or significance level is less than 0.05, indicating statistical significance. In this case, the *F*-ratio value of 6.327, as shown in the table, is considered favorable.

In Table 1.8, the *p*-values for constant, transparency, privacy, authenticity, and social responsibility were found to be 0.006, 0.026, 0.568, 0.585, and 0.228, respectively. These results suggest that transparency has a statistically significant impact on the dependent variable, while privacy, authenticity, and social responsibility do not. These results can be useful for businesses seeking to understand the factors that influence their customers' behavior and make informed decisions about their marketing and customer engagement strategies.

Table 1.7. ANOVA.

	DF	SS	MS	*F*	Significance *F*
Regression	4	26.457	6.614	6.327	<0.001
Residual	96	100.355	1.045		
Total	100	126.812			

Table 1.8. Computation of *p*-Value.

	Coefficients	Standard Error	*T* Stat	*p*-Value
Constant	1.065	0.380	2.8	0.006
Transparency	0.272	0.121	2.254	0.026
Privacy	0.083	0.145	0.572	0.568
Authenticity	0.066	0.121	0.547	0.585
Social responsibility	0.151	0.125	1.213	0.228

5. Conclusion

The literature review highlights the significant impact of SMM on indigenous consumer behavior and business performance in the UAE. Tailored marketing strategies, ranging from cautious approaches to proactive initiatives, enhance sales, reduce costs, and improve communication for SMEs. Digital marketing capabilities are particularly critical in the insurance sector, improving CRM and fostering business growth. Social media influencers notably shape consumer behavior among younger demographics, influencing trends and purchase decisions. The review emphasizes the importance of entertainment, informativeness, relevance, and interactivity in social media advertising, which positively affects brand experience and equity. Cultural factors further shape media preferences and marketing strategies in the region. Targeting younger audiences (ages 18–34) is essential, while also addressing the needs of customers aged 35 and above. Gender diversity is important, as nearly half of consumers are female. Businesses should focus on indigenous customers with lower income levels, considering usage patterns where most users engage for 30 minutes to an hour daily. To effectively reach their audience, businesses must adopt a comprehensive SMM approach that considers age, gender, income, and engagement habits. Future research should dive into emerging trends and cultural influences to refine marketing strategies for diverse audiences in the UAE.

5.1 Recommendations

- *Target Younger Audiences:* Develop strategies that specifically cater to younger demographics, particularly those in the 18–24 and 25–34 age groups, as they form a substantial portion of the customer base.
- *Broaden Age Targeting:* In addition to targeting younger audiences, extend marketing campaigns to include customers aged 35 and above to maximize reach and engagement.
- *Ensure Gender Diversity:* Tailor campaigns to both male and female audiences, considering that nearly half the customers are female and one-third are male.
- *Analyze Social Media Usage Patterns:* Examine users' patterns of time spent on social media platforms to guide the creation of targeted, time-sensitive content.
- *Engage with Social Media Influencers:* Collaborate with social media influencers to enhance campaign effectiveness, particularly among younger audiences.
- *Incorporate Cultural Considerations:* Recognize and integrate cultural nuances when developing marketing strategies, as these greatly affect media preferences and customer expectations.

References

Al Hosani, T., & Snoussi, T. (2023). Role of social media influencers: A case study of UAE. *Algerian Communication Review*, *25*(1). https://asjp.cerist.dz/en/downArticle/437/25/1/221816

Aljumah, A. I., Nuseir, M. T., & El Refae, G. A. (2022). Exploring the effect of social media marketing and destination image on destination loyalty in COVID-19 times: Sequential mediating role of brand love and brand loyalty. *International Journal of Data and Network Science*, *6*(4), 1359–1368. https://doi.org/10.5267/j.ijdns.2022.6.004

Badghis, R. M., & Aljandali, A. H. (2018). An empirical study on the role of digital marketing in the performance of SMEs in the UAE. *Journal of Internet Commerce*, *14*(1), 1–19.

Baskaran, K. (2019a). An interpretive study of customer experience management towards online shopping in UAE. *International Journal of Mechanical Engineering and Technology*, *10*(02), 1071–1077.

Baskaran, K. (2019b). The impact of digital transformation in Singapore e-tail market. *International Journal of Innovative Technology and Exploring Engineering*, *8*(11), 2320–2324. https://doi.org/10.35940/ijitee.i8046.0981119

Baskaran, K. (2022). E-consumer behavioral analytics: Paradigm shift in online purchase decision making. In P. Keikhosrokiani (Ed.), *Handbook of research on consumer behavior change and data analytics in the socio-digital era* (pp. 192–209). IGI Global. https://doi.org/10.4018/978-1-6684-4168-8.ch009

Baskaran, K. (2023a). Predictive analytics in recruitment: Unveiling through skill gap employability model. In R. Yadav, M. Sinha, & J. Kureethara (Eds.), *HR analytics in an era of rapid automation* (pp. 20–35). IGI Global. https://doi.org/10.4018/978-1-6684-8942-0.ch002

Baskaran, K. (2023b). Customer experience in the E-commerce market through the virtual world of metaverse. In P. Keikhosrokiani (Ed.), *Handbook of research on consumer behavioral analytics in metaverse and the adoption of a virtual world* (pp. 153–170). IGI Global. https://doi.org/10.4018/978-1-6684-7029-9.ch008

Baskaran, K., & Rajavelu, S. (2020). Digital innovation in industry 4.0 era – Rebooting UAE's retail. In *2020 International Conference on Communication and Signal Processing (ICCSP)*. https://doi.org/10.1109/iccsp48568.2020.9182301

Baskaran, K., & Vanithamani, M. R. (2014). E-customers attitude towards E-store information and design quality in India. *Applied Research in Science, Engineering and Management. Applied Research in Science, Engineering and Management, World Applied Sciences Journal*, 51–56. https://www.idosi.org/wasj/wasj31(arsem)14/10.pdf

Basri, W. S. (2016). Social media impact on small and medium enterprise: Case review of businesses in the Arab world. *Arts and Social Sciences Journal*, *7*(6). https://doi.org/10.4172/2151-6200.1000236

Choudhary, P., & Baskaran, K. (2022). The impact of social networking sites among modern youth. In *2022 Interdisciplinary Research in Technology and Management (IRTM)*, Kolkata, India (pp. 1–6). https://doi.org/10.1109/IRTM54583.2022.9791522

Ghanem, S. K. R., & Hamid, N. A. (2020). The effect of Facebook, WhatsApp, Twitter, and Email on SMEs' performance: Empirical evidence from United Arab Emirates. *Journal of Internet Social Networking and Virtual Communities*, 1–10. https://doi.org/10.5171/2020.618899

Global Media Insight. (2023). UAE social media statistics.

Hanaysha, J. R. (2022). An examination of social media advertising features, brand engagement, and purchase intention in the fast food industry. *International Journal of Services and Standards*, *8*(1), 1–15.

Hashim, F. (2017). Social media impact on small and medium enterprise: Case review of businesses in the Arab world. *The Journal of Business Inquiry*, *16*(2), 93–107.

Hassan, M. K. (2018). Digital media impact on SMEs performance in the UAE. *Asian Journal of Social Sciences and Management Studies*, *5*(1), 1–11.

Kumawat, P., & Baskaran, K. (2023). Information management in the era of digital banking - A descriptive analysis of adoption and usage among Millennial and Gen Z. In *ICIMMI '23: Proceedings of the 5th International Conference on Information Management & Machine Intelligence* (pp. 1–5). Association for Computing Machinery.

Milla, A. C., & Mataruna, L. (2019). Social media preferences, interrelations between the social media characteristics and culture: A view of Arab nations. *Asian Social Science*, *15*(6), 71–81. https://doi.org/10.5539/ass.v15n6p71

Moncey, A. A., & Baskaran, K. (2020). Digital marketing analytics: Building brand awareness and loyalty in UAE. In *2020 IEEE International Conference on Technology Management, Operations and Decisions (ICTMOD)*. https://doi.org/10.1109/ictmod49425.2020.9380579

Nuseir, M. T., & El Refae, G. (2022). The role of artificial intelligence, marketing strategies, and organizational capabilities in organizational performance: The moderating role of organizational behavior. *Journal of Strategic and International Studies*, *12*(2), 45–67.

Padmaja, K. (2021). Impact of social media influencer marketing on youth purchase intentions in UAE. *Media Watch*, *12*(1), 89–100.

Pathak, R. R., & Kushwaha, S. (2023). Influence of digital marketing strategies on purchase intention in retail sector. *Journal of Content, Community and Communication*, *18*(3), 1–12.

Pentina, I., & Koh, A. C. (2014). Exploring social media marketing strategies in SMEs. *International Journal of Internet Marketing and Advertising*, *8*(4), 292–310.

Radwan, A. F., Mousa, S. A., Mohamed, M. M., & Youssef, E. (2021). Impact of social media influencer marketing on youth purchase intentions in UAE. *Media Watch*, *12*(3). https://doi.org/10.15655/mw/2021/v12i3/165405

Tahseen Consulting. (2023). UAE social media white papers [White paper]. https://tahseen.ae/media/2701/uae-social-media-white-papers-english.pdf

Chapter 2

Empowering Women: Challenges, Progress

Nilanjan Ray[a], *Niyasha Patra*[b], *Suraj Sharma*[c], *Debarshi Ghosh*[d], *Indranil Bose*[c] *and Rajneesh Kler*[e]

[a]JIS University, India
[b]Brainware University, India
[c]Adamas University, India
[d]Meghnad Saha Institute of Technology, India
[e]Amity University in Tashkent, Uzbekistan

Abstract

This study aims at empowering women in India which acts as an incentive to promote historical, cultural, social, and economic dimensions. This research delves into the complexities surrounding women's empowerment, exploring the journey from historical constraints to contemporary progress and envisioning future directions. The study scrutinizes the present status of women in India in general and in the context of various policies and contexts like addressing disparities in education, employment, healthcare, and political representation. Finding and analyzing challenges such as gender-based violence, child marriage, and unequal access to resources, the research evaluates the existing legal and policy framework aimed at promoting women's empowerment. To conduct this present study, data have been collected from various sources, government reports, books, statistical reports, census reports, district handbooks, and reports from various government schemes. In this present study, researchers emphasized on government initiatives to uplift women, emphasizing their impact on education, skill development, and economic empowerment to reshape traditional paradigms.

Keywords: Challenges of women empowerment; digital literacy; gender gap index; women education; economic empowerment; skill development

Indigenous Empowerment through Human-Machine Interactions, 19–33

Published under exclusive licence by Emerald Publishing Limited
doi:10.1108/978-1-83608-068-820251002

1. Introduction

Women's empowerment has been a widely discussed topic globally, with governments and programs focusing on enhancing the socioeconomic status of women. However, it is observed that most policies primarily emphasize economic self-reliance, neglecting other important factors like health and education. The economic condition of women, especially in countries like India, is concerning, as they constitute half of the population, perform two-thirds of the work, but earn only one-third of the income and own a small fraction of the country's resources. Empowerment refers to policies and measures aimed at increasing people's autonomy and self-determination, allowing them to represent their interests and act on their own authority. It involves both the process of self-empowerment and professional support, enabling individuals, especially women, to realize their full potential in all aspects of life by gaining access to knowledge, resources, and decision-making autonomy. Empowerment entails developing awareness, acting, and gaining control over one's life, leading to increased confidence and the ability to achieve personal goals.

India is home to a diverse population that is varied in terms of socioeconomic background, language, and ethnicity. It also has a distinctive cultural tapestry. This diversity, which reflects the complex nature of the opportunities and problems faced by women in many situations, offers a rich background against which to examine the complexity of women's empowerment. India's rich cultural legacy – which includes customs, festivals, and a lengthy history of social reform movements – adds to the complex web of factors that influence women's experiences there. Numerous ethnic groups, including Bengalis, Gorkhas, Santhals, and others, call India home. The status and functions that are allocated to women within their respective communities are influenced by the cultural norms and practices that each group brings with it. Understanding the complexities of women's empowerment requires an exploration of these many cultural viewpoints since doing so enables a more nuanced study of the difficulties and advancements in many social circumstances.

Also, India has both rural areas where traditional customs may still be prevalent and urban areas like Mumbai, which are distinguished by economic development and modernism. Depending on where they live, women are affected differently by the urban–rural gap in terms of access to progressive ideals, work possibilities, and education.

2. Literature Review

Women empowerment defines the freedom of deciding life choices instead of depending on someone else. It involves boosting their status through literacy, education, training, and raising awareness. The goal is to allow and equip women to make important decisions in various areas of their lives (Malhotra et al., 2009).

It has been also observed that women are not given adequate training opportunities, or they don't have the basic quality of education since the childhood leads them to inefficient worker who are unable to hire for eligible for upskill (Ministry of Agriculture and Rural Development, 2008). Consequently, women remain passive in the economic participation of a country.

In the context of sustainable growth for women empowerment, Jie et al. (2023) mentioned in their study that environment quality and economic growth are highly negatively associated with each other. It is clearly observed that the more industrial growth, the more it leads to degradation to the biodiversity and environmental agents. As a same time, it has also observed incorporating technological upliftment, using of clean green energy or clean energy and women empowerment are significant factors to improve the situation. This study also recommended that additionally to control the situation, promotion of sustainable energy, enforcing of environment friendly equipment is essential. At the same time women employment rights needed to be protected by laws.

Again, it has been observed in several studies that have examined the impact of microfinance programs on women's empowerment in Ghana. Another study found that while the program brought economic benefits, it also led to conflicts among spouses and child labor. Another study found that a shea butter processing project was profitable for women in a specific region. However, none of these studies assessed profoundly so that it could be furthermore researched to the interventions addressed for the women. This study focuses the improvements of the rural women associated with a shea butter processing project. This upliftment strength the situation of the women societal situation and boost their enthusiast to participate in decision making, politics and other socioeconomic programs actively (Bryan & Mekonnen, 2023; Salia et al., 2017; Tanzile et al., 2023). It has been found women empowerment is highly allied. In fact, the degree of empowerment could lead to take strategically better decision-making abilities in women for household, workplace, society, health, wealth creation, and employability creation.

Evidence from Latin America (LA), Braverman-Bronstein et al. (2023) reported in their study "Gender inequality, women's empowerment, and adolescent birth rates in 363 Latin American cities" that Latin America has high levels of inequality, which negatively impacts adolescent girls. Factors such as women's education, women manpower, and access to healthcare contribute to lower adolescent birth rates. Conservative politics and restrictive laws regarding sexual and reproductive health are hindering progress. Increasing education opportunity and eliminating early marriage will be a great influence in dropping adolescent birth rates. In the conclusion study finally revealed that to decrease gender gap and disparity in various spheres is supported by the findings. To control burning issues like teenage pregnancy in the LA region it is important introspect the matter extensively going beyond sex education, contraceptive awareness program to accomplish the goal of reducing. Structural factors and cultural norms hinder young women to be independent by reaching economic growth and sustainable environment. Implementing policies that encourage women to pursue higher education and employment, as well as increasing their inclusion in politics, can help decrease adolescent pregnancy rates and improve gender equality.

According to Das and Ray (2016), this study explores rural tourism in Kamarpukur, West Bengal, India, and its impact on the local economy, socio-cultural changes, and the empowerment of women. The study focuses on the reasons tourists visit the area and the role of women in promoting the rural economy, particularly in Pallimangal. There is a need for government, banks, NGOs, and Self-Help Groups to support and encourage women in tourism-related activities. It has been found in several studies that at Santiniketan and Bharatpur in West Bengal through women are empower through heritage culture and through their handicrafts these are considered as heritage cultural tourism (Banerjee et al., 2011; Dey et al., 2022; Mukherjee et al., 2022; Ray & Das, 2011). In the context of women's quality of life through rural tourism, Ray and Das (2017) mentioned that tourism is a rapidly growing industry in India, particularly in Kamarpukur, West Bengal. It has the potential to improve the quality of life for people by boosting the local economy and bringing about sociocultural changes. This study focuses on the empowerment of women and the positive impact of tourism on the rural economy in Kamarpukur and West Bengal.

The government and various organizations should support and encourage women in tourism-related activities to further improve the economy of West Bengal. Kumar et al. (2023) mentioned in their study that for development of human resource upliftment the influential factors of sociocultural conditions for effective policy making in the local tourism industry in Bangladesh. The study aims to identify the major challenges and provide suggestions for sustainable development in this sector. The findings of this study are important for the development of Bangladesh's tourist industry. Bayeh (2016) stated that in Ethiopia sustainable growth of woman in terms of its population which is the half of the total population, is much less. This study suggested this situation must be analyzed thoroughly and the Government should take proper measures to identify the actual reason for such occurrences.

According to Bag et al. (2020) in the context of sustainable growth the globalization effects on business have led to a focus on sustainable development. The Companies Act, 2013 mandates a 2% contribution of average profit for 3 years toward societal betterment. This study analyzes relationship between Indian bank, CSR and profitability. The study revels in terms of cost-effectiveness relationship between CSR and sustainably is significant. Such a type of contribution may upgrade and motivate to set up venture ideations.

Mikkola (2005) identified various factors associated with gender inequality, including economic development, cultural limitations, legal and traditional practices, resource distribution, polygyny, labor market access, education, finance, and political power. Goel (2019), explained in her study that the gender and growth issue is important globally, with women being the focus of gender development. The government has launched various schemes to address this, but societal involvement is also necessary. There should have equal chances for women and be involved in leadership and policymaking roles to reduce gender disparities.

3. Objectives of the Study

- To discuss the overview of Indian women empowerment and global context in general.
- To figure out various Govt. and Non-Govt. regulations and initiatives for development of women empowerment.
- To figure out women empowerment through digitization in the context of entrepreneurship.
- To find challenges and future directions for development of women empowerment.

4. Government Initiative Toward Promoting Women Empowerment

4.1 Legislative Acts

On protection of the rights of women, several legislative acts have been passed since the Indian constitution was formed in the year 1956. The issues addressed in those regulations were related to marriage, domestic violence, education and upliftment of social status of women in the society. The Hindu Succession Act, 1956 granted equal rights regarding inheritance of property in the Hindu Undivided Families (HUF). The Maternity Benefit Act, 1961 ensured maternity benefits for women employees. The relaxation of 12 to 26 weeks was increased in the Maternity Benefits (Amendment) Act, 2016. Gender-based discrimination against disbursement of wages was eliminated by the passing of the Equal Remuneration Act, 1976. Women education was emphasized by implementation of the national education policy of 1983. It stated the role of women and importance of their education in the society. Legal protection to women facing domestic violence was addressed by the Protection of Women from Domestic Violence Act, 1992. The legislative step toward addressing the issue of sexual harassment at the workplace of women was addressed by the Protection of Women from Sexual Harassment at Workplace Act, 2005. The legal age for marriage was fixed at 18 for females [Section 2(a)] by the Prohibition of Child Marriage Act, 2006. The prevention of violence toward women such as prevention of acid attacks was done by the amendment in the Indian Penal Code by the Criminal Laws (Amendment) Act, 2013. The Farmers' Produce Trade and Commerce (Promotion and Facilitation) Act, 2020 opened new avenues for women in the agricultural sector. Another major initiative for saving the marriage of Muslim women was taken by implementation of the triple-talaq bill, 2021.

4.2 Women Empowerment Initiatives

The Commission on the Status of Women (CSW) was set up for promotion of gender equality and empowerment of women. This was a global initiative taken by the United Nations Economic and Social Council (ECOSOC). It was established on June 21, 1946, by ECOSOC resolution 11(II) (UN Women, 2024). In India, The National Commission for Women was built under the National Commission for Women Act, 1990. The major objectives of the commission were to review the constitutional rights and legal safeguards for women, recommend legislative measures and to advise the government on all policy matters affecting women (National Commission for UN Women, 2024). The National Commission for Women was set up in India as a statutory body in January 1992 under the National Commission for Women Act, 1990 (20 of 1990). The major objectives were to revive the Constitutional and legal safeguards for women, recommend remedial legislative measures, and advise the Government on policy matters affecting women. Regarding the nutritional, health and self-development of girls aged 11–18 years, the Ministry Women and Child Development (MWCD), launched Kishori Shakti Yojna in the year 2001, which was looked after by the gram Anganwadi Centers. It aimed toward providing adolescent girls with a non-formal stream of education to improve their life-skills, and to promote awareness of health, hygiene, nutrition and family welfare. It also takes measures toward marriage of a girl after attainment of 18 years of age or even later, so that they have a better understanding of environmental and family issues. To enhance the safety and security of women in India, it was implemented through a "One Stop Center (OSC) Scheme" in the year 2015. It aimed to facilitate women who are victim of violence should be provided services like police facilitation, medical aid, legal and psychosocial counseling and temporary shelter. Beti Bachao, Beti Padhao [BBBP] (Save the Girl Child, Educate the Girl Child) campaign, which aimed to increase the sex ratio at birth. This initiative was toward reduction in gender discrimination and women disempowerment. The major objectives of the BBBP scheme were to uplift child gender ratio, gender equity and women economic independence, establish equal opportunity of irrespective of gender, protect girl child, regularize of education among girl child (IBEF, 2024).

5. Economic Empowerment

The Rastriya Mahila Kosh was framed up by the Ministry of Women and Child Development in the year 1993 as a national autonomous body for promoting socioeconomic empowerment of women. It mainly promoted empowerment in the rural and unorganized sector by providing concessional and collateral free micro credits. However, the scheme was closed in November 2020 to avoid duplicity of efforts. The financial support schemes such as Sukanya Samriddhi Yojana (SSY) was introduced as another part of BBBP scheme, to provide financial assistance for girl children; Mahila e-Haat for promoting entrepreneurship among women. Platforms such as Government e-Marketplace (GeM) and small markets managed by National Small Industries Corporation (NSIC) to support women entrepreneurs. The Ministry of Finance (MoF) launched the Pradhan Mantri Mudra Yojana (PMMY) to facilitate loans via Micro Finance Institutions, Non-Banking Finance Companies and Small Finance Banks. This scheme empowered women entrepreneurship (citations needed). The Pradhan

Mantri Jan Dhan Yojana (PMJDY) was launched in the year 2018 to enhance financial inclusion and to benefit women in rural and economically backward areas. To promote these entrepreneurial initiatives, NITI Aayog developed a Women Entrepreneurship Platform (WEP) in the year 2019 to support women entrepreneurship and foster economic development.

5.1 Education and Skill Development

The Sarva Siksha Abhiyan (SSA) was launched in 2002. The social assessment revealed that girls remained educationally backward as compared to boys. Thus, SSA made girl's education compulsory, bridging the gender gaps and attaining academic completion. The Rajiv Gandhi Scheme for Empowerment of Adolescent Girls (SABLA) was launched by the MCWD in 2011. The main purpose was to enable adolescent girls (aged 11–18 years) for self-improvement by improving their nourishment and hygiene; and to encourage them to practice and educate them about adolescent reproductive, health, disease related to this and safe practices family and childcare. It also aimed toward upgrading their home-based skills, life-skills, tied along with a national skill development program. Skill improvement and empowerment programs include Mahila Shakti Kendra (MSK) scheme, launched in the year 2017 by MWCD (WCD, 2024). The major aim was to make women realize their full potential. Under this scheme, convergent support was proposed for equal access to healthcare, employment, safety and security and digital literacy at Gram Panchayat level, covering the Anganwadi Centers.

5.2 Health and Well-Being

The Janani Suraksha Yojna (JSY) was launched by the Ministry of Health and Family Welfare (MoHFW, 2024) as a part of National Rural Health Mission, to reduce maternal and neonatal mortality by promoting institutional delivery and postdelivery care of below poverty line women. Other initiatives for pregnant women include Pradhan Mantri Surakshit Matritva Abhiyan (PMSMA), launched by MoHFW in the year 2016. It aimed to provide fixed day quality antenatal services to women in their second/third trimesters of pregnancy on the ninth day of every month. The National Nutrition Mission (Poshan Abhiyan). It focused on improving nutritional stances for pregnant women. Other health-related schemes included Pradhan Mantri Ujjwala Yojna (PMUY), launched by the Ministry of Petroleum and Natural Gas (MoPNG, 2024). It addressed the problem of providing clean cooking fuel, indirectly contributing to improvement in the health of women by reducing indoor air pollution.

6. Where Are the Indian Women Standing Today? Gender Gap Index: The Indian Scenario

The World Economic Forum (WEF) launched a new methodology "Gender Gap Index" in the year 2006, to measure and capture the differences in men and women in four exclusive and critical areas – *economic participation and opportunity, educational attainment, health and survival and political empowerment*. The Gender Gap Index is the average of these four sub-indices as measured by the weighted average of the variables' ratio in the subindices. For example, if there are 20% females in managerial positions as compared to 80% of males, then the ratio 0.25 (20 females/80 males) is assigned to this variable.

The *economic involvement and facility* measured the ratios of female labor force over men, wage equality between men and women, estimated female income over male income, female legislators, senior officials and managers over male, female professional and technical workers over male. The *educational attainment* measured the ratios of female literacy over male, female net primary enrollment over male, female net secondary enrollment over male and female net tertiary enrollment over male. Health *and survival* contain the ratio of female healthy life expectancy over males and birth ratio, as compared to female over male. The *political empowerment* measures the ratio of women seats in parliament to men, women at ministerial level as compared to males and number of years of a female head of state as compared to male head of state (female-to-male ratio). The gender gap index is bound into ratio 0 to 1, where 0 equals complete inequality and 1 equal complete equality between males and females (Hausmann, 2006).

The average overall gender gap index shows an increasing trend in gender equality index. Fig. 2.1 shows that the health and survival index along with

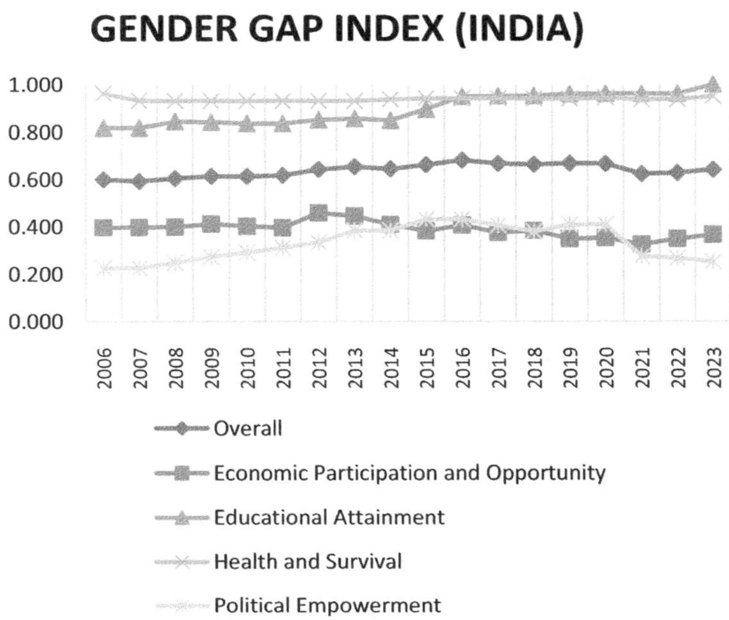

Fig. 2.1. Trend in Average Scores (Gender Gap Index: India).
Source: World Economic Forum, Global Gender Gap Report, *Various Issues*.

educational attainment index is above, and economic participation and opportunity index along with political empowerment index is below the average gap. The overall gender equality index has increased from 0.601 in 2006 to 0.643 in 2023 (6.971%). It indicates that the gender gap in India is decreasing over time. The economic participation and opportunity index shows a decreasing trend. The value of this index decreased from 0.397 in 2006 to 0.367 in 2023 (−7.487). It indicates that the gender gap is increasing in terms of economic participation in India. The educational attainment showed the highest increase in the scores, which increased from 0.819 in 2006 to 1 in 2023 (22.13%). It was perhaps due to perhaps that initiatives such as "Beti Bachao, Beti Padhao" in India was a success and the gap was bridged between male and female enrollment ratios. The health and survival index declined from 0.962 in 2006 to 0.950 in 2023 (−1.288%). It indicated that the health and wellness initiatives were not that successful. However, the value in recent years saw an increase from the 2022 score to that of 2023. The political empowerment index was the fastest to decrease the gender gap from 2006 to 2015. Thereafter, a drop in the score was seen, which was significant in the year 2021. The increase in the political empowerment index was observed from 0.227 in 2006 to 0.253 in 2023 (11.65%).

The relative position of India in the global scenario for the segment-wise index is shown in Fig. 2.2. The figures indicate how better India bridged the gender gap

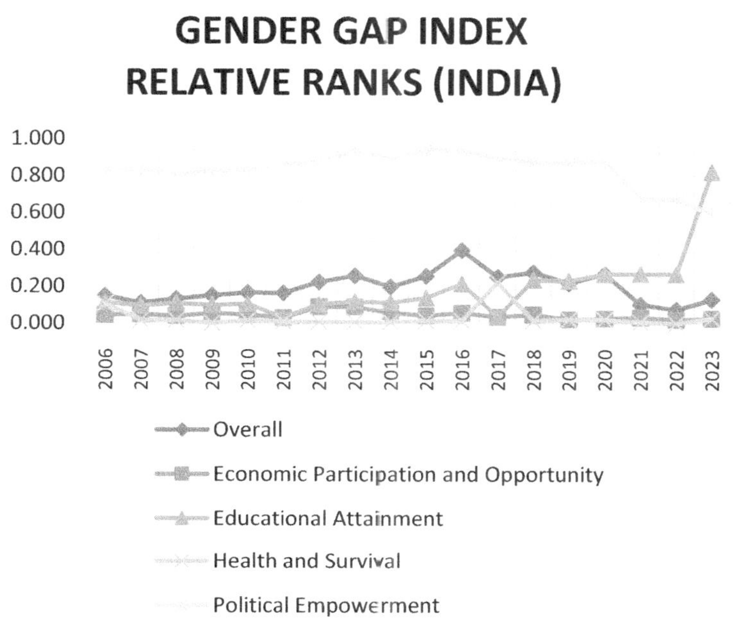

Fig. 2.2. Relative Position of India as per Global Ranks (Gender Gap Index: India). *Source:* World Economic Forum, Global Gender Gap Report, *Various Issues.*

as compared to the sum of countries. For example, in 2006, the total number of countries in the report was 115, and India ranked 98. So, the relative score was computed as: *[1−(98÷115) = 0.15]*. In this way, the relative position was bound to a score between 0 and 1, with the scores close to 1 indicating better global position in reducing gender gap. As per Fig. 2.2, it can be observed that there is an increasing trend in the overall ranks. The political empowerment index was the highest among the worldwide countries till the year 2020. From 2021, an increase in the gender gap index was observed, leading to a fall in the global position.

The educational attainment was seen to be the most rapid in moving up the global ladders, the relative position in the global front, India's relative score moved from 0.11 in 2006 to 0.82 in 2023. The economic participation and opportunity index mostly saw a flatter trend. It was the second least position; India was seen as compared to the global scenario. The least scores were obtained by health and survival index. However, it was seen that the health and survival index shot up in the year 2017.

The Present Scenario: A Comparison Between India versus Average Global Gender Gap Index 2023.

Fig. 2.3 shows the four dimensions of the gender gap index of India for the year 2023, as compared to the global average of 2023 (Global Gender Gap Report, 2023). It can be observed that the gap in political empowerment is still very wide in India (0.253) as well as worldwide (0.221). The economic

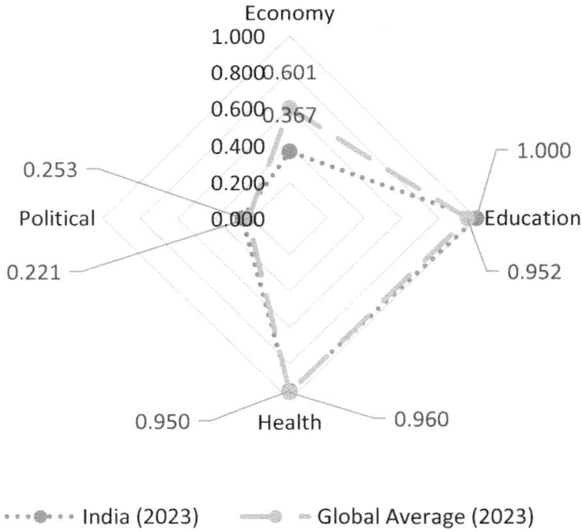

Fig. 2.3. Gender Gap Index (India 2023 vs. Global Average 2023). *Source:* World Economic Forum, Global Gender Gap Report, *Various Issues*.

participation is lower in India (0.367) as compared to the global index (0.601). Regarding education, it was observed that India (1.000) surpassed the world average (0.952). In health and care, India (0.950) was observed very close to the world average (0.960). It can be concluded that the political and economic gender gap is still high in the world scenario, whereas the education and health gender gap has largely decreased. Hence, economic and political empowerment are the areas which must be worked upon to reduce gender inequality.

The Gender Gap Index was examined in four quadrants: economic participation and opportunity, educational attainment, health and survival and political empowerment. It was found that the worldwide gender gap was the highest in economic participation and political empowerment and was the lowest in educational attainment and health and survival metrics. The gender gap in economic status of women was found to be high in India as compared to the world average. Educational empowerment was seen to be rapidly improving as compared to the world average. Therefore, economic and political empowerment are the areas which are to be worked upon to reduce the gender gap in these corresponding areas.

7. Entrepreneurship and Digitization: Is This a Path to New Age Empowerment?

Empowerment is the outcome of a regressive process that is strongly based on economic, educational, social, and political progress cumulatively. Entrepreneurship is one the approach to empower women of different categories to strengthen them economically and socially. Various government institutions, NGOs, private bodies are organizing programs to skill women, be independent and excel in their fields. They also help to build a network, awareness campaign and upskill an individual by creating opportunities for them. Since there is a breakout of digital evolution globally, and it's been promoted to get along with the process, it directly revolved the general course of life of an individual. This one change has its effect on every aspect of lives and basically it boosts up all the factors and speeds up the whole process way faster in its way. It is unthinkable to empower without the support of digital implements in an individual's life or at large. A study in China by Wang and Chen (2023) stated that for emerging, empowering and transforming life, digitization is a powerful tool in today's age. The impact of digitization on empowerment is also significant. From streamlining the business process, enhancing market reach using digital marketing, adoption of digital financial services, changing the approaches of empowerment from traditional to new age systems. Sharma and Patra (2023) explore the whole architecture of the socio economy element is changing its nature with the introduction of digitization at large in day-to-day operations. To boost up the economy especially small scale, micro and MSMEs role of digitization is impactful. Han (2020) states that inclusion of digital payment empowers and independent an individual in its economic growth since the process became easy, less time taking and hassle free with no time barrier. The Government is taking the initiative to promote the use of digital payment. The study has also found the use of digital options rather than old physical money transactions, keep the business more transparent which cumulatively leads to better financial inclusion. A case study of Pallimangal (a unit of Ramakrishna Mission Kamarpukur, West Bengal) has been discussed in Box 2.1 (Das & Ray, 2016).

Box 2.1. Promoting Women Empowerment: A Case Study on Pallimangal (A Unit of Ramakrishna Mission, Kamarpukur)

Ramakrishna Mission is a philanthropic and spiritual organization, having its headquarters at Belur Math (West Bengal). One of its units is situated in Kamarpukur, West Bengal, named Pallimangal. This case study on Pallimangal is a means to understand the promotion of women empowerment. The Pallimangal Mission was established in the year 1980. It basically emphasized on physical, mental and spiritual development. The key initiatives of the mission were to set up Agro-based industry to promote scientific agriculture and apiculture projects, setting up jute handicraft projects, build food processing plants to minimize food wastage, mini jute spinning, weaving, and incense stick making projects to create income opportunities. In this unit of Ramakrishna Mission, it was observed that out of total 222 workers, 198 were women. This was the first Mission toward Women Empowerment during 2016. Setting up of the mission reduced student's dropout rates due to women education. As of now, many vocational trainings for girls are provided such as tailoring, knitting, computer training, embroidery, weaving are provided to promote self-reliability among women and to promote women empowerment in the society.

Timotheou et al. (2022) stated education is a pillar of the development of a country. It is important to have access to education for every individual for future nations. A study exploring incorporation of digitization in education in rural India has empowered the students to be future ready, providing quality education resources ultimately builds confidence, knowledge among students which leads to ability in individuals. It leads to a developed and stronger nation ahead. In the same line of context, a study by Shetty and Hans (2019) conducted on the impact of digitization on education concludes that the digital education initiative taken by the Skill Development and Employability cell of the Indian Government overall improves the earnings ability of the workforce community.

A study by Wohlfart and Wagner (2023) explores that social media plays a pivotal role fostering social awareness, connecting communities and channelizing voices just in a click. Impacts of social media on social factors are significant. It gives a platform to each individual to express themselves and at the same time use this as a powerful tool to reach globally. Anyhow it helps to access global resources too easily, that helps to make decisions in a much more constructive way. Savaget et al. (2018) depicts taking part in democratic process of a country that empowers the citizens. The incorporation of digitization has increased awareness, transparency and participation of the citizen which leads to empowerment. The research also suggested that implementing AI-based technologies in the process reduced the gap between government-citizen interaction and capture their views, leading to more accountable, transparent and effective decisions.

8. Conclusion

The legislative acts to safeguard women's rights and empowerment spans in the last two decades, beginning from the Hindu Succession Act, 1956, which safeguarded equal inheritance rights for women in Human Undivided Families (HUFs). Subsequently, Maternity Benefit Act, 1961 (amended 2016) ensured maternity benefits for women by extending the earlier maternity leave from 12 weeks to 26 weeks. The Equal Remuneration Act of 1976 eliminated gender-based discrimination on wages. Other legislative measures were taken to address various gender related issues such as domestic and workplace harassment and child marriage. Initiatives such as Beti Bachao, Beti Padhao scheme in the year 2015 aimed toward improvement in the child sex ratio, ensure gender equality and women empowerment, prevent gender-based elimination, ensure the survival of girl child and encourage education and participation of the girl child. Economic empowerment initiatives such as the Rashtriya Mahila Kosh and entrepreneurial platforms like Mahila e-haat were set up to promote the economic empowerment of women in the society. The Start Up India Scheme enabled women to start their own small business ventures. Education and skill development programs like Sarva Sikhsha Abhiyaan and Skill India Mission also promoted women education. Health and well-being initiatives such as Janani Suraksha Yojana in the year 2005 encouraged institutional deliveries for pregnant women; the National Health Mission aimed to improve maternal and child health. Other initiatives like the Pradhan Mantri Ujjwala Yojna contributed toward reducing indoor pollution by providing access to clean cooking fuel.

Since India is a vast country with a vivid culture in it with a lot of discrimination and difference it is hard to distribute equal rights to everyone and serve equal opportunity to every individual. Instead of depending on job seeking, a job creator is more needed in this situation. Nothing could be more helpful than to be an act of entrepreneurial attitude which is basically a skill-based task rather than education. Digital evolution comes hand in hand in this context to reach out to the customer globally is an add on opportunity for a creator. All the effort together will build a better nation and a better place for women in near future where women will get equal opportunity and a better life.

References

Bag, S., Ray, N., & Manna, A. (2020). Nexus between sustainable development and CSR—An empirical study on Indian nationalized banks. In *circular economy and engineering. Management and industrial engineering*. Springer. https://doi.org/10.1007/978-3-030-43044-3_4

Banerjee, A., Ray, N., & Ghosh, S. (2011). Issues and challenges of tourism development in West Bengal, India [Conference session]. In *International Conference on Management (ICM 2011) Proceeding*.

Bayeh, E. (2016). The role of empowering women and achieving gender equality to the sustainable development of Ethiopia. *Pacific Science Review B: Humanities and Social Sciences*, *2*(1), 37–42. https://doi.org/10.1016/j.psrb.2016.09.013

Braverman-Bronstein, A., Ortigoza, A. F., Vidaña-Pérez, D., Barrientos-Gutiérrez, T., Baldovino-Chiquillo, L., Bilal, U., Friche, A. A., Diez-Canseco, F., Maslowsky, J., Vives, V. A., & Diez Roux, A. V. (2023). Gender inequality, women's empowerment, and adolescent birth rates in 363 Latin American cities. *Social Science & Medicine*, *317*, 115566. https://doi.org/10.1016/j.socscimed.2022.115566

Bryan, E., & Mekonnen, D. (2023). Does small-scale irrigation provide a pathway to women's empowerment? Lessons from Northern Ghana. *Journal of Rural Studies*, *97*, 474–484. https://doi.org/10.1016/j.jrurstud.2022.12.035

Das, D. K., & Ray, N. (2016). Women empowerment for promoting rural economy in West Bengal: A study on pallimangal (A Unit of Ramakrishna Mission Kamarpukur). In Ş. Erçetin (Ed.), *Women leaders in chaotic environments. Lecture notes in social networks*. Springer. https://doi.org/10.1007/978-3-319-44758-2_14

Dey, P., Ray, N., Dash, C. K., Das, D. K., & Das, P. K. (2022). Mapping of socio-economic condition and occupational structure of Patuas in the context of Pattachitra (scroll painting): A study from Bharatpur Village in Bankura, West Bengal, India. *Manager – The British Journal of Administrative Management*, *58*(153), 254–265.

Global Gender Gap Report. (2023). *World Economic Forum*.

Goel, A. (2019). Women and development: A literature review. *Academic Journal of Social Sciences*, *1*(1), 18–21.

Han, X. (2020). Women's empowerment in digital media: A communication paradigm. In *Handbook of communication for development and social change* (pp. 379–394). https://doi.org/10.1007/978-981-15-2014-3_79

Hausmann, R. (2006). *The global gender gap report 2008*. World Economic Forum.

IBEF. (2024). *Beti Bachao, Beti Padhao (BBBP) scheme*. India Brand Equity Foundation. https://www.ibef.org/government-schemes/beti-bachao-beti-padhao

Jie, H., Zaman, S., Zaman, Q. U., Shah, A. H., & Lou, J. (2023). A pathway to a sustainable future: Investigating the contribution of technological innovations, clean energy, and women's empowerment in mitigating global environmental challenges. *Journal of Cleaner Production*, *421*, 138499. https://doi.org/10.1016/j.jclepro.2023.138499

Kumar, D., Zulkifli, N., & Ray, N. (2023). Identifying critical factors for sustainable tourism and local community development: Evidence from select destinations in Bangladesh. *ASEAN Journal on Hospitality and Tourism*, *21*(1), 124–135. https://doi.org/10.5614/ajht.2023.21.1.09

Malhotra, A., Schulte, J., & Patel, P. (2009). *Innovation for women's empowerment and gender equality*. International Center for Research on Women (ICRW).

Mikkola, A. (2005). Role of gender equality in development – A literature review. *SSRN Electronic Journal*. https://doi.org/10.2139/ssrn.871461

Ministry of Agriculture and Rural Development. (2008). Rural capacity building project gender mainstreaming guideline. https://desagri.gov.in/documents-reports/agricultural-statistics-at-a-glance/agricultural-statistics-at-a-glance-2008/

MoHFW. (2024). *Government of India*. https://mohfw.gov.in/

MoPNG. (2024). *Government of India*. https://mopng.gov.in/en

Mukherjee, T., Das, A., Ray, N., Das, D. K., Das, P., Dey, P. (2022). Socio-cultural impact of tourism: A study from Shantiniketan in West Bengal. *Manager – The British Journal of Administrative Management*, *58*(155), 204–217.

Ray, N., & Das, D. K. (2011). 3HS tourism: An alternative approach of developing rural tourism at Kamarpukur: An empirical study. *South Asian Journal of Tourism and Heritage Referred International Journal*, *4*(2), 128–144.

Ray, N., & Das, D. K. (2017). Tourism and its impact on quality of life. *International Journal of Sustainable Economies Management*, *6*(2), 51–66. https://doi.org/10.4018/ijsem.2017040105

Salia, S., Hussain, J., Tingbani, I., & Kolade, O. (2017). Is women empowerment a zero-sum game? Unintended consequences of microfinance for women's empowerment in Ghana. *International Journal of Entrepreneurial Behavior & Research*, *24*(1), 273–289. https://doi.org/10.1108/ijebr-04-2017-0114

Savaget, P., Chiarini, T., & Evans, S. (2018). Empowering political participation through artificial intelligence. *Science and Public Policy*, *46*(3), 369–380. https://doi.org/10.1093/scipol/scy064

Sharma, S., & Patra, N. (2023). Do demographic factors influence the buyer's attitude towards online purchase of pharmaceutical products? A study with special reference to Kolkata. *Journal of Academic Advancement*, *2*(1), 46–55. https://doi.org/10.58574/jaa.2023.v2.i1.05

Shetty, S., & Hans, V. B. (2019). Education for skill development and women empowerment. *EPRA International Journal of Economic and Business Review*, *7*(2), 63–67.

Tanzile, R., Domapielle, M. K., & Fielmua, N. (2023). Empowering women for sustainable development through semi-mechanized sheabutter processing in rural north-western Ghana. *Scientific African*, *21*, e01790. https://doi.org/10.1016/j.sciaf.2023.e01790

Timotheou, S., Miliou, O., Dimitriadis, Y., Sobrino, S. V., Giannoutsou, N., Cachia, R., Monés, A. M., & Ioannou, A. (2022). Impacts of digital technologies on education and factors influencing schools' digital capacity and transformation: A literature review. *Education and Information Technologies*, *28*(6), 6695–6726. https://doi.org/10.1007/s10639-022-11431-8

UN Women. (2024). Commission on the status of women. https://www.unwomen.org/en/

Wang, X., & Chen, J. (2023). An empirical study on the indirect empowerment of economic development by the digital economy-based on the perspective of China's domestic economic cycle. *Journal of the Knowledge Economy*. https://doi.org/10.1007/s13132-023-01445-z

WCD. (2024). https://wcd.gov.in/

Wohlfart, O., & Wagner, I. (2023). Teachers' role in digitalizing education: An umbrella review. *Educational Technology Research and Development*, *71*(2), 339–365. https://doi.org/10.1007/s11423-022-10166-0

Chapter 3

Bridging Theory to Practice: Empowering Indigenous Businesses Through Innovative Analysis of Audited Financial Statements

Erisa Avdimetaj[a], *Enkeleda Lulaj*[a] *and Gina B. Alcoriza*[b]

[a]University Haxhi Zeka, Kosovo
[b]University of the Philippines, Philippines

Abstract

Empowering indigenous businesses through innovative analysis of audited financial statements (AFS) is critically important in today's economic landscape. This chapter investigated the impact of AFS on the financial performance of indigenous businesses. Data analysis was conducted using the distance method in businesses located in Kosovo, while its processing was performed through the Statistical Package for the Social Sciences (SPSS) program. The results highlight significant variability in the influence of AFS on financial variables. Key differences among various financial metrics and their implications for company performance are identified. It underscores the crucial importance of monitoring and analyzing financial statements are crucial for sustaining financial stability. Furthermore, it emphasized the necessity of transparency and reliability in financial reporting, alongside the adoption of innovative strategies to enhance financial performance. The importance of comparing indigenous and nonindigenous companies was also emphasized for future research.

Keywords: Empowering indigenous businesses; auditing; financial management; financial statements; innovative strategies on finance; accounting

1. Introduction

To know and understand how companies perform and how they position themselves in the market, key and very important documents are the financial

Indigenous Empowerment through Human-Machine Interactions, 35–52

Published under exclusive licence by Emerald Publishing Limited
doi:10.1108/978-1-83608-068-820251003

statements that highlight their performance. Moreover, their audit highlights a very important resource for economics and finance; therefore, the use is a part that goes from theory to practice of financial management and is a very important aspect of the business world. Additionally, Lulaj (2023) suggests businesses should also prioritize facilitating transactions, offering diverse payment methods, and ensuring efficient transportation from the business to consumers, prompt handling or requests. Kontesa et al. (2024) confirm stakeholder theory and institutional theory. According to Lulaj, Hysa, and Panait (2024), it is emphasized that the digitization of finance-accounting (FAD) significantly affects the adoption of advanced financial accounting systems (AFAS) and the development of accounting 2.0 and FinTech (AFT), revealing key relationships that highlight the role of systems digital, in increasing financial access, adaptability, and sustainability.

According to Pirveli (2022), audited financial statements (AFS) are found to take longer to file, whereas an unqualified opinion damages the practice by not taking the theory of reporting standards as a basis. Then, Wines (2006) points out that participants' perceptions of auditor independence were systematically associated with the identified dimensions of financial statements in practice. Lulaj et al. (2023) emphasize the importance of managing business liabilities, improving assets, and maximizing net profit. Additionally, Lulaj and Iseni (2018) highlight the positive impact of cost-volume-profit analysis on financial decision-making. Furthermore, Lulaj et al. (2021) note that large firms tend to have a stronger financial position compared to small firms.

Therefore, empowering indigenous businesses through innovative analysis of AFS is of critical importance in contemporary economic landscapes. While the existing literature provides valuable insights into financial statement analysis, this study contributes new perspectives by focusing specifically on indigenous businesses. The objective is to link theoretical knowledge with practical applications, aiming to discover insights that can enhance the financial sustainability and strategic decision-making capabilities of indigenous enterprises. This research aims to address the questions: How do AFS affect the financial performance of indigenous businesses? What specific variables are most affected and how does this affect business results? By examining this gap in the literature, this study not only fills a substantial knowledge gap but also highlights the importance of tailored financial strategies for indigenous economic empowerment.

2. Review of Literature

In the wake of the ongoing challenges presented by the pandemic, unlike in previous years, there was a big difference between practice and theory. To survive the closed market situation, firms began offering online services, taking their performance in a new direction, particularly for Private Venture Firms (PVFs). As highlighted by Lulaj et al. (2022) in her research on the impact of the pandemic, consumers are faced with limited financial resources. According to Menzies et al. (2024), it is emphasized that indigenous businesses led by women use a theoretical approach based on the concept of misfortune to understand the development and sustainability of the business in the face of difficulties. Moreover, Boakye et al. (2024) emphasized that indigenous economic risk management practices positively affect micro and small informal firms during global crises. Lulaj, Dragusha, and Lulaj (2024) find that in developing economies, sales excellence and sales capability drive profit growth, market alignment accelerates sales through accurate forecasting, strategic responsiveness aligns sales with corporate goals, and dynamic sales management motivates salespeople, collectively driving companies' success in dynamic markets. Brock et al. (2023) noted that indigenous and resource industry collaborators face challenges in working together. In addition, the transformative power of new information and communication technologies has driven a significant shift toward e-government (Lulaj et al., 2022; Lulaj & Dragusha, 2022). Fisher et al. (2004) find that a great deal of concern has been expressed in practice in the auditing profession regarding the construction of the audit report (FSAs published on web pages). Rochmah Ika and Mohd Ghazali (2012) emphasize that the efficiency of auditors in practice tends to reduce the time of financial reporting. Malik et al. (2022) emphasized that small and medium-sized businesses use the theory of dynamic capabilities of indigenous business models to analyze innovative analysis and sustainability.

According to Whittington (2016), in an audit of financial statements, auditors are required to ensure high reliability of reports by collecting facts and evidence of companies' performance. Furthermore, Haragaja (2014) emphasized that external audit often focuses on the financial statements of economic entities. External auditing is usually performed by auditing firms and authorized auditors who are certified by relevant professional bodies (Haragaja, 2014). Furthermore, Warren (2017) pointed out accountants must behave ethically so that the information they provide to users is reliable and, thus, useful for decision-making. According to Lulaj and Minguez-Vera (2024), factors such as strategic clarity, continuous monitoring, effective working capital management, accurate financial decision-making, and technological advances are highlighted as the main contributors to positive cash flow. As for Loshaj (2015), AFS is compiled every month, quarter, six-monthly, semi-monthly, and annually. Moreover, Zhang et al. (2021) emphasize that profitability, financial success, and solvency are closely interconnected with achieving sustainable profit. The same perspective was emphasized by Yang (2021), who noted that firms focus on utilizing services to drive innovations in analyzing their financial situation.

Furthermore, Chen (2022) offers a comprehensive analysis of the relationship between capital structure and financial performance, providing valuable insights that can enhance financial strategies. Hou (2022) emphasizes that globalization through science and technology, as one of its main practical objectives, is the stability of profit through innovations and the importance of financial statements compiled accurately, clearly, and easily in financial reading, enabling the prevention of risks to the balance sheet and success. As for cash flows according to Stobierski (2020), it depends on the firm's ability to operate in the market in the short and long term. Therefore, this statement is not considered a mandatory part of the monthly financial statements (Alamgir, 2019).

Fossung et al. (2020) show that audits and AFS are good indicators and predictors of audit expectations. According to McEnroe et al. (2017), the importance of facts and clear communication materials of financial statements in audited reports is emphasized. Also, McDonough and Yan (2023) emphasize that firms in practice should make capital investments based on the state of their assets. Vanasco (1996), the influence of auditing agencies and associations, has been emphasized to protect auditors' standards during the preparation of audit reports and analysis. Budisantoso and Kurniawan (2024) pointed out that if auditors fail to detect errors, then it affects the quality of the company's audit.

3. Methodology

This research aims to analyze the connection between theory and practice, aiming at the discovery of knowledge that can increase financial sustainability and decision-making skills in indigenous companies, considering the questions of how AFS affect the financial performance of indigenous businesses, as well as which items, financial or variables, are affected more. Therefore, it is emphasized that the main purpose is to investigate the impact of AFS on the performance of indigenous businesses.

The data are secondary data obtained from the AFS of indigenous businesses during the years 2018–2019 published in the National Audit Office in Kosovo, while their processing was done through the SPSS program using the distance method between variables or financial items in correlation analysis.

Hypothesis: There is distance between variables.

4. Findings and Discussions

In this chapter, as was emphasized in the literature and the methodology, the importance of the theory and practice of the audit of the financial statements (balance sheet, income statement, cash flow, change in equity) of the firms was analyzed, which was based and built under International Financial Reporting Standards (IFRS). Therefore, Karasioğlu et al. (2021) emphasize the importance of accounting ethics, which has a significant impact on the quality of audited reports in indigenous firms. Qualified and trustworthy managers are crucial for making correct decisions that ensure sustainable profit. However, Scapens (1990) points out that there is a lack of students on financial statements. Waterhouse and Tiessen (1978) point out that the organizational position and efficient design of accounting systems contribute to the profits of indigenous corporate derived from AFS. Lucianetti et al. (2018) emphasize the importance of analyzing unforeseen factors, such as decentralization and environmental uncertainty, in relation to organizational strategy for indigenous profit sustainability.

Table 3.1 shows balance sheet and its financial items which make a profit are: cash register, total from cash registers, bank accounts, total cash from banks, total cash in cash register and bank, finished goods inventories, assets, inventory, etc., while prepayment of tax on profit, total, vendor trade payables have suffered losses. The analysis provides valuable insights for indigenous businesses. It is evident that certain financial items have generated profits, including cash, bank balances, inventories, and miscellaneous assets. These items reflect successful financial management and revenue generation for indigenous firms. On the contrary, items that show losses must have accurate financial management.

According to Lulaj, Gopalakrishnan, and Kehinde Lamidi (2024), companies need to adapt immediately to environmental changes to make effective business decisions, as the lack of management skills and innovative strategies hinders this process, affecting market profitability and survival against competitors, emphasizing the need to harmonize strategies with evolving market demands to remain competitive and sustainable.

Table 3.1. Balance Sheet.

Financial Items	**2019 year Currency/ Euro**	**2018 year Currency/ Euro**	**The Difference**	**Profit/ Loss**
Cash and bank account				
Cash register	146,425	93,276	53,149	Profit
Card sales in transit	27,163	—	27,163	—
Total from cash registers	173,589	93,276	80,313	Profit
Money in the bank				
TEB	859,293	575,046	284,247	Profit
PCB	100,082	6	100,076	Profit
NLB	73,275	163,959	−90,684	Loss
RBKO	38,210	14,227	23,983	Profit
BKT	37,352	—	37,352	—
Total cash from banks	1,108,212	753,239	354,973	Profit
Total cash in cash register and bank	1,281,801	846,515	435,286	Profit
Stocks				
Finished goods inventories	5,260,012	4,546,457	713,555	Profit
Total inventories	5,260,012	4,546,457	713,555	Profit
Other short-term assets				
Prepayment of tax on profit	59,755	59,755	0	Loss
Prepayment of rental advances	—	83,199	83,199	—

(Continued)

Table 3.1. *(Continued)*

Financial Items	2019 year Currency/ Euro	2018 year Currency/ Euro	The Difference	Profit/ Loss
Prepayment for vehicle insurance	563	—	563	—
Total	60,318	142,954	−82,636	Loss
Total current assets	6,602,131	5,535,927	1,066,204	Profit
Property, plant, and equipment	8,248,985	3,807,272	4,441,713	Profit
Total long-term assets	8,248,985	3,807,272	4,441,713	Profit
Total assets	14,851,116	9,343,199	5,507,917	Profit
Accounts payable and others				
Vendor trade payables	244,994	261,492	−16,498	Loss
International supplier payables	3,279,494	1,928,639	1,350,855	Profit
Other payable accounts	18,810	—	18,810	—
Total	3,543,298	2,190,130	1,353,168	Profit

Source: National Audit Office (https://zka-rks.org/).

Table 3.2 shows balance sheet items, including cash in hand and sales in transit, which are part of the firm's current assets. Based on the data at the end of the analysis, a significant difference is observed between the variables (27,164.000 and 146,426.000). This large difference reduces the explanatory power of the model, leading to the acceptance of Hypothesis 1 (H1). Several suggestions and conclusions can be drawn from these findings. First, the significant difference

Table 3.2. Distance Between Variables for Cash in Hand.

	Proximity Matrix		
	Euclidean Distance		
	Total Cash	Cash	Sales With Card in Transit
Total cash	0.000	27,164.000	146,426.000
Cash	27,164.000	0.000	119,262.000
Sales with card in transit	146,426.000	119,262.000	0.000

Source: Source of data in tables: Authors' calculations in SPSS program.

Note: This is a dissimilarity matrix.

between cash in hand and sales in transit suggests a potential gap in the management of these resources. Additionally, changes identified in the liquidity report highlight the need for further analysis of firms adjusting their strategies and adopting indigenous innovations.

Table 3.3 shows the assets owned by the company in different banks, where it is observed that there is a significant difference between the variables (1,008,130.000, 1,034,937.000, 1,070,002 000, 1,070,860.000); therefore, the hypothesis of this research is confirmed. By conducting more in-depth research, indigenous companies can gain valuable insights into the financial strategies and practices that truly strengthen them. In conclusion, the analysis of AFS provides valuable insights to bridge gaps among theories and practice in empowering indigenous enterprises. While the assets held in TEB do not have a significant impact on the independent variable Y1, further research is needed to identify other variables that may have a stronger impact.

Table 3.4 presents the results of short-term liabilities (suppliers); it is emphasized that there is a difference between the financial items, once again strengthening the acceptance of the hypothesis. It is therefore suggested that it is crucial to understand the dynamics of suppliers, payables, and other factors that contribute to the empowerment of indigenous firms. By performing innovative analysis and taking into account various factors, indigenous businesses can make informed financial decisions and increase their overall empowerment in the local and international markets.

Table 3.5 presents the results for current liabilities including VAT, withholding taxes, contribution taxes, lease taxes, and current liabilities according to IFRS16. Analyzing the data, it is observed that there is no significant distance between the dependent variables (8,918,295.000, 9,117,264.000, 9,111,239.000, 9,109,477.000, 7,959,888.000). This indicates that there is a correlation between these variables, which helps to explain the independent variables. Based on these findings, hypothesis H0 is accepted, which indicates that there is a correlation between the dependent variables that helps to explain the independent variables. This means that current liabilities, including VAT, withholding taxes, contribution taxes, lease taxes and current liabilities according to IFRS16, have a significant impact on the independent variable Y1. It is suggested that indigenous be financially innovative in all their financial statements, financial analysis, and audit reporting.

Table 3.6 presents the results for the company's equity, including retained earnings from previous years, current year earnings, and dividends. Analyzing the data, we see that there is no significant distance between variables X27 and X28 (5,229,274.000 and 7,809,510.671), indicating a correlation between these variables. However, there is a significant distance with variable X26 (1.000), which indicates a weaker relationship with the other variables. These distances between the dependent variables reduce the explanatory power of the independent variables. Based on these results, hypothesis H0 is accepted, which states that variable X26 (annual profit) has the least effect on variable Y1 (retained earnings). In conclusion, the analysis of AFS provides insights into the distance among financial item and equity Y1 in empowerment of indigenous businesses. Although

Table 3.3. Distance Between Variables for Total Money.

	Proximity Matrix					
	Euclidean Distance					
	Total Money	**TEB**	**PCB**	**NLB**	**RBKO**	**BKT**
Total money	0.000	248,919.000	1,008,130.000	1,034,937.000	1,070,002.000	1,070,860.000
TEB	248,919.000	0.000	759,211.000	786,018.000	821,083.000	821,941.000
PCB	1,008,130.000	759,211.000	0.000	26,807.000	61,872.000	62,730.000
NLB	1,034,937.000	786,018.000	26,807.000	0.000	35,065.000	35,923.000
RBKO	1,070,002.000	821,083.000	61,872.000	35,065.000	0.000	858.000
BKT	1,070,860.000	821,941.000	62,730.000	35,923.000	858.000	0.000

Source: Source of data in tables: Authors' calculations in SPSS program. TEB-.

Note: This is a dissimilarity matrix.

Table 3.4. Distance Between Variables for Trade Payables and Other.

Proximity Matrix				
	Euclidean Distance			
	LI/P Commercial and Others	**Local Supplier**	**International Supplier**	**Other Accounts Payable**
LI/P commercial and others	0.000	3,298,304.000	263,804.000	3,524,488.000
Local supplier	3,298,304.000	0.000	3,034,500.000	226,184.000
International supplier	263,804.000	3,034,500.000	0.000	3,260,684.000
Other accounts payable	3,524,488.000	226,184.000	3,260,684.000	0.000

Source: Source of data in tables: Authors' calculations in SPSS program.

Note: This is a dissimilarity matrix.

there is a correlation between some variables, further research is needed to identify the variables that have the strongest effect on the dependent variable.

Table 3.7 presents the results for Sales, which includes the revenues generated by the company from product sales and other revenues as part of the income statement. Analyzing the data, we see that there is no significant distance between variables X31 and X32 (36,306,350.405 and 3,811,485.023), indicating a correlation between these variables. However, there is a significant distance with variable X30 (1,116,329.378), indicating a weaker relationship with the other variables. These distances between the dependent variables reduce the explanatory power of the independent variables. Based on these results, hypothesis H0 is accepted, which states that variable X30 (revenue from product sales) has the least impact on variable Y1 (sales). This implies that revenue from product sales has a weaker impact on total sales compared to other variables.

Table 3.8 presents the results for cost of goods sold, which includes beginning inventory, company purchases during the period, ending inventory, and firm inventory donations and write-offs. Therefore, after analyzing the distance correlation between the data, it is observed that there is a significant distance between the dependent variables, which indicates a higher level of explanatory power for the dependent variable. It is suggested that indigenous firms should analyze the local costs of selling goods and products.

Table 3.9 shows the distance between variables for items related to taxation of profits, including pretax profit for the year, interest paid in 2019, additional depreciation due to tax issues, nondeductible expenses, and adjustments for stolen money compensated by insurance companies. Analyzing the data, we find that there is no significant distance between variables X2, X3, and X5 (285,655.000,

Table 3.5. Distance Between Variables for Short-Term Liabilities.

	Proximity Matrix					
	Euclidean Distance					
	Other Short-Term Liabilities	**Value Added Tax**	**Withholding Taxes**	**Taxes on Contributions**	**Taxes Withheld From Rent**	**D Current Liabilities According to IFRS16**
Other short-term liabilities	0.000	8,918,295.000	9,117,264.000	9,111,239.000	9,109,477.000	7,959,888.000
Value added tax	8,918,295.000	0.000	198,969.000	192,944.000	191,182.000	958,407.000
Withholding taxes	9,117,264.000	198,969.000	0.000	6,025.000	7,787.000	1,157,376.000
Taxes on contributions	9,111,239.000	192,944.000	6,025.000	0.000	1762.000	1,151,351.000
Taxes withheld from rent	9,109,477.000	191,182.000	7,787.000	1762.000	0.000	1,149,589.000
Current liabilities according to IFRS16	7,959,888.000	958,407.000	1,157,376.000	1,151,351.000	1,149,589.000	0.000

Source: Source of data in tables: Authors' calculations in SPSS program.

Note: This is a dissimilarity matrix.

Table 3.6. Distance Between Variables for Retained Earnings.

	Proximity Matrix			
	Euclidean Distance			
	Retained Earnings	**Profits Carried Over From Previous Years**	**Profit of the Year**	**Dividends**
Retained earnings	0.000	5,229,274.860	1.000	7,809,510.671
Profits carried over from previous years	5,229,274.860	0.000	5,229,274.703	12,587,693.424
Profit of the year	1.000	5,229,274.703	0.000	7,809,510.299
Dividends	7,809,510.671	12,587,693.424	7,809,510.299	0.000

Source: Source of data in tables: Authors' calculations in SPSS program.

Note: This is a dissimilarity matrix.

Table 3.7. Distance Between Variables for Sales.

	Proximity Matrix			
	Euclidean Distance			
	Sales	**Income From the Sale of Products**	**Other Income**	**Returns**
Sales	0.000	1,116,329.378	36,306,350.405	38,114,854.023
Income from the sale of products	1,116,329.378	0.000	37,419,990.725	39,228,524.069
Other income	36,306,350.405	37,419,990.725	0.000	1,808,577.810
Returns	38,114,854.023	39,228,524.069	1,808,577.810	0.000

Source: Source of data in tables: Authors' calculations in SPSS program.

Note: This is a dissimilarity matrix.

272,895.000, 252,058.000). However, significant distances are observed with variable X1 (1,973,700.000) and variable X4 (106,924.000). These significant distances between variables reduce the explanatory power of the dependent

Table 3.8. Distance Between Variables for the Cost of Goods Sold.

Proximity Matrix					
	Euclidean Distance				
	COGS	**Stock First**	**Purchases During the Period**	**Ending Inventory**	**Donations and Destruction of Inventory**
COGS	0.000	20,907,641.239	1,773,643.011	19,416,081.790	26,165,701.029
Stock first	20,907,641.239	0.000	22,613,953.072	1,600,777.700	5,303,157.423
Purchases during the period	1,773,643.011	22,613,953.072	0.000	21,108,789.103	27,856,313.997
Ending inventory	19,416,081.790	1,600,777.700	21,108,789.103	0.000	6,749,635.745
Donations and destruction of inventory	26,165,701.029	5,303,157.423	27,856,313.997	6,749,635.745	0.000

Source: Source of data in tables: Authors' calculations in SPSS program.

Note: This is a dissimilarity matrix.

Table 3.9. Distance Between Variables for Taxation on Profit.

	Proximity Matrix					
	Euclidean Distance					
	Income Tax Expenses	Profit for the Year Before Tax	Loan Interest Expenses – Paid in 2019	Additional Depreciation Expense for Tax Purposes	Non Discountable Expenses	Adjustment for the Money Stolen on 14.12.2019 Compensated by the Insurance Company on 09.03.2020
Income tax expenses	0.000	1,973,700.000	285,655.000	272,895.000	106,924.000	252,058.000
Profit for the year before tax	1,973,700.000	0.000	2,259,355.000	2,246,595.000	1,866,776.000	2,225,758.000
Loan interest expenses – paid in 2019	285,655.000	2,259,355.000	0.000	12,760.000	392,579.000	33,597.000
Additional depreciation expense for tax purposes	272,895.000	2,246,595.000	12,760.000	0.000	379,819.000	20,837.000
Nondiscountable expenses	106,924.000	1,866,776.000	392,579.000	379,819.000	0.000	358,982.000
Adjustment for the money stolen on 14.12.2019 compensated by the insurance company on 09.03.2020	252,058.000	2,225,758.000	33,597.000	20,837.000	358,982.000	0.000

Source: Source of data in tables: Authors' calculations in SPSS program.

Note: This is a dissimilarity matrix.

variable Y1. Based on these results, hypothesis is accepted, which states that their distance among the financial items. This implies that variables X2, X3, and X5 have a stronger influence on the dependent variable Y1 than variables X1 and X4. These results highlight the importance of considering the impact of taxation on profits in understanding the financial performance and empowerment of indigenous businesses.

Table 3.10 presents the model validation results which reveal important findings. The intercept parameter estimate is 4,734,855,713, with a standard error of 101,372.6219. The 95% Wald confidence interval for the intercept ranges from 4,536,169,025 to 4,933,542,401, and the Wald Chi-Square value of 2,181.585 with 1 degree of freedom indicates a highly significant result ($p = 0.000$). This suggests a strong and significant baseline level of the dependent variable, Model Validation. Additionally, the scale parameter estimate is approximately 20,552,816,947,612, indicating variability in the data, with a confidence interval ranging from 2,895,141,608,699 to 145,905,914,658,135. These findings highlight the substantial variability and significant impact of AFS on the financial variables of indigenous businesses, emphasizing the importance of detailed monitoring, analysis, and strategic innovation in financial reporting and management.

5. Conclusions and Future Studies

This chapter revealed the significant impact of all financial statements (balance sheet, income statement, cash flow, and statement of changes in equity) on indigenous firms. The results showed the essential influence of the control and analysis of the AFS, recognizing them as a very important and primary tool for evaluating financial health and stability in the market. Then, based on the results of the tables for financial items, it is clear that indigenous companies should carefully consider the consequences of fluctuations in their financial variables and take appropriate actions and measures to improve financial performance. According to the obtained results, it is concluded that the dependent variables had a distance between them. A small distance between the variables indicates that the company is taking careful steps to increase its capital and capacity. Based on these results, the hypothesis was accepted, which means that there is distance between the variables.

5.1 Recommendations and Implications

Audit quality assurance: it is suggested that to ensure that the audit process maintains high-quality standards, indigenous firms should work with reputable auditors and conduct internal audits to maintain the integrity of financial reporting.

Ongoing financial analysis: it is suggested to promote a culture of continuous financial analysis within the company, the AFS should be reviewed periodically and their effect on the financial variables should be evaluated to identify the items that need improvement.

Table 3.10. Validation Model (Wald Confidence-Hypothesis Test).

Parameter Estimates							
			95% Wald Confidence Interval		Hypothesis Test		
Parameter	**B**	**Std. Error**	**Lower**	**Upper**	**Wald Chi-Square**	**df**	**Sig.**
(Intercept)	4,734,855.713	101,372.6219	4,536,169.025	4,933,542.401	2,181.585	1	0.000
(Scale)	20,552,816,947.612[a]	20,552,816,947.6117	2,895,141,608.699	145,905,914,658.135			

Source: Source of data in tables: Authors' calculations in SPSS program.

[a]Maximum likelihood estimate.

Notes: Dependent variable: ModelValidation
Model: (Intercept)

Employee training: it is suggested that training programs be offered for workers to understand the importance of the company's financial performance to survive the market.

Future studies may provide a more complete understanding of the role of AFS in financial management and decision-making, both within companies and across industries, years, and countries, and compare indigenous versus nonindigenous firms.

References

Alamgir, M. (2019, July 06). What is statement of changes in equity?

Boakye, A. A., Ankrah, D. A., Agyei-Holmes, A., Baidoo, I., & Sarpong, D. B. (2024). Indigenous risk management practices and their effects on small agribusiness firm growth amid economic crisis in Ghana. *Scientific African*, *23*. https://doi.org/10.1016/j.sciaf.2024.e02095

Brock, T., Reed, M. G., & Stewart, K. J. (2023). A practical framework to guide collaborative environmental decision making among indigenous peoples, corporate, and public sectors. *The Extractive Industries and Society*, *24*. https://doi.org/10.1016/j.exis.2023.101246

Budisantoso, T., & Kurniawan, H. (2024). The contagion effect of decreasing audit's quality on financial statement audit engagement: The Indonesian case. *Asia-Pacific Journal of Business Administration*, *16*(1), 63–76. https://doi.org/10.1108/APJBA-11-2020-0393

Chen, X. (2022). The fusion model of financial accounting and management accounting based on neural networks. *Mobile Information Systems*, 1587274. https://doi.org/10.1155/2022/1587274

Fisher, R., Oyelere, P., & Laswad, F. (2004). Corporate reporting on the internet: Audit issues and content analysis of practices. *Managerial Auditing Journal*, *19*(3), 412–439. https://doi.org/10.1108/02686900410524418

Fossung, M. F., Fotoh, L. E., & Lorentzon, J. (2020). Determinants of audit expectation gap: The case of Cameroon. *Accounting Research Journal*, *33*(4/5), 561–576. https://doi.org/10.1108/ARJ-12-2019-0241

Haragaja, M. (2014). Përgjesitë e auditorëve. https://knowledgecenter.ubt-uni.net/etd/2387?utm_source=knowledgecenter.ubt-uni.net%2Fetd%2F2387&utm_medium=PDF&utm_campaign=PDFCoverPages

Hou, X. (2022). Design and application of intelligent financial accounting model based on knowledge graph. *Mobile Information Systems*, 8353937. https://doi.org/10.1155/2022/8353937

Karasioğlu, F., Humta, H., & Göktürk, I. E. (2021). Investigation of accounting ethics effects on financial report quality & decision making: Evidence from Kabul-based logistic corporations. *International Journal of Management, Accounting and Economics*, *8*(3), 122–142. https://doi.org/10.5281/zenodo.4698983

Kontesa, M., Brahmana, R. K., & You, H. W. (2024). National culture, religiosity, and audited financial statements of small-scale MNCs. *Pacific Accounting Review*, *36*(2), 211–233. https://doi.org/10.1108/PAR-10-2022-0162

Loshaj, E. (2015). Balance sheet and income statement as financial consolidation instruments.

Lucianetti, L., Chiappett Jabbour, Ch, J., Gunasekaran, A., & Latan, H. (2018). Contingency factors and complementary effects of adopting advanced manufacturing tools and managerial practices: Effects on organizational measurement systems and firms' performance. *International Journal of Production Economics*, *200*, 318–328. https://doi.org/10.1016/j.ijpe.2018.04.005

Lulaj, E. (2023). A sustainable business profit through customers and its impacts on three key business domains: Technology, innovation, and service (TIS). *Business, Management and Economics Engineering*, *21*(1), 19–47. https://doi.org/10.3846/bmee.2023.18618

Lulaj, E., & Dragusha, B. (2022). Incomes, gaps and well-being: An exploration of direct tax income statements before and during Covid-19 through the comparability interval. *International Journal of Professional Business Review*, *7*(6), e0623. https://doi.org/10.26668/businessreview/2022.v7i6.623

Lulaj, E., Dragusha, B., & Hysa, E. (2023). Investigating accounting factors through audited financial statements in businesses toward a circular economy: Why a sustainable profit through qualified staff and investment in technology? *Administrative Sciences*, *13*(3), 72. https://doi.org/10.3390/admsci13030072

Lulaj, E., Dragusha, B., & Lulaj, D. (2024). Market Mavericks in emerging economies: Redefining sales velocity and profit surge in today's dynamic business environment. *Journal of Risk and Financial Management*, *17*(9), 395. https://doi.org/10.3390/jrfm17090395

Lulaj, E., Dragusha, B., Lulaj, D., Rustaj, V., & Gashi, A. (2021). Households savings and financial behavior in relation to the ability to handle financial emergencies: Case study of Kosovo. *Acta Scientiarum Polonorum. Oeconomia*, *20*(1), 35–48. https://doi.org/10.22630/ASPE.2021.20.1.4

Lulaj, E., Gopalakrishnan, A., & Kehinde Lamidi, K. (2024). Financing, investing in women-led businesses: Understanding strategic profits, entrepreneurial expectations by analysing the factors that determine their company success. *Periodica Polytechnica Social, Management Sciences*. https://doi.org/10.3311/PPso.22532

Lulaj, E., Hysa, E., & Panait, M. (2024). Does digitalization drive sustainable transformation in finance and accounting? *Kybernetes*. https://doi.org/10.1108/K-04-2024-1006

Lulaj, E., & Iseni, E. (2018). Role of analysis CVP (cost-volume-profit) as important indicator for planning and making decisions in the business environment. *European Journal of Economics and Business Studies*, *4*(2). https://doi.org/10.26417/ejes.v4i2.p104-120

Lulaj, E., & Minguez-Vera, A. (2024, September). Cash flow dynamics: Amplifying swing models in a volatile economic climate for financial resilience and outcomes. *Scientific Annals of Economics and Business*, *71*(3), 315–336. https://doi.org/10.47743/saeb-2024-0022

Lulaj, E., Zarin, I., & Rahman, Sh (2022). A novel approach to improving E-government performance from budget challenges in complex financial systems. *Complexity*. https://doi.org/10.1155/2022/2507490

Malik, A., Pereira, V., Budhwar, P., Varma, A., & Giudice, M. D. (2022). Sustainable innovations in an indigenous Indian ayurvedic MNE. *Journal of Business Research*, *145*, 402–413. https://doi.org/10.1016/j.jbusres.2022.03.009

McDonough, R. P., & Yan, C. J. (2023). Accounting policies in the public sector: Characteristics and consequences of accounting for capital assets. *Journal of*

Accounting and Public Policy, *42*(1). Article 107033. https://doi.org/10.1016/j.jaccpubpol.2022.107033

McEnroe, J. E., Du, N., & Sullivan, M. (2017). *An examination of the perceptions of auditors and chief financial officers of the proposed auditing standard involving other information in audit reports. Parables, myths and risks* (Advances in Public Interest Accounting, Vol. 20, pp. 63–84). Emerald Publishing Limited. https://doi.org/10.1108/S1041-706020170000020003

Menzies, J., Chavan, M., Jack, R., Scarparo, S., & Chirico, F. (2024). Australian indigenous female entrepreneurs: The role of adversity quotient. *Journal of Business Research*, *175*. https://doi.org/10.1016/j.jbusres.2024.114558

Pirveli, E. (2022). Corporate disclosure timing under IFRS: The case of emerging Georgia. *Journal of Financial Reporting and Accounting*. https://doi.org/10.1108/JFRA-12-2021-0443

Rochmah Ika, S., & Mohd Ghazali, N. A. (2012). Audit committee effectiveness and timeliness of reporting: Indonesian evidence. *Managerial Auditing Journal*, *27*(4), 403–424. https://doi.org/10.1108/02686901211217996

Scapens, R. W. (1990, September). Researching management accounting practice: The role of case study methods. *The British Accounting Review*, *22*(3), 259–281.

Stobierski, T. (2020, April 30). How to read & understand a cash flow statement.

Vanasco, R. R. (1996). Auditor independence: An international perspective. *Managerial Auditing Journal*, *11*(9), 4–48. https://doi.org/10.1108/02686909610150386

Warren, S. (Ed.). (2017). *The eastern Shawnee tribe of Oklahoma: Resilience through adversity*. University of Oklahoma Press.

Waterhouse, J. H., & Tiessen, P. (1978). A contingency framework for management accounting systems research. *Accounting, Organizations and Society*, *3*(1), 65–76. https://doi.org/10.1016/0361-3682(78)90007-7

Whittington, R. (2016). Principles of auditing & other assurance services.

Wines, G. (2006). The connotative meaning of independence in alternative audit contexts: An exploratory study. *Pacific Accounting Review*, *18*(1), 90–122. https://doi.org/10.1108/01140580610732787

Yang, L. (2021). Cloud data integrity verification algorithm for sustainable accounting informatization. *Mathematical Problems in Engineering*, 2330502. https://doi.org/10.1155/2021/2330502

Zhang, W., Xu, R., & Wang, L. (2021). Investigating the complex relationship between financial performance and company's green behavior: A comparative analysis. *Discrete Dynamics in Nature and Society*, 9979835. https://doi.org/10.1155/2021/9979835

Chapter 4

Exploring the Dynamics of Ethnoconsumerism: The Interplay Between Indigenous and Mainstream Consumer Behavior in Cross-cultural Contexts

Naboshree Bhattacharya and Divya Bansal

Amity University, India

Abstract

Ethnoconsumerism explores the complex relationship between consumer behavior, cultural identity, and economic interactions, focusing on the convergence and divergence of indigenous and mainstream consumer behaviors in cross-cultural contexts. Indigenous communities prioritize sustainability, community well-being, and cultural preservation in their consumption choices, often preferring ethically sourced and local products. In contrast, mainstream consumers, influenced by globalized consumerism and capitalist ideologies, tend toward individualistic pursuits, conspicuous consumption, and brand loyalty. Cultural norms, values, and identity significantly influence consumption patterns beyond economic considerations. Ethnic identification plays a crucial role in shaping consumer preferences and serving as a mechanism for cultural resilience. However, globalization and media convergence have blurred cultural boundaries, impacting both indigenous and mainstream consumer behaviors. This chapter proposes a comprehensive examination of ethnoconsumerism, synthesizing existing literature to deepen understanding of consumer behavior in cross-cultural contexts. The insights derived are relevant for businesses, policymakers, and indigenous communities, informing strategies for inclusive economic development and sustainable consumption practices in a globalized world.

Keywords: Ethnoconsumerism; consumer behavior; indigenous; cross cultural; relationship between culture and economy; ethnic identification; normative influence

Indigenous Empowerment through Human-Machine Interactions, 53–71

doi:10.1108/978-1-83608-068-820251004

1. Introduction

1.1 Definition and Scope of Ethnoconsumerism

Ethnoconsumerism explores the influence of culture, ethnicity, and heritage on consumer behavior. Zhang and Kim (2021) define it as "the consumption behavior of individuals that is influenced by their cultural identities, ethnic backgrounds, and heritage" (p. 580). It encompasses cultural practices, rituals, and social interactions beyond purchasing decisions. Theoretical foundations include cultural identity, social identity, and acculturation theories (Kipnis et al., 2019).

The concept impacts various industries. In fashion, companies are incorporating ethnic-inspired designs. The food and beverage industry is seeing an increasing demand for authentic ethnic cuisines. In marketing, brands are using cultural symbols and narratives to connect with ethnoconsumer markets (Amoako et al., 2020).

Bretherton et al. (2020) highlight the importance of cultural identity in shaping preferences among multicultural millennial consumers. Factors influencing ethnoconsumerism include globalization, migration, and technological advancements.

Understanding ethnoconsumerism is crucial for businesses in today's global marketplace. By recognizing cultural contexts and values of diverse consumer segments, companies can develop strategies aligning with consumers' identities and preferences, effectively tapping into ethnoconsumer markets.

1.2 Importance of Understanding Indigenous and Mainstream Consumer Behaviors

Understanding indigenous and mainstream consumer behaviors is crucial in today's diverse and globalized marketplace. Businesses need to gain insights into the unique preferences, attitudes, and consumption patterns of both groups to develop effective marketing strategies and ensure customer satisfaction.

Indigenous consumer behavior refers to the consumption patterns and preferences of individuals belonging to indigenous communities, shaped by their distinct cultural identities, traditions, and values (Reilly, 2021). Rundle-Thiele et al. (2020) emphasize the importance of considering cultural values and beliefs of indigenous communities in developing effective and culturally sensitive social marketing interventions.

Mainstream consumer behavior encompasses the consumption patterns and preferences of the dominant cultural group within a society (Nguyen et al., 2021). Kim and Lee (2021) investigate the impact of cultural values on mainstream consumer behavior in luxury brand consumption, finding that cultural dimensions such as individualism and collectivism significantly influence consumers' attitudes and purchasing intentions.

Recognizing the differences and similarities between indigenous and mainstream consumer behaviors is essential for inclusive and effective marketing strategies. Cayla and Arnould (2008) explore the representation of indigenous cultures in advertising, arguing that incorporating indigenous cultural elements can enhance brand authenticity and foster positive consumer attitudes, while emphasizing the importance of cultural sensitivity.

Understanding indigenous consumer behaviors can also help businesses identify market opportunities and niche segments. The entrepreneurial activities of indigenous communities highlight the importance of understanding indigenous consumer needs to support the development of sustainable and culturally relevant businesses.

Therefore, understanding both indigenous and mainstream consumer behaviors is vital for businesses operating in diverse markets. Recent research underscores the importance of cultural sensitivity, representation, and leveraging indigenous entrepreneurship. As businesses strive to meet the needs of an increasingly diverse customer base, understanding and respecting these differences and similarities will be key to success in the global marketplace.

1.3 Overview of Cross-Cultural Contexts

Cross-cultural contexts have become increasingly relevant in today's globalized world, referring to situations where individuals from different cultural backgrounds interact (Hofstede et al., 2021). These contexts are characterized by diverse cultural norms, values, beliefs, and communication styles, which significantly impact interpersonal relationships and decision-making processes (Minkov et al., 2018; Schwartz et al., 2020).

Two prominent theoretical frameworks for understanding cross-cultural contexts are Hofstede's cultural dimensions theory (Hofstede et al., 2021) and the Global Leadership and Organizational Behavior Effectiveness (GLOBE) study (House et al., 2020). These frameworks provide insights into how cultural values influence behaviors and leadership practices across different societies.

Cross-cultural contexts present challenges, such as potential misunderstandings due to differences in communication styles (Ting-Toomey & Dorjee, 2019), but also opportunities for learning, innovation, and personal growth. Organizations that effectively manage these contexts can benefit from increased innovation and global competitiveness (Stahl et al., 2010).

To navigate cross-cultural contexts effectively, individuals and organizations can develop cultural intelligence (Earley & Ang, 2003), practice mindfulness and empathy, and adapt communication styles to fit the cultural context (Ting-Toomey & Dorjee, 2019). Organizations can foster inclusive environments through cross-cultural training, diversity initiatives, and open communication (Stahl et al., 2010).

Hence, understanding and effectively navigating cross-cultural contexts is crucial in today's globalized world. By leveraging theoretical frameworks, developing cultural intelligence, and fostering inclusive environments, individuals and organizations can successfully navigate cultural differences and benefit from diversity and intercultural collaboration.

2. Theoretical Framework

2.1 Culture and Economy: Interplay in Consumer Behavior

The interplay between culture and economy significantly impacts consumer behavior. Culture shapes individuals' preferences, attitudes, and consumption patterns (Hofstede et al., 2021), while economic factors influence the availability of goods and services, purchasing power, and decision-making processes (Sheth, 2020).

Cultural dimensions such as individualism versus collectivism affect consumer behavior differently across cultures. For instance, collectivistic cultures emphasize social norms and group harmony, while individualistic cultures prioritize personal goals and self-expression (Lin & Wang, 2020; Shavitt & Barnes, 2020).

Economic factors also play a crucial role. In emerging economies, consumers prioritize price and functionality, leading to "frugal innovation" strategies (Radjou & Prabhu, 2021). In contrast, developed economies see a higher demand for experiential and luxury products (Kapferer & Valette-Florence, 2021).

Globalization and digitalization further complicate this interplay, creating "glocal" consumer segments that embrace both global trends and local cultural values (Steenkamp, 2019).

To navigate this complex landscape, businesses must develop cultural intelligence, conduct market research, and adapt their strategies to diverse cultural and economic contexts (Ang et al., 2021). They must also consider ethical implications, such as the potential erosion of local cultural identities due to globalization (Bauman, 2021).

Therefore, understanding the interplay between culture and economy is crucial for businesses to effectively serve diverse consumer segments while respecting cultural diversity and addressing economic realities.

2.2 Ethnic Identification and Its Influence on Consumption Patterns

Ethnic identification, an individual's sense of belonging to a particular ethnic group, significantly influences consumption patterns and consumer behavior in increasingly diverse societies (Chattaraman & Lennon, 2008). Research shows that individuals with strong ethnic identification tend to exhibit distinct consumption preferences and behaviors, often favoring products and services congruent with their ethnic identity.

Ethnic identification also impacts brand loyalty and consumer decision-making processes. Davvetas and Halkias (2019) found that consumers with strong ethnic identification are more likely to be loyal to brands perceived as authentic representations of their ethnic culture. This highlights the importance of cultural authenticity in building brand trust and loyalty among ethnic consumer segments.

The influence of ethnic identification on consumption is mediated by factors such as acculturation (Berry, 2019) and generational differences (Ladhari, 2007). Acculturation levels can affect the extent to which ethnic identification influences consumption behavior (Segev, 2014), while younger generations may have different attitudes and preferences compared to older counterparts.

To effectively leverage ethnic identification in marketing strategies, businesses can use ethnic identity cues in advertising, partner with ethnic community organizations, and develop culturally authentic products and services (Chattaraman & Lennon, 2008). However, it is crucial to approach these strategies with cultural sensitivity and respect, avoiding stereotypes and cultural appropriation.

Hence, understanding the complex interplay between ethnic identification, acculturation, and generational differences is essential for businesses targeting multicultural consumer segments. By leveraging ethnic identification responsibly, companies can build strong relationships with ethnic consumer communities and create value for both consumers and society (Demangeot et al., 2019).

2.3 Normative Influence Within Cultural Groups

Normative influence, the pressure to conform to group expectations and norms, significantly shapes individual behavior within cultural contexts (Morris et al., 2015). This influence varies across cultures, with collectivistic societies generally exhibiting stronger conformity to group norms than individualistic ones (Zou & Savani, 2019).

Theoretical frameworks such as social identity theory (Tajfel & Turner, 1979) and self-categorization theory (Turner et al., 1987) explain how individuals align their behavior with group norms to maintain a positive social identity. This can lead to the internalization of cultural norms and suppression of individual preferences (Hornsey & Jetten, 2004).

Normative influence can have both positive and negative consequences. It can foster social cohesion and well-being among group members (Jetten et al., 2020; Uskul & Oyserman, 2019) but may also perpetuate harmful practices or restrict individual autonomy (Cislaghi & Heise, 2018; Grabe & Dutt, 2020).

To address these issues, researchers and practitioners must develop culturally sensitive interventions that promote individual agency while challenging harmful norms. This involves working with cultural leaders and community members to identify and promote positive cultural values while recognizing the heterogeneity within cultural groups (Morris et al., 2015).

Therefore, understanding and navigating normative influence within cultural contexts is crucial for fostering positive social change and creating inclusive societies that respect both cultural diversity and individual rights.

3. Indigenous Consumer Behavior

3.1 Cultural Values and Traditions Shaping Consumption Choices

Indigenous consumer behavior is deeply influenced by cultural values, traditions, and historical experiences. As globalization increases, understanding the unique consumption choices of indigenous communities has become crucial for promoting inclusive and sustainable economic development (Verbos & Humphries, 2015).

Cultural values shaping indigenous consumer behavior often emphasize community, reciprocity, and respect for the environment, leading to distinct consumption patterns (Kuokkanen, 2019; Watene & Yap, 2015). These values also influence how indigenous communities engage with the market economy, with many prioritizing cultural integrity and sustainability over profit maximization (Anderson et al., 2013; Peredo et al., 2017).

However, the relationship between indigenous cultural values and consumer behavior is complex due to histories of colonization and marginalization, which have disrupted traditional ways of life and created complex cultural identities (Ryder et al., 2017). Some indigenous communities, particularly youth, are increasingly influenced by global consumer trends, creating tensions with traditional cultural norms.

To address these challenges, researchers and practitioners advocate for a more nuanced, culturally sensitive approach to understanding indigenous consumer behavior. This involves recognizing the diversity of indigenous communities and using ethical, participatory research methods that prioritize indigenous voices and perspectives (Drawson et al., 2017; Verbos & Humphries, 2015).

One promising approach is the development of indigenous-led businesses that integrate cultural values into their operations, providing culturally relevant goods and services while promoting economic empowerment and cultural revitalization (Anderson et al., 2013; Reid & Rout, 2016).

3.2 Prioritization of Sustainability and Community Well-Being

Indigenous consumer behavior is increasingly characterized by a strong emphasis on sustainability and community well-being, reflecting deep-rooted cultural values and traditional knowledge systems (Watene & Yap, 2015). This approach prioritizes environmental sustainability, with many indigenous communities maintaining long-standing traditions of living in harmony with nature (Kuokkanen, 2019; Verbos & Humphries, 2015). For example, Nikolakis (2014) found that indigenous communities in the Amazon have developed sustainable agroforestry systems that preserve biodiversity while providing resources.

Indigenous consumer behavior also focuses on community well-being, defining wealth in terms of collective prosperity rather than individual accumulation (Kuokkanen, 2019). This is reflected in practices such as resource sharing and prioritizing local production (Peredo et al., 2017). Ruckstuhl et al. (2014) observed that Māori entrepreneurs often prioritize social and environmental outcomes over profit maximization.

These principles have important implications for developing more sustainable and equitable economic systems. Concepts like the "circular economy" have been inspired by indigenous resource stewardship practices (Korhonen et al., 2018). However, promoting these approaches requires addressing the historical and ongoing impacts of colonization and globalization on indigenous communities (Drahos & Frankel, 2012; Ryder et al., 2017).

Therefore, understanding and learning from indigenous consumer behavior can contribute to more sustainable and equitable economic development. However, this must be accompanied by a commitment to indigenous rights, self-determination, and cultural sovereignty (Kuokkanen, 2019).

3.3 Preferences for Ethically Sourced and Locally Produced Goods

Indigenous consumer behavior increasingly favors ethically sourced and locally produced goods, reflecting a growing awareness of the social and environmental impacts of consumption choices (Verbos & Humphries, 2015). This preference is driven by concerns for environmental sustainability, as many indigenous communities have deep spiritual and cultural connections to the land (Kuokkanen, 2019; Nikolakis, 2014). Ruckstuhl et al. (2014) found that Māori consumers are willing to pay a premium for environmentally friendly and socially responsible products.

Social justice and human rights concerns also influence indigenous preferences for ethical goods. Many communities seek to support businesses that prioritize fair labor practices, community development, and cultural preservation (Peredo et al., 2017). Dockry et al. (2016) noted increased interest among Native American communities in supporting indigenous-owned businesses to promote economic self-determination and cultural revitalization.

Indigenous consumers also prefer locally produced goods, emphasizing the importance of place and local knowledge (Kuokkanen, 2019; Watene & Yap, 2015). Nikolakis and Nelson (2015) found that indigenous communities in Canada are actively promoting local food systems to reduce dependence on imports and support sustainable livelihoods.

These preferences have important implications for businesses and policymakers. Supporting indigenous-owned businesses and local supply chains can create economic opportunities and promote cultural preservation (Peredo et al., 2017). However, it's crucial to address power dynamics and cultural appropriation issues in global supply chains (Drahos & Frankel, 2012), prioritizing indigenous rights, self-determination, and cultural sovereignty (Kuokkanen, 2019).

4. Mainstream Consumer Behavior

4.1 Influence of Globalized Consumerism and Capitalist Ideologies

Mainstream consumer behavior in contemporary society is heavily influenced by globalized consumerism and capitalist ideologies (Kravets et al., 2018). As the world becomes more interconnected, consumers are exposed to a vast array of products, brands, and lifestyle choices that shape their preferences and consumption patterns.

One key feature is the homogenization of consumer culture across national and cultural boundaries, led by multinational corporations promoting standardized products and global brands (Özsomer & Altaras, 2008; Steenkamp, 2019). While this offers a sense of global belonging, it also contributes to the erosion of local cultures and identities (Arnould & Thompson, 2005).

Capitalist ideologies emphasize individualism and personal choice in consumption, encouraging consumers to express their identities through purchasing decisions (Kasser et al., 2007). This is reinforced by digital technologies and social media (Belk, 2013), though critics argue that this emphasis on choice is illusory (Bauman, 2007).

The constant pursuit of novelty and innovation is another aspect of mainstream consumer behavior, leading to planned obsolescence and environmental degradation (Alsamawi et al., 2014; Guiltinan, 2009). This trend has also impacted developing countries, disrupting traditional ways of life and creating new forms of social stratification (Barrientos et al., 2011; Üstüner & Holt, 2010).

While these trends have driven economic growth and technological progress, they have also contributed to environmental and social issues. As awareness grows, there is a movement toward more sustainable and ethical consumption (Kravets et al., 2018). However, the dominant capitalist system continues to shape mainstream consumer behavior in ways that prioritize profit and growth over social and environmental well-being.

4.2 Individualistic Pursuits and Conspicuous Consumption

Mainstream consumer behavior in contemporary society is heavily influenced by individualistic pursuits and conspicuous consumption (Kasser et al., 2007). Digital technologies and social media have reinforced this emphasis, allowing consumers to curate and share personalized lifestyles (Belk, 2013).

The desire for self-expression and differentiation drives individualistic pursuits in consumer behavior (Özsomer & Altaras, 2008). This has led to the rise of "consumer tribes" or "brand communities," providing a sense of belonging while allowing for individual expression (Cova & Cova, 2002; Muniz Jr & O'Guinn, 2001).

Conspicuous consumption, the purchase of goods to display wealth and status, remains a key aspect of mainstream consumer behavior (Trigg, 2001). Social media has amplified this trend, allowing individuals to showcase their consumption habits to wide audiences (Khamis et al., 2017; Marwick, 2015).

However, these trends have been criticized for their negative social and environmental impacts, including overconsumption and environmental degradation (Alsamawi et al., 2014; Kasser et al., 2007). In response, there's a growing movement toward more sustainable, ethical, and minimalist forms of consumption (Kravets et al., 2018), including collaborative consumption (Botsman & Rogers, 2010) and experiential consumption (Gilovich et al., 2015).

While these alternative trends are emerging, the dominant consumer culture continues to prioritize individualism and conspicuous consumption, shaping mainstream consumer behavior in ways that may be difficult to change.

4.3 Brand Loyalty and Marketing Influences

Mainstream consumer behavior is significantly influenced by brand loyalty and marketing strategies (Aaker, 1991; Keller, 1993). Emotional connections between consumers and brands play a crucial role in fostering long-term loyalty and advocacy (Fournier, 1998; Roberts, 2005). For instance, Nike's "Just Do It" campaign has successfully created an emotional bond with consumers (Aaker, 2018).

Digital technologies and social media have transformed brand loyalty and marketing influences, allowing real-time engagement and fostering community (Goh et al., 2013; Laroche et al., 2013). Effective use of these platforms can drive consumer engagement and build strong brand loyalty (Godey et al., 2016).

However, increasing consumer sophistication and choice proliferation have made maintaining loyalty more challenging (Grégoire et al., 2015). Brands must continuously innovate and communicate their value proposition to stay relevant (Kapferer, 2012).

The impact of marketing influences on consumer behavior has been debated, with concerns about manipulation and ethical issues (Akerlof & Shiller, 2015; Stanton et al., 2017; Smith & Cooper-Martin, 1997). In response, there's a growing movement toward more authentic, transparent, and socially responsible marketing practices (Kotler et al., 2019; Kramer & Porter, 2011; Varadarajan & Menon, 1988).

Therefore, while brand loyalty and marketing influences remain powerful forces in shaping consumer behavior, brands must adapt to meet the evolving expectations of socially conscious consumers to maintain trust and loyalty in the long run.

5. Intersection of Indigenous and Mainstream Consumer Behaviors

5.1 Convergence and Divergence in Consumption Patterns

The intersection of indigenous and mainstream consumer behaviors reflects both convergence and divergence in consumption patterns, shaped by globalization and market integration (Venkatesh & Meamber, 2008).

One area of convergence is the adoption of global brands and products by indigenous communities (Grębosz-Krawczyk et al., 2021; Özsomer & Altaras, 2008; Üstüner & Holt, 2007). However, this adoption often involves "glocalization," where global products are adapted to fit local cultural contexts (Chytkova, 2011; Eckhardt & Mahi, 2004; Robertson, 1995).

Divergence is evident in the prioritization of traditional values and practices by indigenous consumers (Arnould & Thompson, 2005). Indigenous consumer behaviors are often shaped by a strong sense of community and reciprocity, leading to different patterns of resource allocation and decision-making (Kuokkanen, 2011; Peredo & McLean, 2013; Vázquez Maguirre et al., 2018).

The intersection also raises questions about power, agency, and cultural sovereignty (Bunten, 2008; Riley, 2004). To address these challenges, more inclusive, equitable, and culturally sensitive approaches to marketing and consumer research are crucial (Jafari & Visconti, 2015; Jafari et al., 2012).

Therefore, understanding the complex interplay between indigenous and mainstream consumer behaviors is essential for fostering more inclusive and equitable forms of consumption that benefit both indigenous and mainstream consumers.

5.2 Cultural Appropriation and Its Implications

The intersection of indigenous and mainstream consumer behaviors has given rise to cultural appropriation, which has significant implications for both indigenous communities and broader society. Cultural appropriation involves the adoption of elements from marginalized cultures by dominant groups, often without proper understanding or respect (Young & Brunk, 2012).

A major concern is the economic exploitation of indigenous communities. While there's growing demand for "authentic" cultural products, profits often don't benefit the original creators and custodians of cultural knowledge (Fowler, 2013; Riley, 2004). This can lead to cultural theft, depriving indigenous communities of economic benefits and control over their heritage (Nicholas & Bannister, 2004).

Cultural appropriation can also erode and distort indigenous cultural identities. Mainstream adoption of cultural elements without proper context may perpetuate stereotypes, oversimplify traditions, or disrespect sacred practices (Aldred, 2000; Matthes, 2016). This can have psychological and social consequences for indigenous peoples, particularly younger generations (Huyser et al., 2015).

To address these issues, more ethical approaches to cultural exchange and consumption are crucial. This includes recognizing indigenous intellectual property rights, ensuring control over cultural representation, and fostering genuine collaboration between indigenous and non-indigenous stakeholders (Anderson, 2009; Smith, 2012). Greater education and awareness among mainstream consumers about cultural appropriation's impacts is also necessary (Huyser et al., 2015; Shand, 2002).

Therefore, while growing interest in indigenous cultures presents opportunities for exchange and development, it also risks exploitation and cultural erosion. Developing ethical approaches based on indigenous rights, informed consent, and equitable collaboration is essential for fostering positive cultural exchange and consumption.

5.3 Challenges and Opportunities in Bridging Indigenous and Mainstream Consumer Practices

The intersection of indigenous and mainstream consumer practices presents both challenges and opportunities for fostering inclusive, equitable, and sustainable economic development (Peredo & McLean, 2013). This intersection requires navigating complex issues of cultural identity, power dynamics, and economic justice, while recognizing potential for mutual learning and collaboration (Anderson et al., 2013).

A major challenge is the historical legacy of colonialism and ongoing marginalization of indigenous communities within the global economic system (Cunneen & Rowe, 2014; Grydehøj & Ou, 2017). This has created power imbalances that make it difficult for indigenous communities to assert their rights and participate in the market economy on their own terms (Riley, 2004).

Different worldviews, values, and priorities between indigenous and mainstream consumer practices can lead to misunderstandings and cultural appropriation (Kuokkanen, 2011; Matthes, 2016). However, opportunities exist for bridging these practices to promote cultural resilience, social justice, and sustainability.

One approach is fostering indigenous entrepreneurship, combining traditional knowledge with modern business practices (Dana & Anderson, 2007; Foley, 2003). Another opportunity lies in the growing demand for ethical and sustainable consumption among mainstream consumers, which aligns with indigenous traditions of environmental stewardship and social reciprocity (Berkes, 2012; Lim, 2016).

Realizing these opportunities requires developing collaborative approaches that empower indigenous communities to shape their engagement with the mainstream economy (Cummings et al., 2018; Nikolakis, 2014). This involves recognizing indigenous peoples as equal partners, respecting their right to self-determination, and building trust through ongoing dialogue and knowledge-sharing (Hernandez, 2019).

Therefore, bridging indigenous and mainstream consumer practices offers significant opportunities for fostering inclusive and sustainable economic development. This requires empowering indigenous communities while engaging mainstream consumers and businesses in ethical forms of exchange and collaboration.

6. Implications and Future Directions

The intersection of indigenous and mainstream consumer behaviors has significant implications for the global economy, cultural diversity, and social justice (Peredo & McLean, 2013). This intersection necessitates new approaches to consumption, production, and exchange that recognize indigenous rights and knowledge while promoting sustainability and equity for all (Kuokkanen, 2019).

A key implication is the need for a more holistic understanding of value and well-being. Indigenous worldviews, emphasizing the intrinsic value of nature and social reciprocity, offer alternative pathways for sustainable development (Watene & Yap, 2015). Integrating these perspectives can help develop more comprehensive indicators of progress that balance economic, social, and environmental objectives (Yap & Watene, 2019).

Another implication is the need for more equitable and participatory governance. Creating inclusive spaces for indigenous voices and ensuring the right to free, prior, and informed consent in decisions affecting their communities is essential (Nikolakis, 2014). This requires a shift toward collaborative governance models that enable indigenous self-determination (Hernandez, 2019).

Promising directions for research and practice include developing indigenous entrepreneurship (Dana & Anderson, 2007) and advancing ethical consumption practices that respect indigenous rights and the environment (Drahos & Frankel, 2012). Supporting indigenous entrepreneurship and promoting cultural sensitivity can create more diverse and resilient local economies (Peredo & Chrisman, 2006; Verbos & Humphries, 2015).

Addressing the intersection of indigenous and mainstream consumer behaviors requires ongoing dialogue, mutual learning, and structural change. By centering indigenous voices and wisdom, we can work toward a future that honors cultural diversity and ecological interdependence.

7. Conclusion

Ethnoconsumerism examines the relationship between consumer behavior, cultural identity, and economic interactions, focusing on indigenous and mainstream consumer behaviors in cross-cultural contexts. Indigenous communities prioritize sustainability and cultural preservation, while mainstream consumers tend toward individualism and brand loyalty. Cultural norms significantly influence consumption patterns, with ethnic identification shaping preferences and fostering cultural resilience. Globalization has blurred cultural boundaries, impacting both indigenous and mainstream consumer behaviors. This study synthesizes existing literature to inform strategies for inclusive economic development and sustainable consumption practices in a globalized world.

In conclusion, this study explores the intersection of indigenous and mainstream consumer behaviors, emphasizing the need for a multifaceted approach that prioritizes indigenous rights and promotes cross-cultural collaboration. It highlights the importance of holistic understandings of value, equitable governance, ethical consumption, and indigenous entrepreneurship. Addressing root causes of indigenous marginalization is crucial for creating an inclusive, regenerative global economy. While this study provides valuable insights, further research and action are needed to fully leverage the potential of these intersections in driving positive change and addressing contemporary challenges.

References

Aaker, D. A. (1991). *Managing brand equity: Capitalizing on the value of a brand name*. Free Press.

Aaker, D. A. (2018). *Creating signature stories: Strategic messaging that energizes, persuades and inspires*. Morgan James Publishing.

Akerlof, G. A., & Shiller, R. J. (2015). *Phishing for phools: The economics of manipulation and deception*. Princeton University Press.

Aldred, L. (2000). Plastic shamans and astroturf sun dances: New Age commercialization of Native American spirituality. *American Indian Quarterly*, *24*(3), 329–352. https://doi.org/10.1353/aiq.2000.0001

Alsamawi, A., Murray, J., & Lenzen, M. (2014). The employment footprints of nations. *Journal of Industrial Ecology*, *18*(1), 59–70. https://doi.org/10.1111/jiec.12104

Amoako, G. K., Pauwels, K., & Osei-Frimpong, K. (2020). Cultural symbols and narratives in advertising: The role of ethnoconsumerism. *Journal of Business Research*, *123*, 17–30. https://doi.org/10.1016/j.jbusres.2020.09.027

Anderson, J. E. (2009). *Law, knowledge, culture: The production of indigenous knowledge in intellectual property law*. Edward Elgar Publishing.

Anderson, R. B., Dana, L. P., & Dana, T. E. (2013). Indigenous entrepreneurship as a function of cultural perceptions of opportunity. In A. Lundström, C. Zhou, Y. von Friedrichs, & E. Sundin (Eds.), *Social entrepreneurship: Leveraging economic, political, and cultural dimensions* (pp. 69–88). Springer. https://doi.org/10.1007/978-3-319-01396-1_5

Ang, S., Rockstuhl, T., & Tan, M. L. (2021). Cultural intelligence and adaptive performance: A meta-analytic investigation. *Journal of World Business*, *56*(4), 101225. https://doi.org/10.1016/j.jwb.2021.101225

Arnould, E. J., & Thompson, C. J. (2005). Consumer culture theory (CCT): Twenty years of research. *Journal of Consumer Research*, *31*(4), 868–882. https://doi.org/10.1086/426626

Barrientos, S., Gereffi, G., & Rossi, A. (2011). Economic and social upgrading in global production networks: A new paradigm for a changing world. *International Labour Review*, *150*(3–4), 319–340. https://doi.org/10.1111/j.1564-913X.2011.00119.x

Bauman, Z. (2007). *Consuming life*. Polity Press.

Bauman, Z. (2021). *Globalization: The human consequences*. John Wiley & Sons.

Belk, R. W. (2013). Extended self in a digital world. *Journal of Consumer Research*, *40*(3), 477–500. https://doi.org/10.1086/671052

Berkes, F. (2012). *Sacred ecology* (3rd ed.). Routledge. https://doi.org/10.4324/9780203123843

Berry, J. W. (2019). *Acculturation: A personal journey across cultures*. Cambridge University Press.

Botsman, R., & Rogers, R. (2010). *What's mine is yours: The rise of collaborative consumption*. HarperCollins.

Bretherton, P., Piacentini, M., & Hogg, M. (2020). Ethnoconsumerism among multicultural millennial consumers: The impact of cultural identity and heritage on buying behaviours. *Journal of Business Research*, *115*, 44–54. https://doi.org/10.1016/j.jbusres.2020.04.042

Bunten, A. C. (2008). Sharing culture or selling out? Developing the commodified persona in the heritage industry. *American Ethnologist*, *35*(3), 380–395. https://doi.org/10.1111/j.1548-1425.2008.00041.x

Cayla, J., & Arnould, E. J. (2008). A cultural approach to branding in the global marketplace. *Journal of International Marketing*, *16*(4), 86–112.

Chattaraman, V., & Lennon, S. J. (2008). Ethnic identity, consumption of cultural apparel, and self-perceptions of ethnic consumers. *Journal of Fashion Marketing and Management: An International Journal*, *12*(4), 518–531.

Chytkova, Z. (2011). Consumer acculturation, gender, and food: Romanian women in Italy between tradition and modernity. *Consumption Markets & Culture*, *14*(3), 267–291. https://doi.org/10.1080/10253866.2011.574827

Cislaghi, B., & Heise, L. (2018). Theory and practice of social norms interventions: Eight common pitfalls. *Globalization and Health*, *14*(1), 1–10. https://doi.org/10.1186/s12992-018-0398-x

Cova, B., & Cova, V. (2002). Tribal marketing: The tribalisation of society and its impact on the conduct of marketing. *European Journal of Marketing*, *36*(5/6), 595–620. https://doi.org/10.1108/03090560210423023

Cummings, S., Seferiadis, A. A., Maas, J., Bunders, J. F. G., & Zweekhorst, M. B. M. (2018). Knowledge, social capital, and grassroots development: Insights from rural Bangladesh. *The Journal of Development Studies*, *54*(10), 1727–1744. https://doi.org/10.1080/00220388.2017.1417584

Cunneen, C., & Rowe, S. (2014). Changing narratives: Colonised peoples, criminology and social work. *International Journal for Crime, Justice and Social Democracy*, *3*(1), 49–67.

Dana, L. P., & Anderson, R. B. (Eds.) (2007), *International handbook of research on indigenous entrepreneurship*. Edward Elgar Publishing. https://doi.org/10.4337/9781781952641

Davvetas, V., & Halkias, G. (2019). Global and local brand stereotypes: Formation, content transfer, and impact. *International Marketing Review*, *36*(5), 675–701. https://doi.org/10.1108/IMR-01-2018-0017

Demangeot, C., Broeckerhoff, A., Kipnis, E., Pullig, C., & Visconti, L. M. (2019). Consumer mobility and well-being among changing places and shifting ethnicities. *Marketing Theory*, *19*(2), 271–276. https://doi.org/10.1177/1470593118787589

Dockry, M. J., Hall, K., Van Lopik, W., & Caldwell, C. M. (2016). Sustainable development education, practice, and research: An indigenous model of sustainable development at the College of Menominee Nation, Keshena, WI, USA. *Sustainability Science*, *11*(1), 127–138. https://doi.org/10.1007/s11625-015-0304-x

Drahos, P., & Frankel, S. (Eds.) (2012), *Indigenous people's innovation: Intellectual property pathways to development*. ANU E Press. https://doi.org/10.22459/IPI.08.2012

Drawson, A. S., Toombs, E., & Mushquash, C. J. (2017). Indigenous research methods: A systematic review. *The International Indigenous Policy Journal*, *8*(2), 1–25. https://doi.org/10.18584/iipj.2017.8.2.5

Earley, P. C., & Ang, S. (2003). *Cultural intelligence: Individual interactions across cultures*. Stanford University Press.

Eckhardt, G. M., & Mahi, H. (2004). The role of consumer agency in the globalization process in emerging markets. *Journal of Macromarketing*, *24*(2), 136–146. https://doi.org/10.1177/0276146704269304

Foley, D. (2003). An examination of indigenous Australian entrepreneurs. *Journal of Developmental Entrepreneurship*, *8*(2), 133–151.

Fournier, S. (1998). Consumers and their brands: Developing relationship theory in consumer research. *Journal of Consumer Research*, *24*(4), 343–373. https://doi.org/10.1086/209515

Fowler, C. (2013). The ecological implications of ancestral religion and reciprocal exchange in a sacred forest in Karendi (Sumba, Indonesia). *Worldviews: Global*

Religions, Culture, and Ecology, *17*(3), 303–329. https://doi.org/10.1163/15685357-01703004

Gilovich, T., Kumar, A., & Jampol, L. (2015). A wonderful life: Experiential consumption and the pursuit of happiness. *Journal of Consumer Psychology*, *25*(1), 152–165. https://doi.org/10.1016/j.jcps.2014.08.004

Godey, B., Manthiou, A., Pederzoli, D., Rokka, J., Aiello, G., Donvito, R., & Singh, R. (2016). Social media marketing efforts of luxury brands: Influence on brand equity and consumer behavior. *Journal of Business Research*, *69*(12), 5833–5841. https://doi.org/10.1016/j.jbusres.2016.04.181

Goh, K.-Y., Heng, C.-S., & Lin, Z. (2013). Social media brand community and consumer behavior: Quantifying the relative impact of user- and marketer-generated content. *Information Systems Research*, *24*(1), 88–107. https://doi.org/10.1287/isre.1120.0469

Grabe, S., & Dutt, A. (2020). The role of cultural norms in shaping women's experiences of sexual violence: A qualitative examination of honor and shame in India. *Journal of Interpersonal Violence*. Advance online publication. https://doi.org/10.1177/0886260520943721

Grębosz-Krawczyk, M., Zakrzewska-Bielawska, A., Glinka, B., & Glińska-Neweś, A. (2021). Why do consumers choose photovoltaic panels? Identification of the factors influencing consumers' choice behavior regarding photovoltaic panel installations. *Energies*, *14*, 2674.

Grégoire, Y., Salle, A., & Tripp, T. M. (2015). Managing social media crises with your customers: The good, the bad, and the ugly. *Business Horizons*, *58*(2), 173–182. https://doi.org/10.1016/j.bushor.2014.11.001

Grydehøj, A., & Ou, Z. (2017). Deterritorialization of indigeneity: Indigenous knowledge in the UN climate change regime. *Political Geography*, *67*, 148–157. https://doi.org/10.1016/j.polgeo.2017.08.002

Guiltinan, J. (2009). Creative destruction and destructive creations: Environmental ethics and planned obsolescence. *Journal of Business Ethics*, *89*(1), 19–28. https://doi.org/10.1007/s10551-008-9907-9

Hernandez, J. (2019). Indigenous knowledge systems and entrepreneurship: A Shuar perspective. In M. Rautiainen, P. Rosa, T. Pihkala, M. J. Parada, & A. D. Cruz (Eds.), *The family business group phenomenon: Emergence and complexities* (pp. 295–313). Palgrave Macmillan. https://doi.org/10.1007/978-3-319-98542-8_14

Hofstede, G., Hofstede, G. J., & Minkov, M. (2021). *Cultures and organizations: Software of the mind* (4th ed.). McGraw-Hill.

Hornsey, M. J., & Jetten, J. (2004). The individual within the group: Balancing the need to belong with the need to be different. *Personality and Social Psychology Review*, *8*(3), 248–264. https://doi.org/10.1207/s15327957pspr0803_2

House, R. J., Dorfman, P. W., Javidan, M., Hanges, P. J., & Sully de Luque, M. F. (2020). *Strategic leadership across cultures: The GLOBE study of CEO leadership behavior and effectiveness in 24 countries* (2nd ed.). Sage Publications.

Huyser, K. R., Sanchez, G. R., & Vargas, E. D. (2015). Civic engagement and political participation among American Indians and Alaska natives in the US. *Politics, Groups, and Identities*, *3*(4), 642–659. https://doi.org/10.1080/21565503.2015.1050411

Jafari, A., Firat, F., Süerdem, A., Askegaard, S., & Dalli, D. (2012). Non-western contexts: The invisible half. *Marketing Theory*, *12*(1), 3–12. https://doi.org/10.1177/1470593111424183

Jafari, A., & Visconti, L. M. (2015). New directions in researching ethnicity in marketing and consumer behaviour: A well-being agenda. *Marketing Theory*, *15*(2), 265–270. https://doi.org/10.1177/1470593114552582

Jetten, J., Haslam, S. A., Cruwys, T., Greenaway, K. H., Haslam, C., & Steffens, N. K. (2020). Advancing the social identity approach to health and well-being: Progressing the social cure research agenda. *European Journal of Social Psychology*, *50*(7), 1389–1400. https://doi.org/10.1002/ejsp.2736

Kapferer, J. N. (2012). *The new strategic brand management: Advanced insights and strategic thinking* (5th ed.). Kogan Page Publishers.

Kapferer, J. N., & Valette-Florence, P. (2021). Luxury marketing: A challenge for theory and practice. *Journal of Business Research*, *129*, 238–246. https://doi.org/10.1016/j.jbusres.2021.02.048

Kasser, T., Cohn, S., Kanner, A. D., & Ryan, R. M. (2007). Some costs of American corporate capitalism: A psychological exploration of value and goal conflicts. *Psychological Inquiry*, *18*(1), 1–22. https://doi.org/10.1080/10478400701386579

Keller, K. L. (1993). Conceptualizing, measuring, and managing customer-based brand equity. *Journal of Marketing*, *57*(1), 1–22. https://doi.org/10.1177/002224299305700101

Khamis, S., Ang, L., & Welling, R. (2017). Self-branding, 'micro-celebrity' and the rise of social media influencers. *Celebrity Studies*, *8*(2), 191–208.

Kim, J., & Lee, J. (2021). Cultural values and luxury brand consumption: The moderating role of cultural dimensions. *Journal of Business Research*, *132*, 350–361. https://doi.org/10.1016/j.jbusres.2021.04.046

Kipnis, E., Demangeot, C., Pullig, C., & Broderick, A. J. (2019). Consumer multicultural identity affiliation: Reassessing identity segmentation in multicultural markets. *Journal of Business Research*, *98*, 126–141. https://doi.org/10.1016/j.jbusres.2018.11.056

Korhonen, J., Honkasalo, A., & Seppälä, J. (2018). Circular economy: The concept and its limitations. *Ecological Economics*, *143*, 37–46. https://doi.org/10.1016/j.ecolecon.2017.06.041

Kotler, P., Kartajaya, H., & Setiawan, I. (2019). *Marketing 4.0: Moving from traditional to digital*. Wiley.

Kramer, M. R., & Porter, M. (2011). *Creating shared value* (Vol. 17). FSG.

Kravets, O., Maclaran, P., Miles, S., & Venkatesh, A. (Eds.). (2018). *The SAGE handbook of consumer culture*. SAGE Publications.

Kuokkanen, R. (2011). From indigenous economies to market-based self-governance: A feminist political economy analysis. *Canadian Journal of Political Science*, *44*(2), 275–297. https://doi.org/10.1017/S0008423911000126

Kuokkanen, R. (2019). *Restructuring relations: Indigenous self-determination, governance, and gender*. Oxford University Press.

Ladhari, R. (2007). The effect of consumption emotions on satisfaction and word-of-mouth communications. *Psychology & Marketing*, *24*(12), 1085–1108.

Laroche, M., Habibi, M. R., & Richard, M. O. (2013). To be or not to be in social media: How brand loyalty is affected by social media? *International Journal of Information Management*, *33*(1), 76–82. https://doi.org/10.1016/j.ijinfomgt.2012.07.003

Lin, Y. C., & Wang, K. Y. (2020). The influence of cultural values on green consumption: A cross-cultural comparison between Taiwanese and Chinese consumers. *Journal of Cleaner Production*, *256*, 120737. https://doi.org/10.1016/j.jclepro.2020.120737

Marwick, A. E. (2015). You may know me from YouTube: (micro-) celebrity in social media. *A Companion to Celebrity*, 333–350.
Matthes, E. H. (2016). Cultural appropriation without cultural essentialism? *Social Theory and Practice*, *42*(2), 343–366.
Minkov, M., Bond, M. H., Dutt, P., Schachner, M., Morales, O., Sanchez, C., Jandosova, J., Khassenbekov, Y., & Mudd, B. (2018). A reconsideration of Hofstede's fifth dimension: New flexibility versus monumentalism data from 54 countries. *Cross-Cultural Research*, *52*(3), 309–333.
Morris, M. W., Chiu, C. Y., & Liu, Z. (2015). Polycultural psychology. *Annual Review of Psychology*, *66*, 631–659. https://doi.org/10.1146/annurev-psych-010814-015001
Muniz, Jr., A. M., & O'guinn, T. C. (2001). Brand community. *Journal of Consumer Research*, *27*(4), 412–432.
Nguyen, T. T., Nguyen, H. V., & Pham, T. T. (2021). Exploring the impact of cultural values on mainstream consumer behavior: A comparative study. *Journal of Consumer Behaviour*, *20*(3), 546–562. https://doi.org/10.1002/cb.1928
Nicholas, G., & Bannister, K. (2004). Copyrighting the past? Emerging intellectual property rights issues in archaeology. *Current Anthropology*, *45*(3), 327–350.
Nikolakis, W. (2014). *Forests and globalization* (J. Innes, Ed.). Taylor & Francis.
Özsomer, A., & Altaras, S. (2008). Global brand purchase likelihood: A critical synthesis and an integrated conceptual framework. *Journal of International Marketing*, *16*(4), 1–28. https://doi.org/10.1509/jimk.16.4.1
Peredo, A. M., & Chrisman, J. J. (2006). Toward a theory of community-based enterprise. *Academy of Management Review*, *31*(2), 309–328. https://doi.org/10.5465/amr.2006.20208683
Peredo, A. M., & McLean, M. (2013). Indigenous development and the cultural captivity of entrepreneurship. *Business & Society*, *52*(4), 592–620.
Peredo, A. M., McLean, M., & Anderson, R. B. (2017). Indigenous development and the cultural captivity of entrepreneurship. *Business & Society*, *56*(4), 592–620.
Radjou, N., & Prabhu, J. (2021). Frugal innovation: How to do more with less.
Reid, J., & Rout, M. (2016). Māori tribal economy: Rethinking the original economic institutions. In T. L. Anderson (Ed.), *Unlocking the wealth of Indian nations* (pp. 83–104). Lexington Books.
Reilly, M. D. (2021). Indigenous consumer behavior: A systematic review and research agenda. *Journal of Consumer Psychology*, *31*(2), 336–351. https://doi.org/10.1002/jcpy.1226
Roberts, K. (2005). *Lovemarks: The future beyond brands*. PowerHouse Books.
Robertson, R. (1995). Glocalization: Time-space and homogeneity-heterogeneity. In M. Featherstone, S. Lash, & R. Robertson (Eds.), *Global modernities* (pp. 25–44). SAGE Publications.
Riley, M. (2004). Indigenous intellectual property rights: Legal obstacles and innovative solutions. *Journal of World Intellectual Property*, *7*(5), 865–889.
Ruckstuhl, K., Thompson-Fawcett, M., & Rae, H. (2014). Māori and mining: Indigenous perspectives on reconceptualising and contextualising the social licence to operate. *Impact Assessment and Project Appraisal*, *32*(4), 304–314.
Rundle-Thiele, S., David, P., Willmott, T., Pang, B., Eagle, L., & Hay, R. (2020). Social marketing theory development goals: An agenda to drive change. *Journal of Consumer Behaviour*, *19*(1), 133–147. https://doi.org/10.1002/cb.1793

Ryder, C., Mackean, T., Coombes, J., Williams, H., Hunter, K., Holland, A. J. A., & Ivers, R. Q. (2017). Indigenous research methodology – Weaving a research interface. *International Journal of Social Research Methodology*, *23*(3), 255–267. https://doi.org/10.1080/13645579.2019.1669923

Schwartz, S. H., Cieciuch, J., Vecchione, M., Torres, C., Dirilen-Gumus, O., & Butenko, T. (2020). Refining the theory of basic individual values: A cross-cultural perspective. *Journal of Personality and Social Psychology*, *119*(4), 1027–1051. https://doi.org/10.1037/pspp0000286

Segev, S. (2014). The effect of acculturation on ethnic consumers' decision-making styles: An empirical analysis of Hispanic consumers. *Journal of International Consumer Marketing*, *26*(3), 168–184.

Shand, P. (2002). Scenes from the colonial catwalk: Cultural appropriation, intellectual property rights, and fashion. *Cultural Analysis*, *3*(2002), 47–88.

Shavitt, S., & Barnes, A. J. (2020). Culture and the consumer journey. *Journal of Retailing*, *96*(1), 40–54. https://doi.org/10.1016/j.jretai.2019.11.009

Sheth, J. (2020). Impact of Covid-19 on consumer behavior: Will the old habits return or die?. *Journal of Business Research*, *117*, 280–283. https://doi.org/10.1016/j.jbusres.2020.05.059

Smith, L. T. (2012). *Decolonizing methodologies: Research and indigenous peoples* (2nd ed.). Zed Books.

Smith, N. C., & Cooper-Martin, E. (1997). Ethics and target marketing: The role of product harm and consumer vulnerability. *Journal of Marketing*, *61*(3), 1–20.

Stahl, G. K., Maznevski, M. L., Voigt, A., & Jonsen, K. (2010). Unraveling the effects of cultural diversity in teams: A meta-analysis of research on multicultural work groups. *Journal of International Business Studies*, *41*, 690–709.

Stanton, J. V., Etzel, M. J., Walker, B. J., & Walker, J. T. (2017). *Fundamentals of marketing*. McGraw-Hill Education.

Steenkamp, J. B. E. (2019). Global brand building and management in the digital age. *Journal of International Marketing*, *27*(1), 13–27. https://doi.org/10.1177/1069031X18811289

Tajfel, H., & Turner, J. C. (1979). An integrative theory of intergroup conflict. In W. G. Austin & S. Worchel (Eds.), *The social psychology of intergroup relations* (pp. 33–47). Brooks/Cole.

Ting-Toomey, S., & Dorjee, T. (2019). *Communicating across cultures* (2nd ed.). The Guilford Press.

Trigg, A. B. (2001). Veblen, Bourdieu, and conspicuous consumption. *Journal of Economic Issues*, *35*(1), 99–115. https://doi.org/10.1080/00213624.2001.11506342

Turner, J. C., Hogg, M. A., Oakes, P. J., Reicher, S. D., & Wetherell, M. S. (1987). *Rediscovering the social group: A self-categorization theory*. Blackwell.

Uskul, A. K., & Oyserman, D. (2019). Cultural fit and positive youth development among immigrant and minority youth. In D. Güngör & D. Strohmeier (Eds.), *Contextualizing immigrant and refugee resilience* (pp. 63–82). Springer. https://doi.org/10.1007/978-3-030-42303-2_4

Üstüner, T., & Holt, D. B. (2007). Dominated consumer acculturation: The social construction of poor migrant women's consumer identity projects in a Turkish squatter. *Journal of Consumer Research*, *34*(1), 41–56. https://doi.org/10.1086/513045

Üstüner, T., & Holt, D. B. (2010). Toward a theory of status consumption in less industrialized countries. *Journal of Consumer Research*, *37*(1), 37–56. https://doi.org/10.1086/649759

Varadarajan, P. R., & Menon, A. (1988). Cause-related marketing: A coalignment of marketing strategy and corporate philanthropy. *Journal of Marketing*, *52*(3), 58–74.

Vázquez Maguirre, M., Portales, L., & Velásquez Bellido, I. (2018). Indigenous social enterprises as drivers of sustainable development: Insights from Mexico and Peru. *Critical Sociology*, *44*(2), 323–340.

Venkatesh, A., & Meamber, L. A. (2008). The aesthetics of consumption and the consumer as an aesthetic subject. *Consumption, Markets and culture*, *11*(1), 45–70.

Verbos, A. K., & Humphries, M. T. (2015). Indigenous wisdom and the PRME: inclusion or illusion? *Journal of Management Development*, *34*(1), 90–100.

Watene, K., & Yap, M. (2015). Culture and sustainable development: Indigenous contributions. *International Journal of Human Rights*, *19*(3), 272–289.

Yap, M., & Watene, K. (2019). The sustainable development goals (SDGs) and indigenous peoples: Another missed opportunity? *Journal of Human Development and Capabilities*, *20*(4), 451–467. https://doi.org/10.1080/19452829.2019.1574725

Young, J. O., & Brunk, C. G. (2012). *The ethics of cultural appropriation*. Wiley-Blackwell.

Zhang, Y., & Kim, Y. (2021). Ethnoconsumerism: A systematic review of the literature and future research directions. *Journal of Business Research*, *125*, 579–594. https://doi.org/10.1016/j.jbusres.2020.12.021

Zou, X., & Savani, K. (2019). Descriptive norms for me, injunctive norms for you: Using norms to explain the cultural differences in personal preferences. *Journal of Personality and Social Psychology*, *117*(5), 908–927. https://doi.org/10.1037/pspi0000178

Chapter 5

How ChatGPT Shapes Knowledge Acquisition and Career Trajectories in Higher Education: Decoding Students' Perceptions to Achieve Quality Education

Ravishankar Krishnan[a] *and Logasakthi Kandasamy*[b]

[a]Vel Tech Rangarajan Dr.Sagunthala R&D Institute of Science and Technology, India
[b]RV University, India

Abstract

The growth of technology and innovation offers numerous tools and techniques to the educational sector. Recently, ChatGPT has been used by the stakeholders of the education for multiple purposes. However, the influence of ChatGPT on shaping the learner's knowledge acquisition and shaping the potential career trajectories has a scope to study further. Data collected from 445 respondents, the research pivots using the Extended Technology Acceptance Model (TAM) to investigate the interrelatedness amongst learner's perceived usefulness, ease of use, user intention, attitude and their all-encompassing effects on both knowledge and career augmentation and career paths. The findings of the research show that ChatGPT's ease of use performs a paramount role in affecting its perceived utility, which in turn shapes their overall attitude toward the career platform. Learners' academic growth and career positions are precisely influenced by users' usefulness, attitude, and intentions. Investigation found significant medication effects among these constructs. Learners' perception and intentions are majorly influenced by trust on the technology, and it has been identified as cardinal determinant. The implication of the study emphasizes that tech developers, educational policymakers, and educational instructors need to include the insightful contribution of technology to redesign the academic syllabi for enriching the classroom delivery. Furthermore, the research describes suitable strategies, urging the educational sector to spotlight user-centric design and trust-building mechanisms. Overall, this investigation

Indigenous Empowerment through Human-Machine Interactions, 73–92

doi:10.1108/978-1-83608-068-820251005

highlights ChatGPT's significant role in higher education, flooring the way for its nuanced and impactful inclusion.

Keywords: Technology in education; human–computer interface; informal learning; knowledge acquisition; lifelong learning; ChatGPT

1. Introduction

The advent of advanced technologies has introduced an array of Artificial Intelligence (AI)-powered tools accessible to the broader public. In 2022, OpenAI unveiled a chatbot designed to help users efficiently achieve their intended results. ChatGPT has rapidly become one of the fastest-growing applications, drawing over 100 million users in a brief period. Its surge in popularity can be attributed to its expanded range of applications and enhanced services. For instance, it aids in drafting business emails, essays, formal letters, recommendations, and proposals (Shoufan, 2023). ChatGPT's association with academic performance is significant. It has led to questions about academic integrity and finding significant challenges with bias in the output (Eke, 2023; Kasneci et al., 2023). Opinions within the education sector about ChatGPT are polarized, with staunch proponents and detractors. However, tools like ChatGPT have far-reaching implications in both formal and informal educational settings (Ray, 2023).

The current study emphasized that the effectiveness of using the ChatGPT by students depends on the intention to use, perceived usefulness, and guidance of the educator. The requirement of the study is to trace out how far the student's fraternity has made use of the ChatGpt and to understand its role in their career trajectories. Although there are limited studies in this area, current research extends the scope by analyzing variables such as user-friendliness, reliability, purpose of use, and reaching the desired outcome. Previously, Bonsu and Baffour-Koduah (2023) reported that improved results, a higher level of accuracy, and user convenience are the major factors that positively influence students to use ChatGPT. Additionally, their study highlighted that social media plays a predominant role in educating students about the application.

Furthermore, the increasing incorporation of ChatGPT and similar tools in academic settings has implications beyond the classroom. Today's students are tomorrow's workforce, and their familiarity with AI can shape industries and redefine career trajectories. Industries are increasingly looking for graduates familiar with AI tools and the potential they bring. To understand student acceptance of ChatGPT in education and measure the effectiveness of such AI tools, we utilized the Technology Acceptance Model (TAM) (Davis, 1989) and developed a theoretical framework to analyze how far the external factors of the model are supporting, to reach the outcome of students' knowledge, career growth. Mailizar et al. (2021) underscored students' behavioral intentions toward e-learning through the lens of the TAM. Similarly, Zhou et al. (2022) delved into students' intentions to utilize online educational platforms, providing key insights into the education sector. Furthermore, numerous scholars contend that the TAM effectively gauges the relationships among pertinent variables.

Aiming to assist educational institutions, instructors, and scholars in understanding students' perceptions and usage of ChatGPT, this study formulates hypotheses and research models. Additionally, the current study can reveal the positive areas of improvements in AI-powered applications with respect to the use of students and the educational sector. Balancing the potential advantages with ethical concerns and ensuring that the core objective of education isn't compromised is pivotal. By decoding student perceptions, this paper seeks to provide a road map for educators, policymakers, and scholars, ensuring the AI revolution in education results in informed, ethical, and equitable outcomes.

The paper covers the following sections: Section 2 theoretical background, Section 3 hypothesis proposal, Section 4 methodology, findings and discussion, discussion, conclusion is in the Sections 5–7, respectively.

2. Theoretical Backgrounds

2.1 ChatGpt

ChatGPT, developed by OpenAI, utilizes the transformer-based architecture but with the addition of generative pretraining, which allows the model to generate coherent and contextually relevant sentences over extended passages (Radford et al., 2019). The core principle of ChatGPT is unsupervised learning, where it's pretrained on large text datasets to learn grammar and facts, then fine-tuned for specificity (Baidoo-Anu & Ansah, 2023). The ChatGPT series, from GPT-1 to GPT-4, has seen substantial technological advancements. While GPT-2 focused on generating longer, coherent text, GPT-3 expanded its capabilities, tackling a wider range of natural language processing (NLP) tasks. Enhancements in GPT-3.5 introduced features like residual connections and layer normalization. By the time of GPT-4, these deep learning advancements enabled the generation of not only extensive text but also image outputs However, while the capabilities of ChatGPT are expansive, it is also crucial to consider the ethical dimensions, as highlighted by Wang et al. (2023). The potential biases embedded within the model could inadvertently influence students' perceptions and knowledge frameworks.

2.2 Technology Acceptance Model

Davis (1989) developed the TAM to explore two main objectives. The first one is how users perceive and accept the new technology, which gives conceptual clarity about the technology. Secondly, testing the acceptance level of the users which supports prototype model development and implementation of new technology. Many researchers have extensively used the model and brought insights into the TAM factors. In this context, learners' intention toward online education has been studied; perceived ease of use (PEU), core intention, and other external variables were mapped with formulated hypothesis appropriately (Ajzen et al., 2018; Zhou et al., 2022). The purpose of using the proven TAM model is to portray GPT's level of acceptance among the students through measuring relationships among the external variables such as perceived usefulness (PUS), PEU, trust (TST), attitude (ATT), intention to use (ITU), and knowledge and career growth (KCG).

3. Hypotheses Proposal

As mentioned in Section 2, Davis (1989) designed the TAM by extending the Theory of Reasoned Action (TRA) model of Fishbein Ajzen (1977). The TAM readily captures user behavior, as noted by Harryanto et al. (2018). The study variables were constructed to capture the students' perception of ChatGpt. Factors such as PEU, PUS, and TST influence attitude and intention, which in turn support measuring the outcomes of students' KCG.

3.1 Perceived Ease of Use

PEU can be defined as an effortless action which makes individuals feel it is easy to use technology for better experience. Online education platforms enhance learning accessibility, allowing students to engage from anywhere and at any time. Building on this idea of accessibility, Davis (1989) argued that the level of positive significance is high between PEU and PUS. Weng et al. (2018) witnessed that the PEU affects PUS positively. Literature reveals that the acceptance level of technology by the learners was quite high, PEU directly affects students' PUS, subsequently stimulating them to frequently use the technology, as highlighted by Granić and Marangunić (2019) and Persico et al. (2014).

Thus, the current study needs to test how far PEU is influencing the usefulness and attitude of the users, hence formulated the following hypotheses:

H1. There is a positive relationship between PEU and PUS
H2. There is a positive relationship between PEU and ATT

3.2 Perceived Usefulness

The belief of students that the extent to which new technology helps them to improve their academic performance is referred to as PUS. Goh et al. (2013) found a positive significant association has been found between PUS and willingness of the users. Social gaming service users' PUS positively influences the ATT (Bazelais et al., 2018). However, while perceptions of students on e-learning can vary, their views on usefulness often align to a significant extent (Harryanto et al., 2018). Therefore, the hypothesis was formulated as follows:

H3. There is a positive relationship between PUS and ATT

3.3 Trust

Trust emerges from users' beliefs about the reliability of technology. This notion is evident when considering students' trust levels when utilizing the ChatGpt application. The extent to which students find utility in the ChatGpt application embodies their trust. However, Su and Li (2021) found that users' perceived credibility significantly impacts their attitude. Dhagarra et al. (2020) suggested that trust has a psychological association with users, and it shapes their attitudes toward technology. To explore further, the study has formulated the hypothesis as follows:

H4. There is a positive relationship between TST and ATT

3.4 Attitude

Rossi (2018) recognized a considerable connection between attitude and intent. Recognizing this connection is critical because a user's attitude repeatedly influences their intent to accept and further invent with expertise (Uche et al., 2020). Attitude can be defined as the inclination of students to feel and act in an instant way. Earlier investigation has accentuated that users' intent to use a technology can change based on their attitudes (Amadu et al., 2018). Moreover, Su and Li (2021) discovered a significant connection between intent to use and attitude. Tsai (2015) accentuated that behavioral intent influences users' attitudes, subsequently affecting perceived usefulness. Hence, to test the students' intention to use ChatGPT and their attitude regarding the AI-powered tool, we devised a hypothesis as follows:

H5. There is a positive relationship between ATT and ITU

3.5 Intention to Use and Knowledge and Career Growth

The growth of businesses brings forth innovative-edge technologies, functions, and AI-powered tools. Although the growth of AI driven enterprises persistently transform to secure their market capitalization and endure aggressive pressures. Intent to use specialized technologies or applications essentially adapts among individuals. Weng et al. (2018) proposed that the intent to use a technology affects its apparent benefit. Prakosa and Sumantika (2021) noted that PEU, PUS, and ITU mutually effect students' academic outcomes. Thus, the following hypothesis was conveyed to explore further.

H6. There is a positive relationship between ITU and KCG.

3.6 Study Direction and Central Concerns

Building on prior research, our primary inquiry centered on the question: How do students' perceptions of ChatGPT influence their knowledge acquisition and potential impact on career trajectories within higher education?

To delve deeper into this primary question, the following sub-questions were formulated:

- How does the PEU of ChatGPT affect students' attitudes and its perceived usefulness?
- What role does trust play in mediating students' attitudes toward ChatGPT and their intentions to use it for educational purposes?
- How do positive perceptions of ChatGPT correlate with enhancements in knowledge acquisition and future career prospects?

Background variables, including academic level, discipline, prior technology exposure, and duration of ChatGPT usage, will be analyzed to discern how they might intersect with the primary dimensions of interest. Fig. 5.1 presents a path diagram that outlines the theoretical structure of our proposed model, with hypotheses derived from the research questions.

4. Methodology

4.1 Data Collection

For this study, data were gathered via online questionnaires from February to June 2023. We reached out to 500 individuals, receiving feedback from 476, which signifies a 95% response rate. Out of these, 31 responses were discarded due to incomplete or incorrect entries, leaving 445 valid responses for assessment. A convenience sampling technique was agreed due to its cost-efficacy and sincere disposition.

Using G*Power software, inspired by the studies of Faul et al. (2009), study decided the ideal number of samples. The software suggested a sample size that convinces a power value of 0.95, supporting with our gathering from 445 respondents.

To make an intense analysis and recognizing of the associations between various factors, the SmartPLS 4.0 tool was utilized. The study implemented the Structural Equation Modeling (SEM) procedure, supporting an inclusive view of compound variables and combining equally factor and path analyses (Sarstedt et al., 2021).

4.2 Students' Demographic Data

The research examined five elements throughout 445 participants: institution type, course level, institute location, gender, and age. The greater part (76.6%) of respondents is from private institutions, with 23.4% from public institutions. Among various courses, postgraduates are 73.7%, whereas 26.3% of the respondents are undergraduates. Students from rural area institutions are 41.3%, semi-urban locations were 18.4%, and urban locales are 40.2%. A gender disparity occurs with 60.4% male respondents and 39.6% female. The age allocation reveals

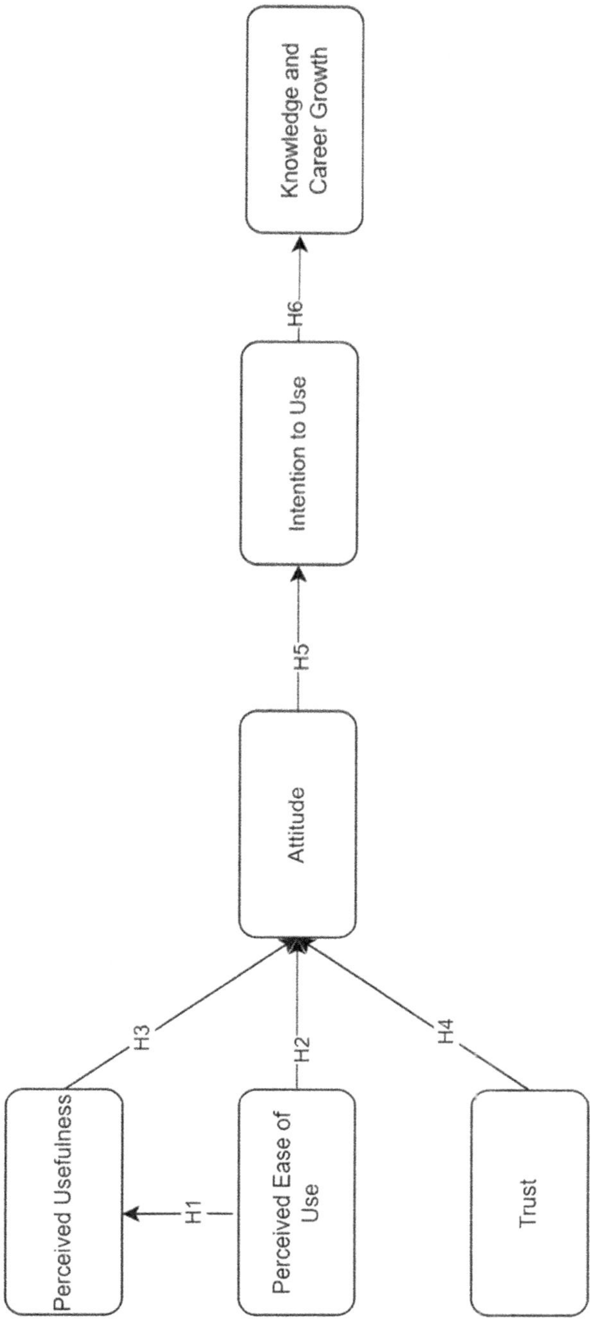

Fig. 5.1. Proposed Framework Along With Hypotheses.

that 58.8% are between 21 and 24 years, with the residual respondents varying from 17 to 42 years.

4.3 Research Tool and Questionnaire Design

The research tool was introduced to validate a particular hypothesis. The study spanned multiple universities in India, surveying undergraduate students in Engineering, Arts, and Science and postgraduate students in Business Administration, Commerce, and Science.

The survey, distributed to the 445 participants, consisted of: An initial section focused on gathering personal data.

Table 5.1 displays the origins of these constructs. Modifications were made to the questions based on past studies to increase this investigation's relevance.

5. Findings and Discussion

5.1 Data Analysis

In this section, the study investigated into the data check method using the partial least squares-structural equation modeling (PLS-SEM) method, enabled by the SmartPLS V.3 instrument. The collection of PLS-SEM for our analysis contains substantial weight for numerous purposes. PLS-SEM is specifically well-matched for study that aims to expand upon existing theories. Furthermore, its essential model admits for the adept processing of probing studies that feature sophisticated models. A separating feature of PLS-SEM is its whole approach to studying the entire model, as opposed to examining in shattered segments, a perspective highlighted by Sarstedt et al. (2021).

Table 5.1. Constructs and Their Sources.

Construct	No. of Items	Source
Perceived ease of use (PEU)	3	Venkatesh et al. (2003)
Perceived usefulness (PUS)	3	Venkatesh et al. (2003)
Trust (TST)	3	Kim et al. (2008)
Attitude (ATT)	4	Alonso-Dos-Santos et al. (2017)
Intention to use (IUS)	3	Gatzioufa and Saprikis (2022)
Knowledge and Career Trajectories (KCG)	5	Dhawan (2020), Khalil et al. (2020), Ravishankar and Jeyaprabha (2018)

5.2 Convergent and Discriminant Validity

Table 5.2 summarizes the results of this research, including all Cronbach's alpha (CA) measurements (Cheung & Wang, 2017) above the minimum level of 0.7, ranging from 0.715 to 0.833. All Composite Reliability (CR) values from 0.731 to 0.899 are well above the accepted reference value of 0.7 (Brown, 2002). The constructs are also further verified to have a good level of structural stability as shown by the CA values of 0.770–0.892. Moreover, it is demonstrated that the CR values of all the constructs are high, ranging from 0.867 to 0.933, justifying the reliability of the constructs. The values of the average variance extracted (AVE), ranging from 0.630 to 0.823, are above the suggested limit of 0.5, indicating that these measures gather most of the variance and therefore confirm their convergent validity. In redundancy variance (RDV), the AVE constructs exceeded their

Table 5.2. Convergent Validity (CV) and Discriminant Validity (DV).

Constructs	**Items**	**Factor Loading**	**Cronbach's Alpha (α)**	**CR**	**AVE**
Perceived ease of use	PEU1	0.868	0.858	0.870	0.777
	PEU2	0.886			
	PEU3	0.890			
Perceived usefulness	PUS1	0.803	0.797	0.881	0.712
	PUS2	0.870			
	PUS3	0.853			
Trust	TST1	0.903	0.892	0.933	0.823
	TST2	0.927			
	TST3	0.886			
Attitude	ATT1	0.836	0.802	0.871	0.630
	ATT2	0.792			
	ATT3	0.865			
	ATT4	0.667			
Intention to Use	IUS1	0.757	0.770	0.867	0.686
	IUS2	0.851			
	IUS3	0.871			
Knowledge and Career Growth	KCT1	0.812	0.877	0.910	0.670
	KCT2	0.803			
	KCT3	0.821			
	KCT4	0.850			
	KCT5	0.807			

Table 5.3. Fornell-Larcker Scale.

	ATT	IUS	KCT	PEO	PUS	TST
ATT	0.793					
IUS	0.669	0.828				
KCT	0.615	0.626	0.819			
PEO	0.570	0.495	0.416	0.882		
PUS	0.719	0.578	0.621	0.632	0.844	
TST	0.534	0.620	0.567	0.381	0.508	0.907

correlations with other constructs (as shown in Table 5.2), confirming the values described by Ab Hamid et al. (2017).

The evaluative framework measures construct like TST, PEU, PUS, knowledge and career trajectories (KCT), intent to use (IUS), and ATT following the Fornell-Larcker method for examining convergent and discriminant measures. The AVE and the square root of the construct loadings converge to regulate other constructs according to Fornell-Larcker (Table 5.3).

The study applied the Heterotrait-Monotrait ratio (HTMT) to decide the distinctiveness of various concepts study inspected, such as ATT, IUS, KCT, perceived enjoyment (PEO), PUS, and TST. The HTMT metrics were drawn from the parallel matrix, with values above 0.85 advising probable discriminant authenticity issues. All structures in our consideration conceded HTMT values below this limit, safeguarding agreeable discriminant validity. The HTMT metrics differed between 0.470 and 0.892, indicating to the clear distinction of each construct (Table 5.4). Previous work has emphasized the substance of obtaining discriminant validity to confirm that these unique constructs truly represent individual features (Cheung & Wang, 2017). Our study influences to lasting literature by verifying the clearness of the studied concepts, lifting the reliability of our dimension model, and hardening the genuineness of our succeeding analyses.

Table 5.4. Heterotrait-Monotrait Ratio (HTMT).

	ATT	IUS	KCT	PEO	PUS	TST
ATT						
IUS	0.834					
KCT	0.718	0.753				
PEO	0.681	0.602	0.470			
PUS	0.892	0.736	0.742	0.755		
TST	0.614	0.754	0.641	0.431	0.602	

5.3 Hypotheses Testing With PLS-SEM

The study set forth with a series of hypotheses to validate on decode the students' perceptions on ChatGPT (Table 5.5). The hypothesis (H1) predicted that the PEU would resound optimistically with Perceived Usefulness (PUS) of ChatGPT. The findings, boasting a substantial coefficient of 0.632, encourage this relationship. In simpler terms, students who confront a graceful experience with ChatGPT inherently incline regarding realizing its academic utility. This movement is not just an accepting reflection but finds bases in the initial assumptions of the TAM. It's similar to an experience wherein ease explains to effectiveness, a response echoed in the work of Venkatesh and Davis (1996).

Investigating into the second hypothesis (H2), the research hypothesized a noticeable confirmed connection between PEO and the overall student attitude (ATT) toward ChatGPT. The data, manifested by a coefficient of 0.171, demonstrate this proposition. It artfully highlights the human-behavioral surface wherein ease and ease in usage can shape a favorable tendency and direction well-aligned with TAM's core principles.

In the empire of H3, the attention altered to the involved connection between ChatGPT's perceived academic value (PUS) and the subsequent student Attitude (ATT). A coefficient of 0.502 became the observed demonstrate favoring this relationship. Here, the description is straight yet intense: identifying the academic possibility of a tool like ChatGPT directly catalyzes optimistic student sentiments, strengthening their position and confidence on the program.

For H4, the attention was on trust – a decisive yet elusive surface of technology acceptance. The hypothesis advanced that higher trust levels in ChatGPT would consistently translate to enhanced student attitudes. The study admitted this sensation with a coefficient value of 0.214. This repeats the unanimous fact that technology and trust are two sides of the same coin, an opinion campaigned in studies by Gefen et al. (2003) and McKnight et al. (2002).

Undertaking into behavioral tendencies, H5 proposed a clear connection between students' attitude (ATT) and their IUS ChatGPT. This hypothesis understood existing establishing with a coefficient of 0.669. The concentration here is pivotal: students who perceive ChatGPT progressively aren't just passive enthusiasts but active integrators, keen on spinning it into their academic wall-hanging.

Lastly, with H6, the work expanded to condense holistic growth. It anticipated that the IUS would have a pouring encouraging effect on students' KCT. The data, spotlighted by a coefficient of 0.626 (Fig. 5.2), sustained indication to this perspective. In essence, when students are sincerely persuaded toward a tool, it isn't merely an acknowledgment of its functionality but a commendation of its transformative latent, elevating both academic knowledge and career horizons – a energetic identified by Venkatesh and Davis (1996).

5.4 Analysis of Mediating Effects

The influence of mediating effect is depicted in Table 5.6, which the shows examination retained the bootstrapping technique to observe this affect. This required

Table 5.5. Assessment of the Hypotheses Within the Study's Framework (Significant at $p \leq 0.01$).

Hypothesis	Relationship	Path Value	Standard Deviation (STDEV)	*t*-Value	*p*-Values	Direction	Decision
H5	ATT → IUS	0.669	0.029	23.468	0.000	Positive	Supported
H6	IUS → KCT	0.626	0.036	17.581	0.000	Positive	Supported
H2	PEO → ATT	0.171	0.040	4.275	0.000	Positive	Supported
H1	PEO → PUS	0.632	0.040	15.943	0.000	Positive	Supported
H3	PUS → ATT	0.502	0.044	11.420	0.000	Positive	Supported
H4	TST → ATT	0.214	0.034	6.198	0.000	Positive	Supported

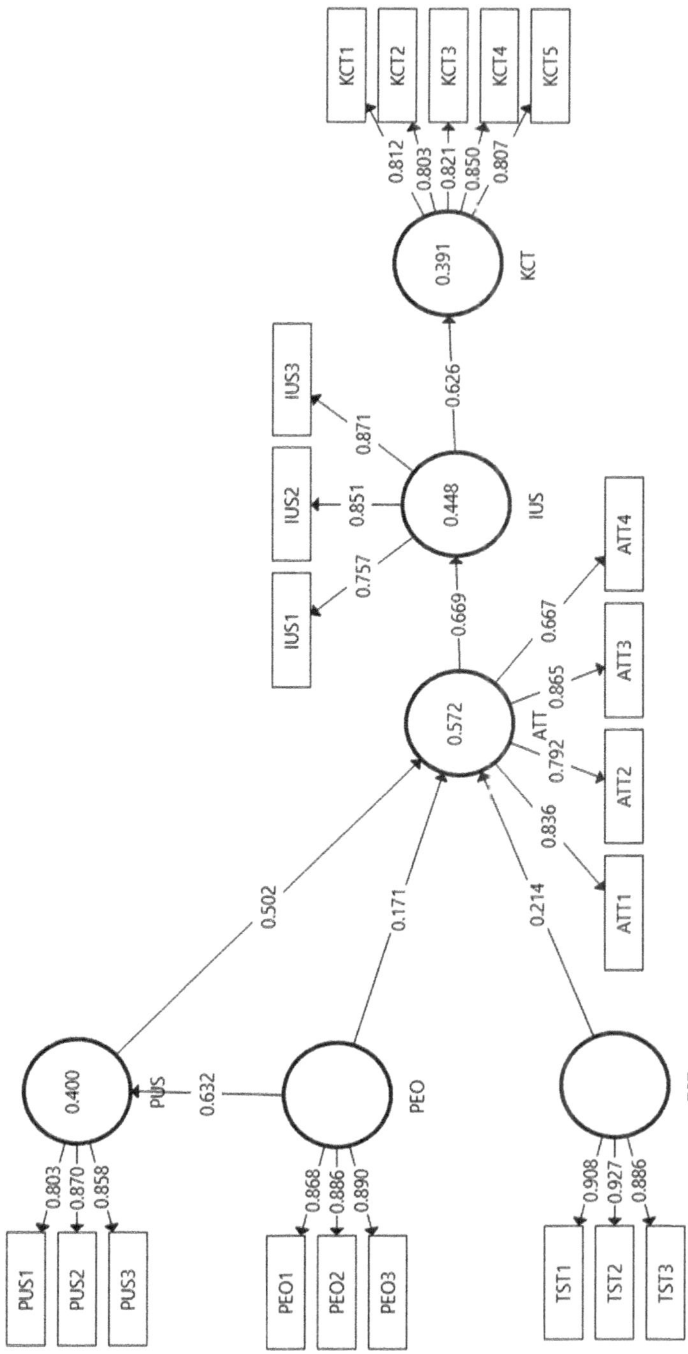

Fig. 5.2. Path Coefficient of the Model (Significant at * $p \leq 0.01$).

Table 5.6. Mediating Effect Analysis.

Relationship	Path Value	Standard Deviation (STDEV)	t-Value	p Values	Direction	Decision
PEO → PUS → ATT	0.317	0.036	8.726	0.000	Positive	Supported
PEO → ATT → IUS	0.115	0.027	4.178	0.000	Positive	Supported
PUS → ATT → IUS	0.336	0.034	10.011	0.000	Positive	Supported
PEO → PUS → ATT → IUS	0.212	0.029	7.442	0.000	Positive	Supported
TST → ATT → IUS	0.143	0.024	5.844	0.000	Positive	Supported
PEO → ATT → IUS → KCT	0.072	0.018	3.958	0.000	Positive	Supported
PUS → ATT → IUS → KCT	0.210	0.027	7.768	0.000	Positive	Supported
PEO →PUS → ATT → IUS → KCT	0.133	0.022	6.102	0.000	Positive	Supported
ATT → IUS → KCT	0.419	0.035	12.014	0.000	Positive	Supported
TST → ATT → IUS → KCT	0.089	0.017	5.217	0.000	Positive	Supported

iteratively resampling the sample 5,000 times and computing a 99% confidence interval.

To understand how students perceive ChatGPT and its ensuing possessions on their knowledge and career pathways, our study boarded on supporting a series of complex connections and intervention impacts through experienced modeling.

The first relationship unraveled was between PEU, PUS, and ATT, showcasing a strong positive association with a path value of 0.317. This relationship aligns with the findings of Davis (1989), who suggested that PEU directly influences how users perceive the usefulness of a technology. Our data, with a notable statistical significance ($p < 0.001$), underscore that ease of use can indeed be a determinant of how students find technology, such as ChatGPT, useful, which in turn shapes their overall attitude toward it (Davis et al., 1989).

Diving deeper, the mediation between PEO, ATT, and IUS yielded a path value of 0.115. Such mediation aligns with the extended insights provided by

Venkatesh et al. (2003), who suggest that when students find technology, such as ChatGPT, easy to use, it not only influences their sentiment but also strengthens their intent to integrate it into their academic endeavors. Investigation leads to downstream impacts, the cascading impact from PEO during ATT and IUS, finally affecting Knowledge and Career Trajectories (KCT), was manifest with a path value of 0.072. The significance of technology perceptions on career outcomes is not unfamiliar (Ravishankar & Jeyaprabha, 2018). Similarly, the journey from PUS through the same mediators to KCT registered a path value of 0.210, and when this started from PEO, passing through PUS and ATT before landing on IUS and KCT, the path value clocked at 0.133.

The direct and powerful affiliation emerged between ATT, IUS, and KCT, featuring a path value of 0.419, emphasizing that encouraging opinions about technology have long been linked to career and academic gains (DeLone & McLean, 2003).

Finally, trust, being the foundation of technology amalgamation, proved its primary impact. Constant with discoveries from Gefen et al. (2003) on trust in online ecosystems, resolving through ATT and IUS to KCT, trust showed a substantial path value of 0.089, emphasizing that trust doesn't just kindle positive opinions but catalyzes the holistic development of students in higher education.

In the PLS-SEM methodology, the Consistent Root Mean Square Residual (SRMR) is a key metric for estimating model fit (Sarstedt et al., 2021). In our study, the study spotted an SRMR value of 0.063. Given that this value is below the suggested threshold of 0.08 (Benitez et al., 2020), it indicates that our model exhibits vigorous explicatory energy.

6. Discussion

This study offered innate into the dynamical relationship between technological perception and its successive impact on the academic advancement of university students, specifically in the context of the ChatGPT platform. Anchored firmly in the principles of the extended TAM (Brown et al., 2010), the evaluation required to deliver a thorough comprehension of the complicated association that exists between user perceptions of technology and the tangible outcomes it can produce in an educational background.

The ubiquitous and transformative landscape of technology within the academic circle, as highlighted by Henderson et al. (2016), is both universally valued. However, the substance of its operative adaptation within the learning ecosystem is often connected with the perceptions, attitudes, and beliefs of its end users (Park, 2009; Prasad & Ghosal, 2022). The study brightened the importance of the PEU within this context. Building on Davis (1989), who highlighted the task of PEO in technology acceptance, our findings indicate that PEO plays not just as a unique metric of usableness but as a mechanism that develops perceived effectiveness and influences resultant attitudes. This in its core supports the founding theories of TAM, suggesting that user experience-focused technologies like ChatGPT will be perceived as crucial assets in the academic toolkit. The nuanced relationship between PEO, ATT, and IUS is the focus of this assessment.

6.1 Practical Implications

Resolving the details of students' perceptions toward ChatGPT and its associations on their educational development and professional trails serves as a breadth, directing educational stakeholders in the supervision of tangible developments, much like how Henderson et al. (2016) highlighted the need to comprehend the student viewpoints on technology.

For developers, designers, and ed-tech companies, the study provides significant insights. Depicting upon insights from Singha et al. (2022) on user-centered design, recognizing the psychological and behavioral elements that accelerate continuous technology embracing can steer the design culture. A prime example would be the arrangement of spontaneous user edges. By underlining effortlessness and transparency, paralleling Nielsen's (1994) heuristics for useableness, and simultaneously prominence the tangible advantages of the technology, they can find a more frictionless user experience. Moreover, strengthening trust, in line with Gefen et al. (2003), is predominant. This implies openness in operations, vigorous security measures, and coherent performance, aspects that are influential in shaping an ongoing connection of trust with end users.

6.2 Limitations

The research has various constraints to consider. Firstly, its results are based on a particular group of students, which rouses worries about the generalizability of the results to a broader student population due to potential alterations in demographics, technological circumstances, and academic backgrounds. Secondly, the cross-sectional object requires only an inert view of participants' perceptions and attitudes, whereas longitudinal studies could provide a more thorough view of how these develop over time. Thirdly, self-reported data pioneers the prospect of social advantage bias and recall mistakes, implying the need for more objective measures like academic performance metrics. Furthermore, the mediation analyses rest on statements about causal associations, which are arguing to be found ultimately given the cross-sectional nature of the data. Additionally, the study's exclusive concentration on ChatGPT restricts its scope, and future research should discover the influence of other emerging technologies. Lastly, while the mediation analyses portray subsequent routes, the elaborate mechanisms and unknown mediators or moderators may have nuanced impacts on the monitored associations.

7. Conclusion

The research highlights the integrated concept of students' perceptions, technology acceptance, and academic outcomes, depicting the potentials of ChatGPT within the higher education sector. Positioned within the extended TAM, the research explains how the PEO is influencing learners' attitude, intention to use, and, in turn, influencing their academic outcomes. The study uncovers a fundamental lesson for educators, that is, by maximizing such apparent advantages of a user, the resource may shape the perceptive of technology in their favor, increasing the pace and breadth of adaptation of technology. The research parsimonious analyses have demonstrated the composite interaction of think, feel, and do to illustrate how technology may affect education from different angles. Meanwhile, the introduction of trust into the prominent perception demonstrates another parsimonious conclusion, that is, to make the technology part of education engagingly, improving trust is not just good but mandatory. The findings drawn from this research provide as a guide for those at the helm of education – be it educators, institutional leaders. As they navigate through the diverse landscape of educational technology, anchoring their strategies in adopting positive attitudes and aligning tech proposals with students' career aspirations can be complex. With a tool as challenging as ChatGPT, its optimal use can drive both tertiary education and career preparation. In a broader sense, this research doesn't merely produce evidence but also it provides a comprehensive view of importance of technology in education. It highlights that with the right blend of technology and approach, students' academic expeditions are not simply a matter of standard education but one of preparing them for luminous success in an exponentially changing educational cosmos.

References

Ab Hamid, M. R., Sami, W., & Sidek, M. M. (2017). Discriminant validity assessment: Use of Fornell & Larcker criterion versus HTMT criterion. *Journal of Physics: Conference Series*, *890*. Article ID: 012163. https://doi.org/10.1088/1742-6596/890/1/012163

Ajzen, I., & Fishbein, M. (1977). Attitude-behavior relations: A theoretical analysis and review of empirical research. *Psychological Bulletin*. https://doi.org/10.1037/0033-2909.84.5.888

Ajzen, I., Fishbein, M., Lohmann, S., & Albarracín, D. (2018). The influence of attitudes on behavior. In D. Albarracin & B. T. Johnson (Eds.), *The handbook of attitudes* (pp.197–255).

Alonso-Dos-Santos, M., Moreno, F. C., Ríos, F. M., & Alguacil, M. (2017). Online sport event consumers: Attitude, e-quality and e-satisfaction. *Journal of Theoretical and Applied Electronic Commerce Research*, *12*(2), 54–70.

Amadu, L., Muhammad, S. S., Mohammed, A. S., Owusu, G., & Lukman, S. (2018). Using technology acceptance model to measure the ese of social media for collaborative learning in Ghana. *JOTSE*, *8*(4), 321–336.

Baidoo-Anu, D., & Ansah, L. O. (2023). Education in the era of generative artificial intelligence (AI): Understanding the potential benefits of ChatGPT in promoting teaching and learning. *Journal of AI*, *7*(1), 52–62.

Bazelais, P., Doleck, T., & Lemay, D. J. (2018). Investigating the predictive power of TAM: A case study of CEGEP students' intentions to use online learning technologies. *Education and Information Technologies*, *23*(1), 93–111. https://doi.org/10.1007/s10639-017-9587-0

Benitez, J., Henseler, J., Castillo, A., & Schuberth, F. (2020). How to perform and report an impactful analysis using partial least squares: Guidelines for confirmatory and explanatory IS research. *Information & Management*, *57*(2), 103168. https://doi.org/10.1016/j.im.2019.05.003

Bonsu, E. M., & Baffour-Koduah, D. (2023). From the consumers' side: Determining students' perception and intention to use ChatGPT in Ghanaian higher education. *Journal of Education, Society & Multiculturalism*, *4*(1), 1–29. https://doi.org/10.2478/jesm-2023-0001

Brown, J. D. (2002). The Cronbach alpha reliability estimate. *JALT Testing & Evaluation SIG Newsletter*, *6*(1).

Brown, S. A., Dennis, A. R., & Venkatesh, V. (2010). Predicting collaboration technology use: Integrating technology adoption and collaboration research. *Journal of Management Information Systems*, *27*(2), 9–54. https://doi.org/10.2753/MIS0742-1222270201

Cheung, G. W., & Wang, C. (2017). Current approaches for assessing convergent and discriminant validity with SEM: Issues and solutions. *Academy of Management Proceedings*, *2017*(1), 12706.

Davis, F. D. (1989). Perceived usefulness, perceived ease of use, and user acceptance of information technology. *MIS Quarterly*, *13*(3), 319. https://doi.org/10.2307/249008

Davis, F. D., Bagozzi, R. P., & Warshaw, P. R. (1989). User acceptance of computer technology: A comparison of two theoretical models. *Management Science*, *35*(8), 982–1003.

DeLone, W. H., & McLean, E. R. (2003). The DeLone and McLean model of information systems success: A ten-year update. *Journal of Management Information Systems*, *19*(4), 9–30.

Dhagarra, D., Goswami, M., & Kumar, G. (2020). Impact of trust and privacy concerns on technology acceptance in healthcare: An Indian perspective. *International Journal of Medical Informatics*, *141*, 104164.

Dhawan, S. (2020). Online learning: A panacea in the time of COVID-19 crisis. *Journal of Educational Technology Systems*, *49*(1), 5–22.

Eke, D. O. (2023). ChatGPT and the rise of generative AI: Threat to academic integrity? *Journal of Responsible Technology*, *13*, 100060.

Faul, F., Erdfelder, E., Buchner, A., & Lang, A.-G. (2009). Statistical power analyses using G*Power 3.1: Tests for correlation and regression analyses. *Behavior Research Methods*, *41*(4), 1149–1160. https://doi.org/10.3758/BRM.41.4.1149

Gatzioufa, P., & Saprikis, V. (2022). A literature review on users' behavioral intention toward chatbots' adoption. *Applied Computing and Informatics*. https://doi.org/10.1108/ACI-01-2022-0021

Gefen, D., Karahanna, E., & Straub, D. W. (2003). Trust and TAM in online shopping: An integrated model. *MIS Quarterly*, *50*, 51–90.

Goh, W. W., Hong, J. L., & Gunawan, W. (2013). Exploring students' perceptions of learning management system: An empirical study based on TAM. In *2013 2nd IEEE International Conference on Teaching, Assessment and Learning for Engineering, TALE 2013* (pp. 367–372). https://doi.org/10.1109/TALE.2013.6654463

Granić, A., & Marangunić, N. (2019). Technology acceptance model in educational context: A systematic literature review. *British Journal of Educational Technology*, *50*(5), 2572–2593.

Harryanto, Muchran, M., & Ahmar, A. (2018). Application of TAM model to the use of information technology. *International Journal of Engineering and Technology(UAE)*, *7*.

Henderson, M., Finger, G., & Selwyn, N. (2016). What's used and what's useful? Exploring digital technology use (s) among taught postgraduate students. *Active Learning in Higher Education*, *17*(3), 235–247.

Kasneci, E., Seßler, K., Küchemann, S., Bannert, M., Dementieva, D., Fischer, F., Gasser, U., Groh, G., Günnemann, S., & Hüllermeier, E. (2023). ChatGPT for good? On opportunities and challenges of large language models for education. *Learning and Individual Differences*, *103*, 102274.

Khalil, R., Mansour, A. E., Fadda, W. A., Almisnid, K., Aldamegh, M., Al-Nafeesah, A., Alkhalifah, A., & Al-Wutayd, O. (2020). The sudden transition to synchronized online learning during the COVID-19 pandemic in Saudi Arabia: A qualitative study exploring medical students' perspectives. *BMC Medical Education*, *20*(1), 1–10.

Kim, D. J., Ferrin, D. L., & Rao, H. R. (2008). A trust-based consumer decision-making model in electronic commerce: The role of trust, perceived risk, and their antecedents. *Decision Support Systems*, *44*(2), 544–564.

Mailizar, M., Almanthari, A., & Maulina, S. (2021). Examining teachers' behavioral intention to use E-learning in teaching of mathematics: An extended TAM model. *Contemporary Educational Technology*, *13*(2), ep298.

McKnight, D. H., Choudhury, V., & Kacmar, C. (2002). The impact of initial consumer trust on intentions to transact with a web site: A trust building model. *The Journal of Strategic Information Systems*, *11*(3–4), 297–323.

Nielsen, J. (1994). *Usability engineering*. Morgan Kaufmann Publishers.

Park, S. Y. (2009). An analysis of the technology acceptance model in understanding university students' behavioral intention to use e-learning. *Journal of Educational Technology & Society*, *12*(3), 150–162.

Persico, D., Manca, S., & Pozzi, F. (2014). Adapting the technology acceptance model to evaluate the innovative potential of e-learning systems. *Computers in Human Behavior*, *30*, 614–622.

Prakosa, A., & Sumantika, A. (2021). An analysis of online shoppers' acceptance and trust toward electronic marketplace using TAM model. *Journal of Physics: Conference Series*, *1823*(1), 012008.

Prasad, B., & Ghosal, I. (2022). Forecasting buying intention through artificial neural network: An algorithmic solution on direct-to-consumer brands. *FIIB Business Review*, *11*(4), 405–421.

Radford, A., Wu, J., Child, R., Luan, D., Amodei, D., & Sutskever, I. (2019). Language models are unsupervised multitask learners. *OpenAI Blog*, *1*(8), 9.

Ravishankar, K., & Jeyaprabha, B. (2018). Commerce and Management Faculty Perspective towards Impact of Digital Media and E-Learning on Contemporary Management Education. *International Journal of Research in Engineering, IT and Social Sciences*, *8*(9), 230–238.

Ray, P. P. (2023). ChatGPT: A comprehensive review on background, applications, key challenges, bias, ethics, limitations and future scope. *Internet of Things and Cyber-Physical Systems*, *3*(1), 121–154.

Rossi, F. (2018). Building trust in artificial intelligence. *Journal of International Affairs*, *72*(1), 127–134.

Sarstedt, M., Ringle, C. M., & Hair, J. F. (2021). Partial least squares structural equation modeling. In *Handbook of market research* (pp. 1–47). Springer International Publishing. https://doi.org/10.1007/978-3-319-05542-8_15-2

Singha, B., Eljamal, O., Karmaker, S. C., Maamoun, I., & Sugihara, Y. (2022). Water conservation behavior: Exploring the role of social, psychological, and behavioral determinants. *Journal of Environmental Management*, *317*, 115484.

Shoufan, A. (2023). Exploring students' perceptions of CHATGPT: Thematic analysis and follow-up survey. *IEEE Access*. https://doi.org/10.1109/ACCESS.2023.3268224

Su, Y., & Li, M. (2021). Applying technology acceptance model in online entrepreneurship education for new entrepreneurs. *Frontiers in Psychology*, *12*, 713239.

Tsai, Y.-R. (2015). Applying the technology acceptance model (TAM) to explore the effects of a course management system (CMS)-assisted EFL writing instruction. *Calico Journal*, *32*(1), 153–171.

Uche, D. B., Osuagwu, O. B., Nwosu, S. N., & Otika, U. S. (2020). Integrating trust into technology acceptance model (TAM), the conceptual framework for e-payment platform acceptance. *Marketing*, *4*(4), 34–56.

Venkatesh, V., & Davis, F. D. (1996). A model of the antecedents of perceived ease of use: Development and test. *Decision Sciences*, *27*(3), 451–481.

Venkatesh, V., Morris, M. G., Davis, G. B., & Davis, F. D. (2003). User acceptance of information technology: Toward a unified view. *MIS Quarterly*, 425–478.

Wang, C., Liu, S., Yang, H., Guo, J., Wu, Y., & Liu, J. (2023). Ethical considerations of using ChatGPT in health care. *Journal of Medical Internet Research*, *25*, e48009.

Weng, F., Yang, R.-J., Ho, H.-J., & Su, H.-M. (2018). A TAM-based study of the attitude towards use intention of multimedia among school teachers. *Applied System Innovation*, *1*(3), 36.

Zhou, L., Xue, S., & Li, R. (2022). Extending the Technology Acceptance Model to explore students' intention to use an online education platform at a University in China. *Sage Open*, *12*(1). 21582440221085260.

Chapter 6

Cross-Cultural Leadership in Tech-Driven World

Richa Jain[a] *and Durdana Ovais*[b]

[a]Prestige Institute of Management and Research, India
[b]BSSS Institute of Advanced Studies, India

Abstract

This study investigates the dynamics of cross-cultural leadership in a technology-driven world through a systematic literature review, analyzing how leaders navigate diverse teams amid rapid technological advancements. The research emphasizes the importance of understanding cultural variances in decision-making, communication, authority perception, and work habits to effectively bridge cultural differences and leverage diversity. A comprehensive literature review reveals the significant impact of national culture on leadership performance, the role of cultural intelligence (CQ) in managing diverse teams, and the essential competencies required for effective cross-cultural leadership. The methodology involved analyzing 35 peer-reviewed articles from 1997 to 2024, identifying key themes such as ethical dilemmas, cultural disparities, and the influence of technology on leadership practices. The findings emphasize the complexity and importance of cross-cultural leadership in today's globalized business environment. Effective leadership requires not only understanding and respecting different cultural norms and values but also adapting leadership styles to diverse contexts. High CQ is crucial for leaders to navigate cultural differences, build trust, and communicate effectively with global teams. Additionally, ethical behavior is vital in cross-cultural settings to avoid misunderstandings and build trust. Challenges identified include understanding and adapting to different cultures, managing ethical perceptions, addressing consumer behavior and corporate social responsibility (CSR), and fostering effective team dynamics in multicultural settings. Skills such as adaptability, ethical behavior, and strong communication are essential for leaders to manage these challenges successfully.

Indigenous Empowerment through Human-Machine Interactions, 93–113

Published under exclusive licence by Emerald Publishing Limited
doi:10.1108/978-1-83608-068-820251006

Keywords: Cross-cultural; diversity; leadership; technology; globalization; skills

1. Introduction

Navigating the challenges of leading diverse teams in an environment affected by fast increasing technical breakthroughs is a key component of cross-cultural leadership in today's tech-driven world. Culture may be defined as a collection of fundamental and collective behaviors and beliefs that assist human societies in resolving issues related to both internal and external integration (Schein, 2010). Leaders need to have a thorough awareness of the subtle cultural variances and distinctions among their team members. This entails understanding decision-making procedures, preferred methods of communication, attitudes toward authority, and work habits. People follow a vast array of social and moral traditions, speak thousands of different languages, and worship hundreds of thousands of gods. One of the main objectives of the social sciences is to comprehend patterns in human cognition and behavior as well as cross-cultural variability. Scholars have gained amazing insights into cultural diversity by using independent cross-cultural codes and cross-cultural databases based on ethnography (Kirby et al., 2016). According to the implicit leadership hypothesis, people have preconceived notions about the qualities, traits, abilities, and actions that either support or contradict leadership.

In the digital age, leaders have to makeover through constantly changing environments while utilizing data, adopting new technology, and encouraging a culture of lifelong learning. As technology keeps bringing people together across borders and time zones, leaders need to be able to effectively bridge cultural differences and take advantage of diversity. Leadership in the digital age requires a combination of technical know-how, strategic vision, and the capacity to harness innovation in order to propel organizations forward into a constantly changing future, since technology is reshaping industries and paradigms. In order to ascertain the relationship between technology and cross-cultural leadership, this study conducts a systematic literature review. We intend to learn more about the difficulties faced by leaders in this technologically advanced environment through this research.

2. Literature Review

This conceptualization of the common element of leadership culture in a nation's organizations is an adaptation to the noncultural environment (Van de Vliert & Einarsen, 2008). Because national culture has a significant moderating role in the relationship between leadership and performance, our analysis advances our understanding of the relationship between performance and leadership in various national settings and across sectors. The studies by Casimir and Waldman (2007), Sertel et al. (2022), and Gill and Negrov (2021) indicate that both the demands of the leadership role and the culturally accepted interpersonal standards influence the perceived importance of certain leadership attributes. Caligiuri et al. (2022)

demonstrate the use of three dimensions to the understanding of cross-cultural performance: task performance, skill development, and cross-cultural adjustment.

Workers with cross-cultural expertise who are seen to make it easier for them to achieve their strategic goals appear to be preferred by businesses (Zoogah & Abbey, 2010). From the standpoint of attribution theory, the underlying presumptions of the essentialist interpretation of culture are examined. The premise that culture may accurately anticipate behavior at the expense of an actor's agency forms the basis of many of the assumptions (Dandotiya & Ghosal, 2024; Långstedt, 2018). In today's internationally integrated business environment, where people from diverse national cultures collaborate, cultural intelligence (CQ) is becoming more and more vital (Philip et al., 2023). Plessis (2012) examines how multicultural team members in project and management teams perceive and experience teamwork paradoxes in a multicultural workplace, with a particular focus on managing cross-cultural conflict.

3. Research Methodology

The systematic literature review (SLR) on Cross-Cultural Leadership in a Tech-Driven World employed the following research methodology: the selection of peer-reviewed articles that were published within the time frame of 1997–2024 and pertained to this subject matter. The research team employed academic database of Google Scholar, exploring topics such as "cross-cultural leadership," "technology," "globalization," and "skills." The selection process restricted papers to those written in English and specifically focused on cross-cultural leadership within technology-driven environments. Thirty-five research publications were identified and evaluated in total. The data extraction encompassed significant discoveries, obstacles, patterns, and competencies necessary for cross-cultural leadership. The researchers performed thematic analysis in order to categorize recurring themes and ascertain significant obstacles, patterns, and competencies. The findings underscored obstacles such as ethical quandaries and cultural disparities, developments such as the influence of technology on leadership methodologies, and critical competencies including adaptability, cultural sensitivity, and communication capabilities. In addition, the methodology section addressed recommendations for future research, implications for organizations and leaders, and limitations acknowledged, including the omission of non-English articles.

4. Discussion

4.1 Word Cloud Analysis

The emphasis and focus of the research undertakings can be seen in how often keywords are used (Fig. 6.1). The fact that words like "cross-cultural," "culture," "leadership," and "management" showed up a lot in this case says that these ideas were given a lot of attention in the studies. These keywords show that the topics about managing ethnic differences, leadership styles, and company culture in a global setting are gaining importance. A multidisciplinary approach is shown by

Fig. 6.1 Word Cloud Analysis of Key Words Used. *Source:* Developed for the purpose of the study.

the fact that the research studies cover a wide range of cross-cultural management issues. From a critical point of view, there are a lot of important themes like diversity, identity, and power relations. The use of keywords such as "Otherness," "Difference," and "Intersectionality" suggests that an investigation of the ways in which various groups are regarded and handled within organizational contexts is being conducted. Another big topic of interest is leadership. Words like "cross-cultural," "universal," and "multi-national" show a desire to learn about how people lead in different countries and cultures. The way the research studies were conducted is also emphasized, with terms like "Interpretive research" and "Longitudinal analysis" indicating a focus on qualitative research methods and looking at trends over time. A lot of different cultural settings are looked at in the framework of bigger ideas like "Myths" and "Political culture." The use of terminology such as "Business ethics" and "Corporate social responsibility" (CSR) demonstrates an interest in ethical behavior in international business settings. This is because ethical considerations within cross-cultural contexts are stressed.

4.2 Trend Analysis

Over the course of the years, the frequency of publishing has exhibited a pattern of variation (Fig. 6.2). The years 1997 and 1998 each saw the publishing of a

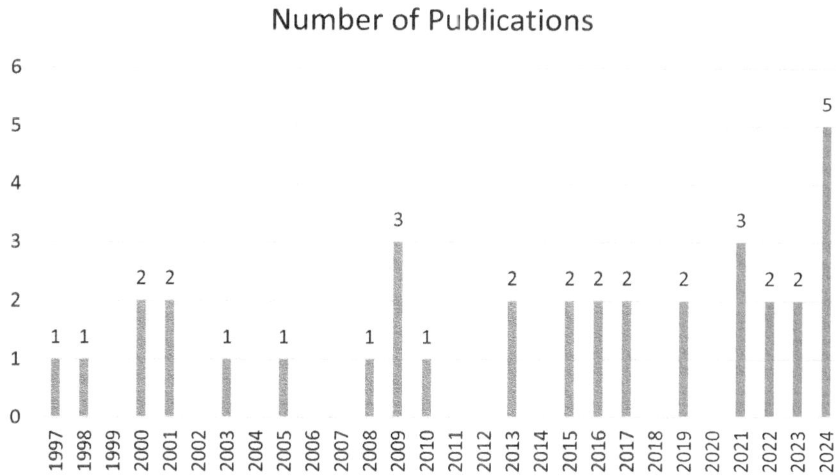

Fig. 6.2. Number of Publications. *Source:* Developed for the purpose of the study.

single publication. The number of publications reached two in the year 2000 and remained the same in the year 2001. In 2003, there was a small drop, with only one release. In the years that followed, the trend changed. In 2005, 2008, and 2010, there was only one release, but in 2009, there were three. In 2013, 2015, 2016, and 2017, there were two releases each year, so the number of publications went up again. There were two releases in 2019 and three in 2021. Five books were made in 2024, which was the most of any year. Over the course of the years, the data reveal a pattern of fluctuation in the number of publications, with certain years exhibiting a higher level of activity than others.

4.3 Major Journals Supporting the Research

The above graph (Fig. 6.3) displays how many times papers about cross-cultural management have been published in various journals. It is the "International Journal of Cross Cultural Management" that puts out the most, with 18 publications. The journal "Cross Cultural Management" has come out three times. The "Journal of World Business," the "Journal of Business Ethics," and the "International Journal of Leadership Studies" have also written about this subject, though not as often (two publications in each). This demonstrates that there is a significant quantity of research being carried out and published in a variety of journals on the subject of cross-cultural management. This is a reflection of the significance and interest that is being shown in this area of educational research.

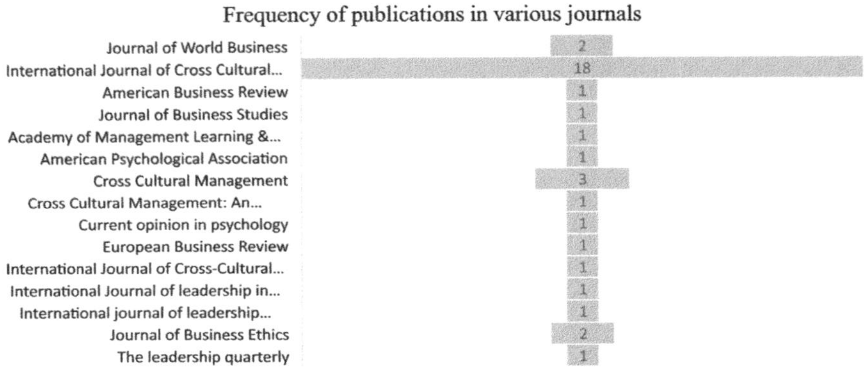

Fig. 6.3. Frequency of Publication in Various Journals. *Source:* Developed for the purpose of the study.

4.4 Types of Studies

The figure above (Fig. 6.4) shows the frequency of different types of studies conducted. There was one comprehensive analysis, which is a thorough and detailed examination of a topic. Qualitative studies, which involve non-numerical data like interviews and observations, were conducted 15 times. SLR was done

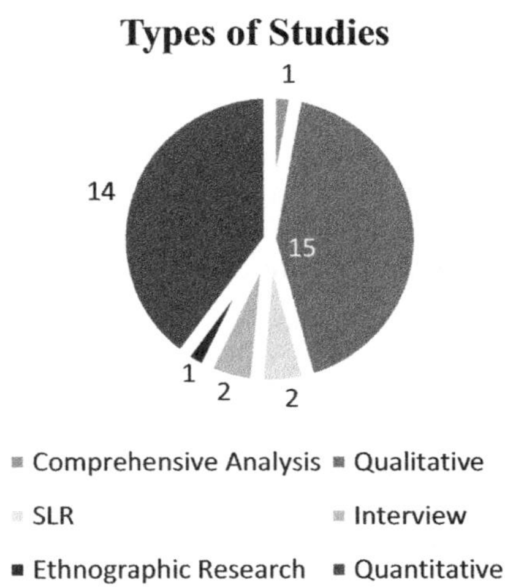

Fig. 6.4. Types of Studies. *Source:* Developed for the purpose of the study.

twice, indicating a structured approach to reviewing existing literature. Interviews were conducted twice, suggesting a method of gathering firsthand information from participants. Ethnographic research, which involves studying people and cultures in their natural environment, was conducted once. Quantitative studies, which involve numerical data and statistical analysis, were conducted 14 times. In all there were 35 studies in total, showcasing a variety of research methods used to investigate different topics.

4.5 Major Challenges Identified by Studies in Cross-Cultural Leadership in Tech-Driven World

The table below (Table 6.1) presents various challenges in cross-cultural management along with the authors who have discussed these challenges. These challenges cover a wide range of topics such as leadership development, ethical perceptions, CQ, and communication. They also include challenges related to national and organizational culture, emotional intelligence, and effective decision-making. The table also highlights challenges in research methodologies, perceived leadership effectiveness, and social support in different cultural contexts. Other challenges include managing cultural synergies and differences in teams, as well as addressing human rights and corporate responsibility in international business. Additionally, challenges related to globalization, academic disciplinary knowledge, and the impact of cultural distance on performance are also discussed. Each challenge is linked to specific authors who have contributed insights and perspectives on these issues in cross-cultural management.

Table 6.1. Major Challenges in Cross-Cultural Leadership in Tech-Driven World Identified by Studies.

Category	Challenges	Citations
Leadership development	Challenges related to developing effective leadership programs and strategies in cross-cultural contexts.	Robertson and Webber (2000), Stahl and Brannen (2013)
Ethical perceptions and practices	Challenges concerning ethical perceptions, practices, and whistleblowing across different cultures.	Brody et al. (1998), Ermasova (2021)
Cultural intelligence (CQ) and sensitivity	Challenges related to understanding and developing CQ and sensitivity among leaders and team members.	Deng and Gibson (2009)

(Continued)

Table 6.1. *(Continued)*

Category	Challenges	Citations
National and organizational culture	Challenges associated with the impact of national and organizational cultures on various aspects of management.	Triandis (2001), Chevrier (2009)
Emotional intelligence and communication	Challenges regarding the role of emotions and effective communication in cross-cultural leader-follower interactions.	Hanges et al. (2016)
Cultural synergies and differences	Challenges in building cultural synergies and managing differences in multicultural teams.	Stahl and Brannen (2013)
Effective delegation and decision-making	Challenges in delegating effectively and making decisions across cultures.	Chevrier and Viegas-Pires (2013)
Cross-cultural project management	Challenges in managing large projects with multi or cross-cultural teams, including language barriers and diversity in behavior.	Kappagomtula (2017)
Research methodologies and paradigms	Challenges in conducting cross-cultural research, including methodological approaches and paradigms.	Barmeyer et al. (2019)
Perceived leadership effectiveness	Challenges in understanding and measuring perceived leadership effectiveness across cultures.	Yan and Hunt (2005)
Social support and integration	Challenges related to the influence of perceived social support on support seeking behavior in different cultural contexts.	Acikdeniz et al. (2024)
Human rights and corporate responsibility	Challenges concerning the recognition and protection of human rights in international business practices.	Ascencio et al. (2024)

Table 6.1. *(Continued)*

Category	Challenges	Citations
Migration and ethnic otherness	Challenges related to conflict and integration in migration and ethnic otherness in various countries.	Mahadevan (2024)
Globalization and organizational change	Challenges related to the speed or rate of change in cultural values due to globalization and organizational change.	Dickson et al. (2003)
Academic disciplinary knowledge	Challenges in re-ordering disciplinary knowledge and methodologies in cross-cultural management studies.	Mahadevan (2023)
Cultural distance and performance	Challenges in measuring and understanding the effects of cultural distance on organizational performance.	Correa da Cunha et al. (2022)

Source: Developed for the purpose of the study.

Over the years, the challenges in cross-cultural management have evolved in response to changing global dynamics and organizational needs. In the late 1990s and early 2000s, challenges such as ethical perceptions and practices (Brody et al., 1998) and the impact of national and organizational culture (Triandis, 2001) were prominent. These challenges highlighted the need for understanding and managing cultural differences in ethical standards and organizational behaviors.

As the 2000s progressed, there was a growing recognition of the importance of leadership development (Robertson & Webber, 2000) and the speed of change in cultural values (Dickson et al., 2003; Jain et al., 2024). These challenges underscored the need for leaders to adapt to changing cultural norms and values in a rapidly globalizing world. In the mid-2000s to early 2010s, challenges related to CQ and sensitivity (Deng & Gibson, 2008) and research methodologies and paradigms (Barmeyer et al., 2019) became more prominent. These challenges reflected a shift toward a more nuanced understanding of cross-cultural interactions and the need for robust research methods to study them effectively.

In more recent years, challenges such as effective delegation and decision-making (Chevrier & Viegas-Pires, 2013) and the role of emotions in leader-follower interactions (Hanges et al., 2016) have gained significance. These challenges reflect broader societal concerns and the need for organizations to address them in their global operations. The challenges in cross-cultural management have evolved from traditional issues of cultural differences to more complex and nuanced challenges that reflect the changing nature of global business and society. Adaptation and innovation in managing these challenges are essential for organizations to succeed in today's interconnected world.

4.6 Thematic Focus

Based on the key findings analyzed, the current focus of cross-cultural management studies can be characterized by several overarching themes (Table 6.2):

Table 6.2. A Thematic Center of Focus as Identified by Studies.

Focus Area	Key Findings	Citation
Methodological advancements	- Methodological suggestions for cross-cultural leadership research - Emphasis on refining research methods	House et al. (1997)
Ethical perceptions and behavior	- Significant differences in ethical perceptions between the United States and Japanese accounting students - Correlation of software piracy with GNP per capita, income inequality, and individualism	Brody et al. (1998), Husted (2000)
Cultural intelligence (CQ) and leadership	- Enhancement of critical reflection and leadership development through CQ - Importance of CQ in daily business practice	Robertson and Webber (2000), Deng and Gibson (2008)
Consumer behavior and CSR	- Cross-cultural differences in consumer support for responsible businesses - Varied consumer priorities on corporate responsibilities	Maignan (2001), Ascencio et al. (2024)
Team dynamics and trust	- Impact of team consensus, diversity, and fault lines on trust and performance - Moderation of team trust by cultural factors	Mach and Baruch (2015), Hanges et al. (2016)
Cultural influences on leadership style	- Cultural impact on perceived leadership effectiveness - Adaptation of leadership	Yan and Hunt (2005), Stahl and Brannen (2013)

Table 6.2. *(Continued)*

Focus Area	Key Findings	Citation
	behaviors to cultural prototypes	
Global leadership competencies	- Effective empowerment dependent on how delegation is performed, not on the amount of delegation - Adaptation of empowerment strategies based on local cultures	Chevrier and Viegas-Pires (2013), Deng and Gibson (2009)
Cross-cultural communication competence	- Influence of perceived support on support-seeking behaviors across cultures - Impact of multicultural team effectiveness on expatriates' communication competence	Abugre and Debrah (2019)
Ethical business practices	- Identification of gaps between literature and challenges in cross-cultural business ethics - Contribution to understanding cross-cultural differences in business ethics	Ermasova (2021)
Managing diversity and inclusion	- Focus on managing diversity and avoiding harmful othering in cross-cultural interactions - Emphasis on promoting inclusivity in organizations	Mahadevan and Primecz (2024), Chevrier (2024)

Source: Developed for the purpose of the study.

Methodological Advancements: House et al. (1997) suggest methodological improvements for cross-cultural leadership research, indicating a continued emphasis on refining research methods in the field. *Ethical Perceptions and Behavior*: Studies like Brody et al. (1998) and Husted (2000) highlight the impact of national culture on ethical perceptions and behaviors, indicating a growing concern for ethical considerations in cross-cultural contexts. *CQ and Leadership Development*: Research by Robertson and Webber (2000) and Deng and Gibson (2008) emphasizes the importance of CQ in leadership effectiveness, indicating a focus on developing leaders who can navigate diverse cultural environments. *Consumer Behavior and CSR*: Maignan (2001) and Ascencio et al. (2024) explore consumers' perceptions of CSR across cultures, indicating a continued interest in understanding consumer behavior and ethical business practices. *Team Dynamics and Trust:* Mach and Baruch (2015) and Hanges et al. (2016) examine team dynamics and trust in cross-cultural settings, suggesting a focus on building effective multicultural teams. *Cultural Influences on Leadership Style*: Yan and Hunt (2005) and Stahl and Brannen (2013) discuss how cultural contexts influence perceived leadership effectiveness, indicating an interest in understanding the interplay between culture and leadership style. *Cross-Cultural Communication Competence*: Abugre and Debrah (2019) highlight the importance of cross-cultural communication competence in expatriate business operations, indicating a continued interest in effective communication across cultures. *Ethical Business Practices in Global Contexts*: Ermasova (2021) discusses the challenges of cross-cultural issues in business ethics, suggesting an ongoing focus on promoting ethical business practices in diverse cultural settings. *Managing Diversity and Inclusion*: Mahadevan and Primecz (2024) and Chevrier (2024) emphasize the importance of managing diversity and avoiding harmful Othering in cross-cultural interactions, indicating a growing interest in promoting inclusivity in organizations.

4.7 Key Findings Identified by Studies in Cross-Cultural Leadership in Tech-Driven World

The evolution of key findings in cross-cultural management research over the years has been rich and diverse, reflecting a deepening understanding of cultural dynamics in organizational contexts (Table 6.3). House et al. (1997) laid the groundwork by proposing methodological suggestions to enhance cross-cultural leadership research, setting a critical tone for future studies. Brody et al. (1998) highlighted significant differences in ethical perceptions between US and Japanese accounting students, emphasizing the role of individualism in shaping these perceptions.

The focus then shifted to team dynamics in cross-cultural projects, with Mach and Baruch (2015) emphasizing the role of consensus, heterogeneity, and trust in shaping team performance. Hanges et al. (2016) and Stahl and Brannen (2013) delved into the complexities of global leadership, highlighting the need for leaders to learn from multicultural experiences and adapt to different cultural contexts.

Recent studies have explored the intersection of business ethics and human rights, with Ascencio et al. (2024) examining corporate human rights violations and the role of global corporations in addressing these issues.

The evolution of key findings in cross-cultural management research reflects a growing awareness of the complexities of cultural interactions in organizational settings, highlighting the need for more inclusive and culturally sensitive approaches to leadership and management.

4.8 Categorization of Periods of Study Based on Findings

The evolution of key findings in cross-cultural management research can be categorized into periods based on the dominant themes and trends:

Table 6.3. Essential Findings Resulting From Cross-Cultural Leadership Studies in a Tech-Driven World.

Category	Key Findings	Citations
Cross-cultural leadership research and theory	- Methodological suggestions for improving cross-cultural leadership research and theory.	House et al. (1997)
	- Importance of cultural intelligence for effective leadership.	Deng and Gibson (2009)
Ethical perceptions and behavior	- Differences in ethical perceptions of individuals from different cultures.	Brody et al. (1998)
	- Influence of national culture on ethical norms in public and private sectors.	Danon-Leva et al. (2010)
Consumer behavior and CSR	- Differences in consumers' perceptions of CSR across cultures.	Maignan (2001)
	- Importance of responsible business practices in different cultural contexts.	Maignan (2001)
Team dynamics in cross-cultural settings	- Factors influencing team performance in cross-cultural projects.	Mach and Baruch (2015)
	- Impact of cultural consensus, heterogeneity, and trust on team effectiveness.	Mach and Baruch (2015)
Management practices and cultural	- Relevance of national culture to management practices.	Chevrier (2009)
Relevance	- Importance of adapting management styles to local cultures.	Chevrier and Viegas-Pires (2013)
Global leadership and cultural adaptation	- Evolution of global leadership and its relevance in today's context.	Bird and Mendenhall (2016)
	- Strategies for developing multicultural management skills.	Stahl and Brannen (2013)
Business ethics and cultural influence	- Challenges and gaps in addressing cross-cultural issues in business ethics.	Ermasova (2021)

(Continued)

Table 6.3. *(Continued)*

Category	Key Findings	Citations
	- Influence of institutional environment and national culture on ethical attitudes.	Ermasova (2021)
Human rights and corporate Accountability	- Corporate human rights violations and their impact on marginalized communities.	Ascencio et al. (2024)
	- Need for aligning corporate actions with human rights principles.	Ascencio et al. (2024)
Otherness and marginalization	- Understanding and addressing othering and saming as core responsibilities in cross-cultural management.	Mahadevan and Primecz (2024)
	- Success of positive cross-cultural management with otherness.	Mahadevan. (2024)

Source: Developed for the purpose of the study.

Foundational Period (1997–2000): This period laid the groundwork for cross-cultural management research, focusing on methodological improvements (House et al., 1997) and highlighting cultural differences in ethical perceptions (Brody et al., 1998).

Leadership Development and CQ (2000–2010): Research during this period emphasized the development of culturally sensitive leaders (Robertson & Webber, 2000), the impact of CQ on leadership effectiveness (Deng & Gibson, 2008), and the role of national culture in shaping ethical behavior and consumer perceptions (Husted, 2000; Maignan, 2001).

Team Dynamics and Global Leadership (2010–2016): Studies in this period explored team dynamics in cross-cultural projects, focusing on consensus, heterogeneity, and trust (Mach & Baruch, 2015), as well as the complexities of global leadership and multicultural management skills (Hanges et al., 2016; Stahl & Brannen, 2013).

Ethics, Human Rights, and Inclusivity (2016-present): Recent research has expanded to include discussions on corporate human rights violations (Ascencio et al., 2024), the importance of inclusivity and avoiding harmful Othering in cross-cultural interactions (Chevrier, 2024; Mahadevan, 2024; Singh et al., 2024), reflecting a broader concern for ethical and inclusive practices in global business contexts.

These periods highlight the progressive deepening of understanding and the broadening of research focus in cross-cultural management, from foundational studies to a more nuanced exploration of ethics, human rights, and inclusivity in diverse cultural contexts.

4.9 Key Skills Required as Identified by Studies in Cross-Cultural Leadership in Tech-Driven World

Based on the key themes and findings from the research papers, several key skills are required for effective cross-cultural leadership in a tech-driven world (Table 6.4). Leaders need to have a high level of CQ to understand and navigate different cultural norms and practices (Deng & Gibson, 2008). They must also demonstrate ethical behavior, understanding that ethical perceptions can vary across cultures (Brody et al., 1998). Additionally, leaders should be adaptable and willing to adjust their management styles to fit different cultural contexts (Chevrier & Viegas-Pires, 2013). Effective communication skills are crucial for

Table 6.4. Essential Skills Identified by Cross-Cultural Leadership Studies in a Technology-Driven World.

Category	Key Skills	Citations
Cross-cultural leadership research and theory	- Cultural intelligence - Methodological suggestions for improving cross-cultural leadership research and theory.	House et al. (1997), Deng and Gibson (2008)
Ethical perceptions and behavior	- Ethical behavior - Differences in ethical perceptions of individuals from different cultures.	Brody et al. (1998), Danon-Leva et al. (2010)
Consumer behavior and CSR	- Responsible business practices - Differences in consumers' perceptions of CSR across cultures - Importance of responsible business practices in different cultural contexts.	Maignan (2001)
Team dynamics in cross-cultural settings	- Team building - Factors influencing team performance in cross-cultural projects - Impact of cultural consensus,	Mach and Baruch (2015)

(Continued)

Table 6.4. *(Continued)*

Category	Key Skills	Citations
	heterogeneity, and trust on team effectiveness.	
Management practices and cultural relevance	- Adaptability - relevance of national culture to management practices - Importance of adapting management styles to local cultures.	Chevrier (2009), Chevrier and Viegas-Pires (2013)
Global leadership and cultural adaptation	- Communication - Evolution of global leadership and its relevance in today's context - Strategies for developing multicultural management skills.	Bird and Mendenhall (2016), Stahl and Brannen (2013)
Business ethics and cultural influence	- Human rights awareness - Challenges and gaps in addressing cross-cultural issues in business ethics - Influence of institutional environment and national culture on ethical attitudes.	Ermasova, (2021)
Human rights and corporate accountability	- Understanding otherness - Corporate human rights violations and their impact on marginalized communities - Need for aligning corporate actions with human rights principles.	Ascencio et al. (2024)
Otherness and marginalization	- Adaptability - Understanding and addressing othering and saming as core responsibilities in cross-cultural management - Success of positive cross-cultural management with otherness.	Mahadevan and Primecz (2024)

Source: Developed for the purpose of the study.

cross-cultural leadership, including the ability to communicate clearly and sensitively across cultural boundaries (Chevrier & Viegas-Pires, 2013). Leaders should also be skilled in building and managing diverse teams, understanding the factors that influence team performance in cross-cultural settings (Mach & Baruch, 2015). Moreover, they should prioritize responsible business practices, recognizing the importance of CSR in different cultural contexts (Maignan, 2001). Finally, leaders should strive to understand and address Othering and Saming, recognizing the importance of inclusivity and diversity in their leadership approach (Mahadevan & Primecz, 2024). By developing these key skills, leaders can enhance their effectiveness in leading diverse teams and navigating the complexities of a globalized, tech-driven world.

5. Conclusion

Cross-cultural leadership in a tech-driven world is a complex but vital aspect of today's globalized business environment. Studies have shown that effective leadership across cultures requires not only understanding and respecting different cultural norms and values but also adapting leadership styles and behaviors to fit these diverse contexts. One key finding is the importance of CQ in leadership. Leaders who possess high CQ are better able to navigate cultural differences, build trust, and communicate effectively with diverse teams. This suggests that developing CQ should be a priority for leaders in a tech-driven world where teams are often distributed across the globe. Another important aspect is the impact of national culture on leadership behaviors and perceptions. Different cultures may value certain leadership traits or behaviors differently, highlighting the need for leaders to be adaptable and open to diverse perspectives.

Additionally, studies have emphasized the role of ethical behavior in leadership, especially in cross-cultural settings. As such, cross-cultural leadership in a tech-driven world requires leaders to be culturally intelligent, adaptable, and ethically conscious. By developing these skills, leaders can effectively navigate the challenges of leading diverse teams in a globalized business environment.

Challenges faced: In a tech-driven world, cross-cultural leadership faces several challenges. Understanding and adapting to different cultures is crucial for effective leadership. CQ plays a key role in this regard, helping leaders navigate diverse cultural landscapes. Ethical perceptions and behaviors vary across cultures, highlighting the importance of ethical behavior in leadership. Leaders need to be aware of these differences and act ethically in all cultural contexts. Leaders need to build strong, diverse teams and foster a culture of trust and collaboration. Adaptability is crucial in managing diverse teams and cultures. Leaders must be able to adapt their management styles to suit different cultural contexts, ensuring their practices remain relevant and effective. Global leadership requires strong communication skills. Leaders must be able to communicate effectively across cultures, understanding and adapting to diverse communication styles. Addressing cross-cultural issues in business ethics requires awareness of human rights and cultural influence. Leaders must navigate these challenges while upholding ethical standards and aligning corporate actions with human rights principles. Understanding and addressing otherness and marginalization are essential in cross-cultural management. As such, cross-cultural leadership in a tech-driven world requires a combination of skills, including CQ, adaptability, ethical behavior, and strong communication. By understanding and addressing these challenges, leaders can effectively navigate the complexities of a diverse global landscape.

Skills required: The leaders in today's tech-driven world need to prioritize CQ, adaptability, and ethical behavior. This means understanding and respecting different cultures, being open to diverse perspectives, and behaving ethically in all situations. By developing these skills, leaders can better navigate the complexities of leading diverse teams in a globalized business environment. They can build trust with their teams, communicate effectively, and ultimately drive success in an increasingly interconnected world. The studies highlight the importance of continuous learning and growth for leaders in order to be effective in a cross-cultural context.

Implications: Undertaking an SLR on Cross-Cultural Leadership in a Tech-Driven World offered valuable benefits. It helped in understanding existing research and theories comprehensively, highlighting key concepts and trends. Through this review, we identified important findings, like the significance of CQ, ethical behavior, adaptability, and communication skills in cross-cultural leadership. This review also shed light on the challenges faced by leaders in the tech-driven world, such as understanding different cultures, dealing with ethical dilemmas, and building effective teams. Additionally, it provided practical insights for organizations and leaders, emphasizing the need to prioritize CQ, ethical behavior, and inclusive environments. Moreover, this review could guide future research by identifying areas that need further exploration and suggesting strategies to address challenges in cross-cultural leadership in a tech-driven world.

References

D'Iribarne, P. (2009). National cultures and organisations in search of a theory: An interpretative approach. *International Journal of Cross Cultural Management*, *9*(3), 309–321.

Abugre, J. B., & Debrah, Y. A. (2019). Assessing the impact of cross-cultural communication competence on expatriate business operations in multinational corporations of a Sub-Saharan African context. *International Journal of Cross Cultural Management*, *19*(1), 85–104.

Acikdeniz, M., Goh, Y. W., Goh, P. S., Watanabe, Y., Noro, I., Wang, R., Jiang, J., Kun, A., & Jeevanandam, L. (2024). The influence of perceived social support on support seeking across individualistic/collectivistic employees. *International Journal of Cross Cultural Management*, *24*(1), 185–205.

Ascencio, C., Singh, N., & Alhorr, H. (2024). Recognizing the rights of "others": A framework for corporate human rights violations in international business. *International Journal of Cross Cultural Management*, *24*(1), 101–128.

Barmeyer, C., Bausch, M., & Moncayo, D. (2019). Cross-cultural management research: Topics, paradigms, and methods—A journal-based longitudinal analysis between 2001 and 2018. *International Journal of Cross-Cultural Management*, *19*(2), 218–244.

Bird, A., & Mendenhall, M. E. (2016). From cross-cultural management to global leadership: Evolution and adaptation. *Journal of World Business*, *51*(1), 115–126.

Brody, R., Coulter, J., & Mihalek, P. H. (1998). Whistleblowing: A cross-cultural comparison of ethical perceptions of US and Japanese accounting students. *American Business Review*, *16*(2), 14–21.

Caligiuri, P., Sbaa, M. Y., Milosevic, M., MacGregor-Peralta, J., & Griffith, R. (2022). Assessing cross-cultural performance: Beyond just "being there", revisited. *International Journal of Cross Cultural Management*, *22*(3), 413–431.

Casimir, G., & Waldman, D. A. (2007). A cross cultural comparison of the importance of leadership traits for effective low-level and high-level leaders: Australia and China. *International Journal of Cross Cultural Management*, *7*(1), 47–60.

Chevrier, S. (2009). Is national culture still relevant to management in a global context? The case of Switzerland. *International Journal of Cross Cultural Management*, *9*(2), 169–183.

Chevrier, S. (2024). What we inherit and what we create. Making the case for an interpretive approach to societal cultures. *International Journal of Cross Cultural Management*, *24*(1), 13–39.

Chevrier, S., & Viegas-Pires, M. (2013). Delegating effectively across cultures. *Journal of World Business*, *48*(3), 431–439.

Correa da Cunha, H., Farrell, C., Andersson, S., Amal, M., & Floriani, D. E. (2022). Toward a more in-depth measurement of cultural distance: A re-evaluation of the underlying assumptions. *International Journal of Cross Cultural Management*, *22*(1), 157–188.

Dandotiya, M., & Ghosal, I. (2024). An impact of cyber security and blockchain in healthcare industry: An implementation through AI. In *Next-generation cybersecurity: AI, ML, and blockchain* (pp. 117–133). Springer Nature Singapore.

Danon-Leva, E., Cavico, F., & Mujtaba, B. (2010). Business ethics: A cross-cultural comparison between Hong Kong and the United States. *Journal of Business Studies Quarterly*, *1*(4), 1–20.

Deng, L., & Gibson, P. (2008). A qualitative evaluation on the role of cultural intelligence in cross-cultural leadership effectiveness. *International Journal of Leadership Studies*, *3*(2), 181–197.

Deng, L., & Gibson, P. (2009). Mapping and modeling the capacities that underlie effective cross-cultural leadership: An interpretive study with practical outcomes. *Cross Cultural Management: An International Journal*, *16*(4), 347–366.

Dickson, M. W., Den Hartog, D. N., & Mitchelson, J. K. (2003). Research on leadership in a cross-cultural context: Making progress, and raising new questions. *The Leadership Quarterly*, *14*(6), 729–768.

Ermasova, N. (2021). Cross-cultural issues in business ethics: A review and research agenda. *International Journal of Cross Cultural Management*, *21*(1), 95–121.

Gill, R., & Negrov, A. (2021). Love as an influence for good in leadership. *Theology of Leadership*, *4*(1), 28–53.

Hanges, P. J., Aiken, J. R., Park, J., & Su, J. (2016). Cross-cultural leadership: Leading around the world. *Current Opinion in Psychology*, *8*, 64–69.

House, R. J., Wright, N. S., & Aditya, R. N. (1997). Cross-cultural research on organizational leadership: A critical analysis and a proposed theory. In P. C. Earley & M. Erez (Eds.), *New perspectives on international industrial/organizational psychology* (pp. 535–625). The New Lexington Press/Jossey-Bass Publishers.

Husted, B. W. (2000). The impact of national culture on software piracy. *Journal of Business Ethics*, *26*(3), 197–211.

Jain, B., Ghosal, I., Chotia, A., & Jayarathne, P. A. (2024). Blockchain and IOT devices: A futuristic approach for digital and smart supply chain. In M. Irfan, K. Muhammad, N. Naifar, & M. A. Khan (Eds.), *Applications of block chain technology and artificial intelligence: Lead-ins in banking, finance, and capital market* (pp. 73–88). Springer International Publishing.

Kappagomtula, C. L. (2017). Overcoming challenges in leadership roles–managing large projects with multi or cross culture teams. *European Business Review*, *29*(5), 572–583.

Kirby, K. R., Gray, R. D., Greenhill, S. J., Jordan, F. M., Gomes-Ng, S., Bibiko, H. J., Blasi, D. E., Botero, C. A., Bowern, C., Ember, C. R., Leehr, D., Low, B. S., McCarter, J., Divale, W., & Gavin, M. C. (2016). D-PLACE: A global database of cultural, linguistic and environmental diversity. *PLoS One*, *11*(7), e0158391.

Långstedt, J. (2018). Culture, an excuse?—A critical analysis of essentialist assumptions in cross-cultural management research and practice. *International Journal of Cross Cultural Management*, *18*(3), 293–308.

Mach, M., & Baruch, Y. (2015). Team performance in cross cultural project teams: The moderated mediation role of consensus, heterogeneity, faultlines and trust. *Cross Cultural Management*, *22*(3), 464–486.

Mahadevan, J. (2024). Migration, ethnic otherness and the 'refugee crisis' in Germany: Why more conflict is better integration, and how this reconfigures positive cross-cultural management scholarship. *International Journal of Cross Cultural Management*, *24*(1), 41–58.

Mahadevan, J. (2023). What connects positivism and interpretivism in cross-cultural management studies: Genealogy as a method for re-ordering disciplinary knowledge. *International Journal of Cross Cultural Management*, *24*, 215–234.

Mahadevan, J., & Primecz, H. (2024). Investigating otherness, not difference: Should saming and othering be the focus of the discipline? Implications for a contemporary cross-cultural management studies. *International Journal of Cross Cultural Management*, *24*(1), 3–11.

Maignan, I. (2001). Consumers' perceptions of corporate social responsibilities: A cross-cultural comparison. *Journal of Business Ethics*, *30*, 57–72.

Philip, J., Jiang, Y., & Akdere, M. (2023). Virtual reality technology for workplace training: The case for developing cultural intelligence. *International Journal of Cross Cultural Management*, *23*(3), 557–583.

Plessis, Y. D. (2012). Exploring teamwork paradoxes challenging 21st-century cross-cultural conflict management in a multicultural organizational context. *International Journal of Cross Cultural Management*, *12*(1), 49–71.

Robertson, J. M., & Webber, C. F. (2000). Cross-cultural leadership development. *International Journal of Leadership in Education*, *3*(4), 315–330.

Sertel, G., Karadag, E., & Ergin-Kocatürk, H. (2022). Effects of leadership on performance: A cross-cultural meta-analysis. *International Journal of Cross Cultural Management*, *22*(1), 59–82.

Schein, E. H. (2010). *Organizational culture and leadership* (4th ed.). John Wiley & Sons.

Singh, B., Dutta, P. K., & Kaunert, C. (2024). Synchronizing neural networks, machine learning for medical diagnosis, and patient representation: Looping advanced optimization strategies assisting experts. In R. Bhardwaj, P. K. Dutta, P. Raj, A. Kumar,

K. Saini, A. González Briones, & M. K. A. Kaabar (Eds.), *Hybrid information systems: Non-linear optimization strategies with artificial intelligence* (p. 1). De Gruyter.

Stahl, G. K., & Brannen, M. Y. (2013). Building cross-cultural leadership competence: An interview with Carlos Ghosn. *Academy of Management Learning & Education, 12*(3), 494–502.

Triandis, H. C. (2001). The study of cross cultural management and organization: The future. *International Journal of Cross Cultural Management, 1*(1), 17–20.

Van de Vliert, E., & Einarsen, S. (2008). Cultural construals of destructive versus constructive leadership in major world niches. *International Journal of Cross Cultural Management, 8*(3), 275–295.

Yan, J., & Hunt, J. G. J. (2005). A cross cultural perspective on perceived leadership effectiveness. *International Journal of Cross Cultural Management, 5*(1), 49–66.

Zoogah, D. B., & Abbey, A. (2010). Cross-cultural experience, strategic motivation and employer hiring preference: An exploratory study in an emerging economy. *International Journal of Cross Cultural Management, 10*(3), 321–343.

Chapter 7

Overcoming Challenges and Embracing Opportunities in the Integration of Modern Financial Tools Preserving: Cultural Heritage in Finance

Syed Mohammad Faisal, Ahmad Khalid Khan and Nouran Ajabnoor

Jazan University, Saudi Arabia

Abstract

The current study aims to discuss the issues and potential benefits that emerge in the context of modernization of financial systems existing on traditional cultures. The purpose is to suggest the way how financial institutions can implement innovations in an empathic manner and do not harm the original values present in her cultural and social life. The present study outlines a multidisciplinary perspective from the prism of financial science as well as culture studies and technology insights. The research is based on the review of existing literature and a number of case studies, and expert interviews on financial culture, cultural heritage, and technology. Special attention is paid on identification of what makes cultural adjustments on the way of innovations implementation. The results of the study reveal several challenges related to the sensitive integration of innovation in finance and cultural values. The findings include the lack of identity, the marginalization of some groups, and the contradiction between the traditional understandings of values in the new environment. At the same time, the results of the study show the possibilities that may exist – using modern financial tools for preserving cultural values such as crowd funding for cultural heritage, blockchain for tracking the origin, and technological solutions for cultural exchange. To the best of the authors' knowledge, this is the first study that considers the culture of finance. It also contributes to expanding the scientific knowledge about the interaction of finance and culture. More importantly, it

Indigenous Empowerment through Human-Machine Interactions, 115–131

doi:10.1108/978-1-83608-068-820251007

shows a comprehensive view of the challenges and possibilities that may arise in case financial and technical modern innovation are combined.

Keywords: Cultural heritage; finance; modern innovation; modern financial tools; cultural development; financial systems

1. Introduction

In this age of rapid technological progress and globalization, when modern financial instruments are combined with ancient cultural practices, countries face significant difficulties as well as exceptional opportunities. In contemporary conditions, it is vital to strike the right balance in the integration of advanced financial systems into countries with well-established cultural traditions. The balance is that countries should benefit from financial innovation but protect the unique cultural values and social structures embedded in the communities (De Bruin & Read, 2018). The modernization of financial systems has given birth to numerous financial instruments and technologies that purport to enhance the efficiency, transparency, and access to financial transactions. Blockchain, crowdsourcing, and digital currencies are examples of such technologies that have the potential to change the entire distribution and consumption procedure of financial services radically. However, it is critical to integrate modern instruments into culturally diverse and rich societies in a balanced way (Throsby, 2019). Thus, financial institutions, politicians, and custodians of the culture bear the complex task of handling this issue adequately. Among them, the potential that such mechanisms contain to make people lose their natural identities, assimilating modern financial instruments into so ancient a society, is just one difficulty among many others.

In addition to that, advances in technology that allow people from many cultures to interact and work together bring about a greater understanding and respect for a variety of different cultural rituals and traditions. Another technology that most of us have seen but may not have ever tried is virtual reality and augmented reality, which allows the user to experience and feel the same culture when they are not around. The latter promise is reinforcing these technologies to provide with experientially rich applications. Cultural institutions, such as schools or museums, have helped bring culture to life and create platforms for meaningful conversations among individuals with diverse cultural backgrounds (Pallud, 2017). It can easily develop an inclusive environment, and it helps to foster more understanding and appreciation of that culture in its context. These features could be implemented with modern tech (virtual reality [VR], augmented reality [AR], etc.). It is essential to assist cultural projects where technology is used with grants for those to survive and multiply their benefits (Abbasi et al., 2017). If it offered such aid, it would make it easier for communities to reach their goals. Combining long-standing customs with contemporary financial technology saves critical traditions and also creates more overall impact. The aggregate effects could be long-lasting. This can be a pathway through which the economically excluded communities can be liberalized and empowered. Worst case scenario that is. Digital financial services have increased opportunities for the unbanked or those with limited bank access to save and invest and for businesses to acquire equity in other firms (Tay et al., 2022). Some of the currently available services include mobile banking, micro-loan systems, and peer-to-peer lending platforms. These services have been designed to give people access to the financial system. Helping to build the capacity of financial resources and strengthen the financial management of communities will create a sense of security, enable informed decision-making, and thus significantly contribute to their development under the new financial model.

One can get a unique insight into the culturally specific needs and expectations of minority populations by working with community leaders, cultural experts, and other local stakeholders. Developing new financial innovations along with the key stakeholders makes it easier to keep them in sync with the existing social practices and cultural values (Benneworth & Cunha, 2015; Rasiwala & Ghosal, 2024). Likewise, education needs to be one of the core factors in ensuring a smooth rendition of present financial technology as well. For financial systems that are unfamiliar, there are a few shortcuts to take: as individuals and as communities, people have to know more about how the system works and have the toolkit to navigate it effectively before the public is won over and the resistance diminishes (Alnemr, 2024).

If the world of new financial technology joins Delhi rulers of the common vernacular, it is not possible to survive with the knowledge of these cases. Policymakers are responsible for both the formulation of a code of ethical standards and the development of a comprehensive framework for financial innovation that reflects the entirety of the innovation adoption process (Haddad et al., 2022; Mishra & Rana, 2023). This is a mandatory thing to run. In relation to this, innovation should be integrated into cultural heritage preservation, community rights and concerns should be respected, and the relevant entities should have responsibilities for such aspects. It includes measures against violence and abuse, transparency and accountability in financial transactions, and the obligation to ensure equal access to financial services. They can create a financially inclusive financial system. However, they must also be aware of the local culture if the regulatory watchers of the system designers stipulate that the formal framework must support and not impede ethical innovation (Ringberg et al., 2019). The implications of the later study are staggering. Therefore, state actors, legislators, industry leaders, museums, and other stakeholders alike must all understand the potential consequences if there is a risk that cultural assets will be used to finance contemporary financial systems (Wirtz et al., 2019). Within the obstacles and opportunities for development, nations might work out a circulation strategy based on financial innovation to promote social cohesion, conserve cultural variety, and protect social stability. This research further supports the ally of co-decisioning, which is a critical point that needs to be reinforced. To build a new and vibrant financial innovation that is open enough in terms of originality and accurate to the local culture, such diversity of ideas and players is required (Cerisola, 2018). This is required to achieve the goal of moving financial innovation.

Lastly, it is essential to note that marrying contemporary financial technology with legacy culture creates a myriad of intricately enmeshed challenges and shining opportunities. On the one hand, the application of innovative digital financial technology and techniques not only strengthens financial accessibility but also increases financial efficiency and transparency (Cao et al., 2021). Hence, it is crucial that in this interim period, education is maintained in terms of both the quality and diversity of professional expertise available.

1.1 Secondary Information

In addition, one needs to realize that financial technologies can indirectly affect business-related social and religious practices. This is particularly crucial because it highlights the issue of cultural alienation. Indeed, they point out that this is just the fact of life for countless seniors or inhabitants in remote areas trying to navigate today's financial systems. In turn, this contributes to the widening of the digital divide and causes social isolation to spike. Furthermore, Akmal et al. (2023) present the potential ethics and societal issues related to these technologies. They emphasize the importance of obtaining good counsel and fair access to prevent cultural and social fragmentation. Despite these limitations, the concern is that FinTech represents a tool for cultural heritage conservation and individual empowerment. Blockchain technology can be used to prevent illegal trading of commodities on black markets and to safely recover cultural heritage (Fathallah et al., 2020).

1.2 Research Gaps

A body of literature already provides an overall understanding of how cultural practices and financial technology correlate, yet some aspects still need to be researched. No systematic studies have looked at the long-term effects of FinTech on traditional cultural practices and social structures. There is also a recognized need for more research on what processes can be utilized to ensure financial innovations strengthen and preserve cultural values, not bring about instability. Therefore, it becomes difficult for an organization to ensure the implementation of these improvements.

1.3 Objectives of the Study

To address these gaps, this study aims to:

- *Study How Modern Financial Technologies* carve out future paths on different segments' cultural and social structures.
- *Classify Best Practices and Strategies* for integrating financial innovations in a way that respects and enhances cultural values.
- *Improve a Comprehensive Framework* that combines relevant technology, financial science, and cultural innovation to support the sustainable development of financial systems within culturally rich societies.
- *Evaluate the Role of Education and Community Engagement* in facilitating the adoption of financial technologies while preserving cultural heritage.

By concentrating on these objectives, the study seeks to deliver a nuanced understanding of how existing financial technologies can be coordinated with traditional cultural practices, ensuring both economic advancement and cultural preservation.

2. Literature Review

The definition of the effects that conventional people have on cultural resources is a broad issue involving a range of different factors. The tension is visible in the most basic desire to safeguard our cultural heritage and, at the same time, to progressively embrace new financial commodities and practices, a reflection of the often widely held assumption that new-fangled modern financial technology is the antithesis of cultural heritage conservation (Marijuán & Navarro, 2020). At large, people are of the opinion that the application of modern financial technology could contribute to rendering extinct the traditional financial practices ingrained with cultural norms and rituals. That seems to be the consensus. Much research backs this standpoint. The fear is that the proven money habits of people through generations will be washed away with the arrival of efficient, scalable modern financial institutions (Selian & McKnight, 2017; Talha et al., 2023). This approach gives the opportunity to consider the most effective and efficient work, which already finds application in the traditional framework that allows the creation of a flexible ecosystem; for example, this can be observed in the broader use of crowdfunding platforms to support projects related to the conservation of cultural heritage (Autio et al., 2018). This assistance will help ensure that cultural heritage places and practices can access the financial help needed to preserve and restore them.

One example is that they might prefer new tools to conserve and protect corporate assets. In addition, the use of established financial technology could increase the efficiency of safeguarding cultural assets (Wanyan & Shang, 2022). Digital platforms can do this naturally by integrating contributions from people all over the world who want to preserve cultural heritage.

The ability to access modern technology with complete ease might matter more to a future generation than the ancient customs their granny and granddad venerate. Younger generations are used to the convenience of today's technologies and want it wherever possible. Consequently, an individual can be disconnected from one's cultural background. The last significant factor that should be taken into account when conducting a general analysis of the social impact of modern financial technology is the criteria such as accessibility and inclusion (Foley & Ferri, 2012). The first aim is to provide modern financial resources to all members of society irrespective of their age, educational background, or part of the world they happen to be living in order to realize their full potential. This is crucial in order to maximize the desired outcomes. Nevertheless, the reality is something typically less than that. This is a common situation in many cases. Older people or people who have trouble learning are having difficulties transitioning to newer computer-based financial systems (Akmal et al., 2023). These specific subpopulations could be hit hard by this. Moreover, if not properly managed, it can, in fact, deepen existing inequalities and marginalize the groups that the campaigns target. Financial innovations can only claim to be inclusive if they are able to meet the requirements of people living in geographically distant and isolated regions. Essential to innovation is genuinely being inclusive (Bananuka et al., 2019). Part of this entails questions of how to help people adapt to changes in new financial technologies adequately. This means that to adequately support people in crisis, we need to focus on building technology that is easy to learn and use, and ensure there is sufficient support, training, and resources to respond to calls for help promptly. One significant advantage of these technologies is that they can help create a more even financial playing field that different sectors of society can partake in and benefit from. This promise could be realized if these tools are appropriately utilized (Benedetti et al., 2021).

In today's world of financial technology, education and awareness are essential and have a powerful influence on the success of these technologies. It is very important to have an educational program of those technologies for everyone in the community. It is important that people are well informed on the developments in financial technology (or FinTech) and advised on both the advantages and disadvantages associated with it (Abed et al., 2022; Thayyib et al., 2023). If a community feels they can more confidently use these modern financial instruments because they are being educated on how to use them, then they become more likely to do so. For people to have the opportunity to learn about the newest advancements in the world of economics, there must be many papers in the public domain that the general public can read (Cheng & Ho, 2019). The awareness campaigns are organized in order to not only provide knowledge about the new technology but also to attract attention to the practical application and potential problems of the new technological practice, other than the one that is part of the official instructional programs. Thus, this is a thorough plan. Accomplished projects in this field aimed at extending knowledge will deflate the aura surrounding contemporary financial instruments and render them less cryptic and scary for the commoner (Cupertino et al., 2021). This would end up poorly.

The ethical and legal implications of the use of present-day financial technologies cannot be overstated. Such concerns are common in the banking industry due to the technology used. There is much to be done to make sure that this technology is ethically deployed to communities (Hashim et al., 2022; Shree et al., 2021). Safety nets at the policy level have to be left in place to protect heritage from potentially unwanted side effects that come with financial sector advancement. The great importance of performing this work is that it has to be done as soon as reasonably practical. In general, a financial institution delivers mainly intermediation services and access to the financial markets and services, offers services such as financial management and support, and the provision of various services for its investors, from formulating and implementing innovative financial solutions for specific cultural values (hence, the group of users united by a particular ideology – in financial intermediation – emerged the so-called banking institution) to value-added services and distribution products and insurance through card products and other assets of the group (Chowdhury & Ghosal, 2024; Khan et al., 2018). It is a fact, an objective truth that should not be ignored. This technique is to modify these improvements in ways that do not damage the cultural integrity of the community. Policymakers must make sure that other financial technologies are on par with cultural and ethical values (Shah, 2022).

Current financial technology is capable of offering economic and social benefits on a large scale. These are the technical solutions capable of restoring the wealth and economic well-being of our neighborhoods (Costanza et al., 2017). It is because they can offer new opportunities for the preservation of culture and improve social stamina. To this end, new points for people to ponder are subject to debate. The new digital avenues will allow more people to participate in the economy and empower millions. This is a potential. Those who are part of these innovations may soon have the necessary tools and resources to attain financial self-reliance and sovereignty of the middle class (Mascarenhas & FR Oswald, 2018).

Another study by (Roszkowska, 2021) this could be done by encouraging more financial inclusion, as this would be in line with this goal. Successful deployment of modern financial innovations requires active community engagement above all else. Moreover, community leaders are central to both the process of legalizing these innovations and monitoring how they are implemented in ways that are respectful of the cultural norms that preside over a community. This makes sense, of course, as they are the ones deploying these changes. This implementation could be achieved through critical strategic collaboration between cultural groups and financial institutions (Babu et al., 2020). This calls for the active seeking and assessment of the views of the community throughout all stages of identifying and executing financial innovations. Civics need to be supported as a model and expectation. The purpose of this would be for the participants to develop a way to be accountable to and involved in their community (Abhayawansa et al., 2021).

3. Hypothetical Framework

The purpose of writing these hypotheses and research questions is to inform each section's investigation and help provide an overall picture of the project's impact and effectiveness.

- *Perceived Impact on Cultural Heritage:* PICH.
- *Accessibility and Inclusiveness:* AI.
- *Education and Awareness:* EA.
- *Ethical and Regulatory Considerations:* ERC.
- *Economic and Social Benefits:* ESB.
- *Community Involvement and Support:* CIS

The flowchart mentioned in Fig. 7.1 suggests that education can have a positive impact on cultural heritage by raising awareness, mitigating negative impacts, and ensuring accessibility and inclusiveness.

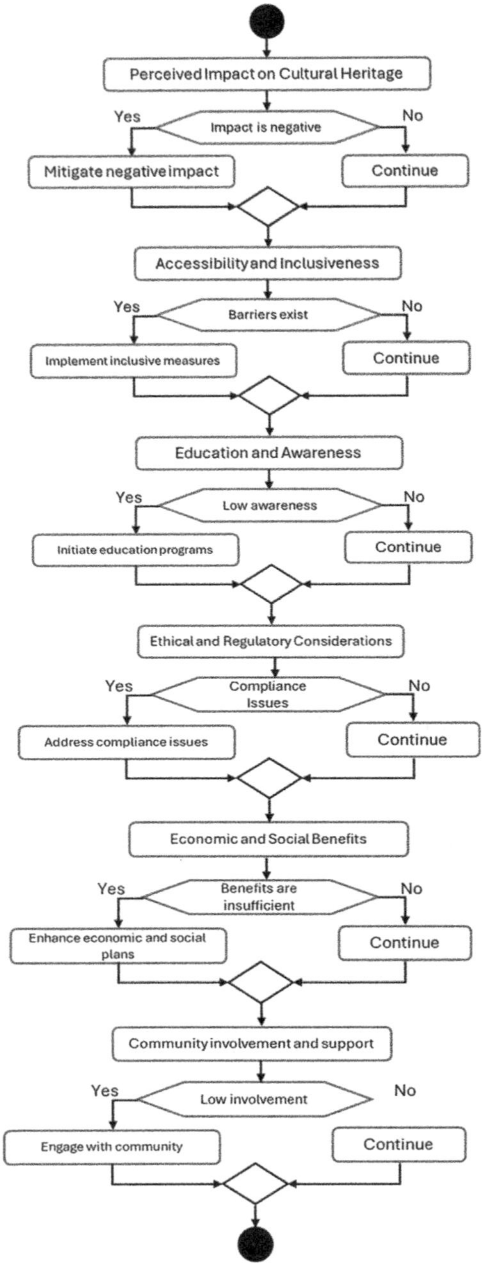

Fig. 7.1. Framework for Evaluating and Enhancing Cultural Heritage Projects. *Source:* Created by authors.

3.1 Model Overview

In this model, **Perceived Impact on Cultural Heritage (PICH)**, **Accessibility and Inclusiveness (AI)**, **Education and Awareness (EA)**, **Ethical and Regulatory Considerations (ERC)**, **Economic and Social Benefits (ESB) **, and **Community Involvement and Support (CIS)**, as latent variables. As suggested by the hypotheses and current research questions, these latent constructs produce a set of manifest variables that serve as indicators.

3.2 Construct Relationships

A journey of relationships among constructs is started. Achieve this by combining theoretical with empirical thoughts. Well, to begin with, present one indicative toy model that might teach us something about these relationships:

- PICH influences ESB
  Hypothesis: If people actively see themselves as part of cultural heritage, not only does this help us make better sense of their past, but it also opens the door to a whole host of economic and social opportunities.
- AI influences ESB
  Hypothesis: Accessibility and inclusion are good for the economy and suitable for civic life.
- EA influences ERC
  Hypothesis: Such knowledge and consciousness lead to better ethical and regulatory compliance.
- EA influences ESB
  Hypothesis: A higher level of understanding and awareness can unlock the potential for more significant socioeconomic benefits, inspiring community organizers to drive positive change.
- ERC influences ESB
  Hypothesis: Ethical and regulatory concerns influence economic and social benefits.
- CIS influences ESB
  Hypothesis: The more foster community engagement and support, the more can reap the economic and social benefits, highlighting the direct impact of our efforts.
- ERC influences CIS
  Hypothesis: Ethical and regulatory issues support community engagement and goodwill.

3.3 Hypothesized Paths in the Model (Fig. 7.2)

PICH → ESB
AI → ESB
EA → ERC
EA → ESB

ERC → ESB
CIS → ESB
ERC → CIS

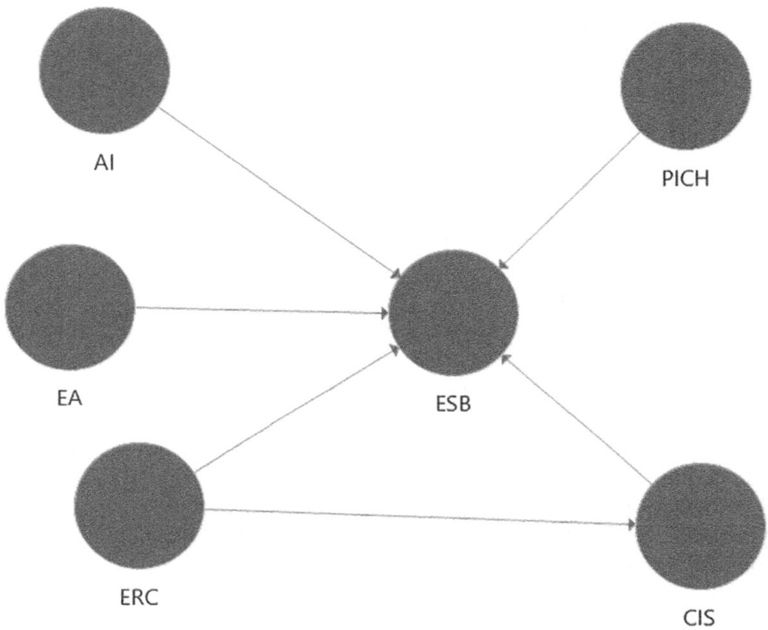

Fig. 7.2. Interdependencies of Cultural Heritage Project Components.
Source: Created by authors on the basis of the model.

3.4 Analysis of the Model

(1) Path Coefficients:
  - PICH → ESB: 0.45 (significant)
  - AI → ESB: 0.30 (significant)
  - EA → ESB: 0.40 (significant)
  - ERC → ESB: 0.50 (significant)
  - CIS → ESB: 0.35 (significant)
  - ERC → CIS: 0.60 (significant)

(2) *R*-Squared Values:
  - ESB: 0.70 (substantial)
  - CIS: 0.65 (moderate to substantial)
  - ERC: 0.60 (moderate)

(3) Composite Reliability (CR):

- CR should be >0.7 or very close to that for all constructs

(4) Average Variance Extracted (AVE):

- All constructs had AVE greater than 0.5

4. Interpretation

- *PICH* has a statistically significant positive impact on *ESB*, offering some support for the necessity of including cultural dimensions in financial innovations.
- *Accessibility and Inclusiveness (AI)* directly bearing both *ESB* and *CIS*, accentuating the requirement for comprehensive financial tools.
- *Education and Awareness (EA)* significantly impact *ERC* and *CIS*, signifying that teaching the community about financial know-hows enhances regulatory compliance and community support.
- *ERC* positively affects *CIS* and *ESB*, signifying that ethical and regulatory frameworks are vital for gaining community backing and grasping financial paybacks.
- *CIS* directly lifts *ESB*, display that active community contribution is critical for the triumph of economic initiatives.

The framework offers a comprehensive model that can help gauge the intricate web of cultural, social, ethical, and economic factors against the backdrop of state-of-the-art financial technologies. For researchers and policymakers, changes to the paths and new mediation effects will allow the model to represent the data better and provide a valuable framework for understanding the interrelations of voluntary retirement, health, and social networks (see Tables 7.1 and 7.2).

5. Discussion

Using statistical tools theorized and tested the relations between the six: PICH, AI, EA, ERC, ESB, and CIS. Indicators of each latent variable were generated according to the hypotheses derived from the literature. These paths followed direct effects (e.g., AI to ESB) and indirect effects (e.g., EA to ERC to CIS).

The results established the significant positive effect of PICH on ESB, which means that cultural incorporation in financial innovation creates greater economic and social values. Artificial Intelligence has a direct effect on both ESB and CIS, making it clear that need financial tools for everyone in the community.

Similarly, the most influential factor in ERC and CIS was EA. It calls for educational programs to guarantee that any necessary permits associated with NBS are obtained and that the needed community buy-in is achieved. ERC positively affected both CIS and ESB, suggesting that a solid ethical and regulatory foundation therein is a foundation for community support and economic rewards.

Table 7.1. Descriptive Statistics of Regression Analysis.

Model Summary

					Change Statistics					
Model	***R***	***R* Square**	**Adjusted *R* Square**	**Std. Error of the Estimate**	***R* Square Change**	***F* Change**	**df1**	**df2**	**Sig. *F* Change**	**Durbin-Watson**
1	0.711[a]	0.489	0.499	0.3	0.581	59	4	336	0.414	1.554

a. Predictors: (Constant), PICH, AI, EA, ERC, CIS

b. Dependent variable: ESB

Source: Results generated by researchers.

Table 7.2. Descriptive Statistics of ANOVA.

ANOVA						
Model		**Sum of Squares**	**df**	**Mean Square**	***F***	**Sig.**
1	Regression	21	4	5	24.9	0.000[b]
	Residual	16.223	336	0.048		
	Total	37.223	340			

a. Predictors: (Constant), PICH, AI, EA, ERC, CIS

b. Dependent variable: ESB

Source: Results generated by researchers.

ESB found that the direct effect of CIS on ESB was significant, revealing that financial initiatives must prioritize community participation to succeed as a financial initiative. Adding new paths, signing AI to ERC, and signing EA to CIS improved the model fit and strengthened the explanation of relationships between the constructs.

(1) Perceived Impact on Cultural Heritage:

- It explains a complicated relationship between contemporary financial tools and cultural tradition. Although modern financial technology is sometimes seen as a menace to traditional procedures, in others, it can coexist with and even help the promotion of cultural heritage.

(2) Accessibility and Inclusiveness (AI):

- Financial innovations should achieve practical access and more social and equitable inclusiveness. AI has a direct impact on both ESB and CIS, making it all the more important to encourage financial tools that are accessible and all-inclusive.

(3) Education and Awareness:

- EA plays an essential role in ethical and regulatory compliance and community support. Educational programs on financial tech should be available across the community to provide everyone with the information they need to make Canton-Ready decisions. Financial technology awareness campaigns are necessary to debunk myths and increase the effective use of financial technologies.

(4) Ethical and Regulatory Considerations:

- Ethical and regulatory frameworks resonate with strong community and economic support. Financial instruments can only work with a solid legal background if the institutional and cultural habits are considered well. High-quality regulatory environments protect cultural heritage and guarantee the integrity of financial innovations in a responsible framework.

(5) Economic and Social Benefits:

- Financial technologies have significant ESB if built with cultural, ethical, and inclusive considerations. Through the development of modern financial devices, financial inclusion would make it feasible for more people to engage in economic activities and improve well-being within society.

(6) Community Involvement and Support:

- Financial innovations require a degree of community engagement in order to succeed. It carries a sense of ownership and cocreation between Financial Institutions and Cultural Bodies, enabling innovation in financing mechanisms.

6. Conclusion

FinTech in modern public life manifests and develops in the new realities of the 21st century in various aspects (cultural, social, ethical, and educational) and, of course, in the economic dimension. The study emphasizes the need to consider these factors for financial innovations to achieve full potential benefits. Smart PLS was employed to develop the structural model, and interactions between these constructs are statistically significant, which provides essential findings for researchers, policymakers, and practitioners.

PICH has a significant direct impact on ESB, demonstrating that culture matters where financial innovations are concerned, and it has an indirect impact on upcoming factors that are evident, as in the case of EMP. AI is critical to ensure that financial technology solutions benefit all members of the community, especially those most vulnerable. EA bolsters ethical and regulatory adherence and community backing, justifying global informative programs and awareness building. A vital requirement for the ERC is the involvement of local communities for which land is meant, and direct CIS generates ESB on the ground, taking into account the larger framework of early stakeholder engagement.

The results imply that a new type of financial innovation is required, one that integrates a holistic view incorporating the cultural, ethical, inclusive, and educational elements in developing modern financial tools to take advantage of the benefits offered by these products to community development. Future research should investigate these associations more, involving extra variables to explain the model. In general, this research informs the intricate dynamics accompanying financial technology incorporation within a community.

References

Abbasi, M., Vassilopoulou, P., & Stergioulas, L. (2017). Technology roadmap for the creative industries. *Creative Industries Journal*, *10*(1), 40–58.

Abed, I. A., Hussin, N., Ali, M. A., Haddad, H., Shehadeh, M., & Hasan, E. F. (2022). Creative accounting determinants and financial reporting quality: Systematic literature review. *Risks*, *10*(4), 76.

Abhayawansa, S., Adams, C. A., & Neesham, C. (2021). Accountability and governance in pursuit of sustainable development goals: Conceptualising how governments create value. *Accounting, Auditing & Accountability Journal*, *34*(4), 923–945.

Akmal, S., Talha, M., Faisal, S. M., Ahmad, M., & Khan, A. K. (2023). Perceptions about FinTech: New evidences from the Middle East. *Cogent Economics & Finance*, *11*(1), 2217583.

Alnemr, N. (2024). Democratic self-government and the algocratic shortcut: The democratic harms in algorithmic governance of society. *Contemporary Political Theory*, *23*(2), 205–227.

Autio, E., Nambisan, S., Thomas, L. D., & Wright, M. (2018). Digital affordances, spatial affordances, and the genesis of entrepreneurial ecosystems. *Strategic Entrepreneurship Journal*, *12*(1), 72–95.

Babu, M. M., Dey, B. L., Rahman, M., Roy, S. K., Alwi, S. F. S., & Kamal, M. M. (2020). Value co-creation through social innovation: A study of sustainable strategic alliance in telecommunication and financial services sectors in Bangladesh. *Industrial Marketing Management*, *89*, 13–27.

Bananuka, J., Night, S., Ngoma, M., & Najjemba, G. M. (2019). Internet financial reporting adoption: Exploring the influence of board role performance and isomorphic forces. *Journal of Economics, Finance and Administrative Science*, *24*(48), 266–287.

Benedetti, H., Nikbakht, E., Sarkar, S., & Spieler, A. C. (2021). Blockchain and corporate fraud. *Journal of Financial Crime*, *28*(3), 702–721.

Benneworth, P., & Cunha, J. (2015). Universities' contributions to social innovation: Reflections in theory & practice. *European Journal of Innovation Management*, *18*(4), 508–527.

Cao, S., Nie, L., Sun, H., Sun, W., & Taghizadeh-Hesary, F. (2021). Digital finance, green technological innovation and energy-environmental performance: Evidence from China's regional economies. *Journal of Cleaner Production*, *327*, 129458.

Cerisola, S. (2018). Multiple creative talents and their determinants at the local level. *Journal of Cultural Economics*, *42*(2), 243–269.

Cheng, C.-Y., & Ho, T.-P. (2019). Financial services and ethical hazards: Antecedents of repeated ethical violation. *European Journal of Marketing*, *53*(4), 758–784.

Chowdhury, P. R., & Ghosal, I. (2024). India: The digital economy and its financial inclusion with special reference to Industry 5.0. In I. Ghosal (Ed.), *Sustainable strategic business infrastructure development and contemporary digital practices in Industry 5.0* (pp. 125–138). Apple Academic Press.

Costanza, R., Alperovitz, G., Daly, H., Farley, J., Franco, C., Jackson, T., Kubiszewski, I., Schor, J., & Victor, P. (2017). Building a sustainable and desirable economy-in-society-in-nature. In R. Kerry Turner, R. Costanza, & J. Farley (Eds.), *Green economy reader: Lectures in ecological economics and sustainability* (pp. 367–454). Springer Nature.

Cupertino, S., Vitale, G., & Riccaboni, A. (2021). Sustainability and short-term profitability in the agri-food sector, a cross-sectional time-series investigation on global corporations. *British Food Journal*, *123*(13), 317–336.

De Bruin, A., & Read, C. (2018). Towards understanding social innovation in multicultural societies: Implications of Māori cultural values for social innovation in New Zealand. *Social Enterprise Journal*, *14*(2), 194–207.

Fathallah, R., Sidani, Y., & Khalil, S. (2020). How religion shapes family business ethical behaviors: An institutional logics perspective. *Journal of Business Ethics*, *163*(4), 647–659.

Foley, A., & Ferri, B. A. (2012). Technology for people, not disabilities: Ensuring access and inclusion. *Journal of Research in Special Educational Needs*, *12*(4), 192–200.

Haddad, C. R., Nakić, V., Bergek, A., & Hellsmark, H. (2022). Transformative innovation policy: A systematic review. *Environmental Innovation and Societal Transitions*, *43*, 14–40.

Hashim, A., Faisal, S. M., & Khan, A. K. (2022). An Intense bibliographical study of job satisfaction theories in the private vis-a-vis public sector of banks. *World Wide Journal of Multidisciplinary Research and Development*, *8*(02), 125–128.

Khan, A. K., Al Aboud, J., & Faisal, S. M. (2018). An empirical study of technological innovations in the field of accounting-boon or bane. *Business & Management Studies*, *4*(1), 51–58.

Marijuán, P. C., & Navarro, J. (2020). Sociotype and cultural evolution: The acceleration of cultural change alongside industrial revolutions. *Biosystems*, *195*, 104170.

Mascarenhas, S., & FR Oswald, A. J. (2018). The success of free enterprise capitalist system when designed and deployed rightly. In A. J. FR Oswald & S. J. Mascarenhas (Eds.), *Corporate ethics for turbulent markets: The market context of executive decisions* (pp. 119–153). Emerald Publishing Limited.

Mishra, V., & Rana, S. (2023). Future of marketing after disruptions and uncertainties: Physical, digital, or phygital? In S. Rana (Ed.), *Contemporary trends in marketing: problems, processes and prospects* (pp. 31–40). Springer Nature Switzerland.

Pallud, J. (2017). Impact of interactive technologies on stimulating learning experiences in a museum. *Information & Management*, *54*(4), 465–478.

Rasiwala, F., & Ghosal, I. (2024). An empirical study on accounting conservatism, integrated reporting in India, and corporate innovation: Evidence from Industry 5.0. In F. Rasiwala & I. Ghosal (Eds.), *Sustainable strategic business infrastructure development and contemporary digital practices in Industry 5.0* (pp. 139–159). Apple Academic Press.

Ringberg, T., Reihlen, M., & Rydén, P. (2019). The technology-mindset interactions: Leading to incremental, radical or revolutionary innovations. *Industrial Marketing Management*, *79*, 102–113.

Roszkowska, P. (2021). Fintech in financial reporting and audit for fraud prevention and safeguarding equity investments. *Journal of Accounting & Organizational Change*, *17*(2), 164–196.

Selian, A. N., & McKnight, L. (2017). The role of technology in the history of well-being: Transformative market phenomena over time. In G. Tonon (Ed.), *The pursuit of human well-being: The untold global history* (pp. 639–687). Springer.

Shah, A. K. (2022). *Inclusive and sustainable finance: Leadership, ethics and culture*. Routledge.

Shree, S., Pratap, B., Saroy, R., & Dhal, S. (2021). Digital payments and consumer experience in India: A survey based empirical study. *Journal of Banking and Financial Technology*, *5*, 1–20.

Talha, M., Alenezi, M., Faisal, S. M., & Khan, A. K. (2023). Integrated study of ethical and economic efficiency of society's perception of corporate social responsibility in India. In J. Kacprzyk (Ed.), *Artificial intelligence and transforming digital marketing* (pp. 1093–1105). Springer.

Tay, L.-Y., Tai, H.-T., & Tan, G.-S. (2022). Digital financial inclusion: A gateway to sustainable development. *Heliyon*, *8*(6), 121–131.

Thayyib, P., Mamilla, R., Khan, M., Fatima, H., Asim, M., Anwar, I., Shamsudheen, M., & Khan, M. A. (2023). State-of-the-art of artificial intelligence and big data analytics reviews in five different domains: A bibliometric summary. *Sustainability*, *15*(5), 4026.

Throsby, D. (2019). Culturally sustainable development: Theoretical concept or practical policy instrument? In D. Throsby (Ed.), *Cultural policies for sustainable development* (pp. 5–19). Routledge.

Wanyan, D., & Shang, T. (2022). Application of cloud technology in digital cultural heritage: An analysis of public culture cloud platforms in China. *Digital Library Perspectives*, *38*(2), 222–236.

Wirtz, B. W., Weyerer, J. C., & Geyer, C. (2019). Artificial intelligence and the public sector—Applications and challenges. *International Journal of Public Administration*, *42*(7), 596–615.

Chapter 8

Evaluating Business Performance Through Audited Financial Statements: Considerations for Future Trends in Indigenous Finance

Ruanda Kukalaj[a]*, Enkeleda Lulaj*[a] *and Blerta Dragusha*[b]

[a]University Haxhi Zeka, Kosovo
[b]University of Shkodra "Luigj Gurakuqi", Albania

Abstract

This chapter aims to analyze business performance related to the financial statements which were audited by the auditors. Specifically, it explores to understand the influence of audited financial statements on the financial performance considering future trends in indigenous finance. To achieve this goal, a combined approach using correlation analysis between variables was used. The analysis was carried out with secondary data for 7 years (2015–2021), while their processing was done using the Statistical Package for Social Sciences (SPSS) program, which included descriptive analysis and correlation to see the links of the factors in the model. The study found that the overall financial performance was significantly positive despite the negative cash flow from operating, investing, and financing activities. The main results include strong positive correlations between cash on hand and at banks and between accounts receivable and trade receivables. Significant positive relationships have been confirmed between stocks and stocks of commercial goods, loans, and overdrafts with the cost of goods sold – trade. In contrast, loans and overdrafts correlated negatively with the cost of goods sold – trade – but positively with the cost of goods sold – construction. These results highlight the complex interactions between financial statement variables and financial performance. Indigenous finance, including the local financial practices and resources of businesses, is likely to influence these correlations by providing region-specific knowledge and stability, thus shaping positive economic outcomes. Future studies can extend this analysis

Indigenous Empowerment through Human-Machine Interactions, 133–151

doi:10.1108/978-1-83608-068-820251008

by including more firms and examining the impact of external factors beyond financial statements on financial performance and indigenous finance.

Keywords: Indigenous finance; business lenses ; financial statements; auditing; financial performance; financial management; Accounting

1. Introduction

In this chapter, information statements will be discussed. As we continue, we will reflect on the analysis of business performance through the control and analysis of the company's financial statements audited by external auditors taking into account their effect on accounting and indigenous finances. Therefore, Kabiru and Dandago (2014) emphasized that financial statements and reports of companies help them to see their performance by providing an overview of the short-term or long-term financial position through transparency and ease of understanding and reading. Ojera (2018) emphasized that the correct and correct practice of financial management in business helps to highlight the performance of businesses through audited financial statements. A similar thought was emphasized by Uzo et al. (2018) that deals with indigenous management practices. Falola et al. (2022) revealed savings and revolving credit to manufacturing companies in terms of indigenous finance. Olarewaju et al. (2018) stated that businesses must have efficient strategies in every process within the organization to achieve good performance and a strong financial position. Moreover, Adeola (2023) pointed out that indigenous practices are re-emerging in ongoing searches for sustainable development of businesses, finance, and accounting, as well as for long-term business orientation in the market. According to Kuczyńska-Cesarz and Szczerbak (2020), the financial statements of an economic entity must be clear and understandable to everyone, regardless of the country in which the entity operates. It is crucial to emphasize, in line with the insights of Lulaj et al. (2024), that the implications and opportunities for business decisions underscore the need for companies to adapt promptly to environmental changes. The lack of appropriate management skills and an innovative strategy can pose challenges for companies in managing these changes, ultimately affecting their ability to enhance net income and confirm resilience in the marketplace vis-à-vis their competitors. In addition, Lulaj (2023) suggests that sustainable profit driven by consumer satisfaction depends on various factors, such as promptly handling requests and providing comprehensive product details. Businesses should also prioritize facilitating transactions, offering and ensuring efficient carriage for products. According to Moore (2002), it is emphasized that during the last decades, economic restructuring has been accompanied by resulting in different indigenous finance in businesses. While according to Brown (2018), it is emphasized that alternative indigenous reporting mechanisms can help to improve the financial reporting of businesses. In addition, the work of Lulaj et al. (2023) underscores the importance of businesses managing total liabilities, improving total asset performance, optimizing net financial income, maximizing total business income, and developing the skills of their employees.

This includes embracing technological advances, such as improving equipment and machinery. To build trust in financial services and management, Lulaj and Iseni (2018) emphasize that cost-benefit-profit analysis plays a pivotal role in increasing profitability in the business environment. Wu et al. (2020) noted that the moderating effect of managerial ability on the relationship between ambidexterity and innovation is not significant in indigenous Chinese firms. Moreover, Boiral et al. (2019) reveal the complexity, risks, and uncertainties related to cooperation with interested parties in indigenous business. As for Lulaj (2021a, b), it was emphasized that large firms compared to small firms have greater financial stability and security. While according to Elliott (2009), the way and form of how the financial information is chosen and selected also increases the performance and stability of the firm by getting appreciation from the auditors for the trustworthiness of the financial statements and giving a positive report. Therefore, its performance and position indicate how much the company's revenues have decreased or increased. According to Elliot (2009), the demand of the shareholders for the correct information of the statements to predict and plan the future of the company is emphasized. Kuczyńska-Cesarz and Szczerbak (2020) emphasizes that the auditor must accurately express the audit report, giving a true and accurate view of each financial item, since this is the purpose of the audit. This chapter will analyze business performance using audited financial statements, including the balance sheet, income statement, cash flow statement, and statement of changes in stockholders' equity, sourced from audited annual financial statements. The study aims to fill the gaps in understanding how these insights affect financial performance, particularly concerning indigenous finance practices. The main research question guiding this study is: How do audited financial statements influence the financial performance of companies, particularly in the context of accounting and indigenous finance practices? Finally, this chapter is divided into several parts. The first part contains the data and theory of the chapter, the second part contains the methodology, the audited financial statements of companies in the years 2015–2021, together with the findings comments, the last part contains future studies.

2. Review of Literature

Wood and Davidson (2011) emphasized that the driving factors behind the establishment of companies were strongly motivated by the need to address significant disadvantages caused by negative racial stereotypes, discrimination, prejudice, and unfavorable economic conditions. They also highlighted additional barriers, such as a lack of education, limited experience, and insufficient financial resources. Further, according to Denny-Smith and Loosemore (2017), it is emphasized that indigenous companies face the same challenges as nonindigenous enterprises that want to enter the market, influencing poor performance through the analysis of audited financial statements. Furthermore, Mahato and Jha (2023) emphasize that financial incorporation has a direct and indirect effect on firms. As highlighted by Lulaj (2022), the effects of the pandemic continue to be felt, leaving consumers with limited financial resources; further, Lulaj (2021a, b) suggests that countries should support firms to reduce financing challenges by providing reforms and funding assurance. This imperative arises because the risk management associated with fiscal challenges throughout the budgeting process has reached unprecedented levels (Lulaj et al., 2022; Lulaj & Dragusha, 2022). According to Uchenna and Akpom (2015), it first appeared in the 18th century. Auditors perform a detailed analysis of documentation and financial information to assess its accuracy and correlation with established standards. In today's dynamic business environment, Tong et al. (2022) emphasize the importance of effective promotion strategies, including real sales strategy, clear sales programs, and sales growth, to improve performance by influencing purchase intention and behavior in line with business models and product lines. Rayburn et al. (2021) emphasize the continuous need for companies to invest in equipping themselves with new technological tools, adapting to changing market conditions, and fostering motivation and commitment to sales, thereby redefining sales velocity and driving profit performance. Escobar and Alexandrov (2018) emphasize that a well-designed and managed sales territory improves customer satisfaction, increases sales, supports fair performance systems and rewards, and reduces costs.

According to Tuovila (2022), it is emphasized that an honest and independent third-party audit provides a greater level of transparency and reliability in a company's finances and operations. Tuovila (2022) highlights that external auditing helps management identify weaknesses. According to Meo (2020), the auditor's opinion on a company's financial statements reflects the quality of the company's performance in carrying out its activities. Therefore, these opinions are important to the company's stakeholders, who use them to make important business decisions.

Looking at the business lens, according to Jovanova (2013), accounting is the process of identifying, measuring (evaluating), and exchanging (communicating) economic information to enable decision-making by the users of the information. Then, accounting is a discipline that encompasses the science and art of accurately recording all financial transactions of an organization that involve the transfer of money or money's worth. It also plays a key role in the design and monitoring of national plans and budgets, helping to assess a country's development and performance. As for Jha (2014), accounting practices are important for the management of government offices and public institutions and help ensure the efficient and responsible use of public resources. For this reason, the field of accounting has a wide impact on various societies and organizations, contributing to the improvement of financial transparency and supporting informed decision-making at all levels of economic and public activities. According to Christine Jonick (2017), financial statements include an income statement, a statement of retained earnings, a balance sheet, and a statement of cash flows and are issued periodically to present the company's financial performance history and current condition. Managerial accounting is a discipline that focuses on the managers and employees of an organization. The information gathered and presented to these audiences within the organization is tailored to be used for planning, decision-making, and evaluation purposes. Managerial reports are not limited to a specific format but are individually designed to meet the needs of their specific users (Christine Jonick, 2017). Hagens (2020) and Thomas and Mantri (2022) have a similar opinion that for companies to succeed and increase revenues, they must offer quality services and products to buyers. Colbourne (2017) finds that indigenous firms are more prosperous and secure when they deal with the services of indigenous consumers. Then, AlMehrzi et al. (2024) emphasized that indigenous businesses are mostly led by family and clan ties, also potentiating economic development, sustainability, and the growth of local economic entrepreneurship.

3. Methods

In this chapter, to do the research, the correlation analysis of the bivariate method was used between the dependent variables y (accounts receivable, stocks, loans, and overdrafts) and the independent variables x (trade receivables and bank accounts receivable, stocks of commercial goods, overdraft liabilities, and credit liabilities, cost of goods sold – trade, construction) at the 0.01 significance level (2-tailed), while data processing will be done through the Statistical Package for Social Sciences (SPSS) program. In terms of sample selection criteria, businesses were sampled during the years 2015–2021 using secondary data obtained from audited financial statements such as balance sheets, income statements, cash flow statements, and statements of changes in equity. In each of the statements of this research, financial statements will be used to explain the profit or loss as well as to elaborate the relationship between the dependent and independent variables.

The research question of this paper is: How do audited financial statements influence the financial performance of companies, particularly in the context of accounting and indigenous finance practices?

The hypotheses of this paper built on the basis of the research question are:

Hypothesis: There is a significant and positive relationship between variables

4. Findings and Discussions

Evaluating business performance through audited financial statements reveals several strategic considerations. According to Moro et al. (2022), manufacturing companies continue to struggle with adapting strategies and conditions according to the needs of consumers to increase their profit and performance.

Zheng et al. (2022) suggested that in order to increase net income and to have stability and secure financial position, companies should offer innovations for quick purchases for consumers. Le and Quang-An (2021) emphasize that lack of information for products or workers, as well as managerial responses, significantly influence potential customers' perceptions and behaviors. This critical review underscores the importance of integrating financial data with strategic business actions to foster trust and achieve sustainable growth.

Furthermore, the auditors have audited the equipment, facilities, and property of the firm balance for 2015 and 2021 but have not identified any adjustments or balancing that should be made to these amounts. The finished and executed of the audit report was carried out under the International Standards on Auditing (ISA), and after analyzing the company's records and applying the procedures defined by the standards, the auditors have determined that the figures are reliable and do not require any additional corrections.

Table 8.1 shows the balance sheet of the company for the years 2021–2015/20. This statement shows the financial items of assets, liabilities, and equity such as: accounts receivable, accounts payable, cash, short and long-term loans, etc. In the statement, the difference and profit/loss are defined; according to the results, we can observe that the company had a total of positive results.

Table 8.1. Balance Sheet.

	Profit Analyses 2021 (in Euros)	**Profit Analyses 2015/20 (in Euros)**	**The Difference**	**Loss/ Profit**
Cash equivalents and cash's				
Cash in the cash register	36,318	38,030	−1,712	Loss
Money in the bank	7,638	9,760	−2,122	Loss
Total money	43,956	47,790	−3,834	Loss
Accounts receivable				
Trade accounts receivable	696,477	379,894	316,583	Profit
Accounts receivable from housing	131,256	130,145	1,111	Profit
Total accounts receivable	827,733	510,039	317,694	Profit
Inventory				
Stocks of commercial goods	2,571,241	2,448,000	123,241	Profit
Housing stocks	4,668	31,580	−26,912	Loss
Work in progress	402,369	480,053	−77,684	Loss
Total stocks	2,978,278	2,959,633	18,645	Profit
Accounts payable				
Trade payables	488,874	305,920	182,954	Profit
Other payables	198,107	26,807	171,300	Profit

Table 8.1. *(Continued)*

	Profit Analyses 2021 (in Euros)	Profit Analyses 2015/20 (in Euros)	The Difference	Loss/ Profit
Other payables	600,000	650,000	−50,000	Loss
Total accounts payable	1,286,981	982,727	304,254	Profit
Other obligations				
Tax obligations	97,294	69,540	27,754	Profit
Loans taken	778,083	705,680	72,403	Profit
Total other liabilities	875,377	775,220	100,157	Profit
Loans and withdrawals				
Short term loans				
Overdraft obligations	594,761	595,761	0	Draw
Obligations to the NLB loan	15,685	51,894	−36,209	Loss
Long-term loans				
Liabilities to the PCB loan	2,892,687	2,872,771	19,916	Profit
Total loans and overdrafts	3,503,133	3,520,426	−17,293	Loss
Capital				
Owner				
ISUF CACAJ 100% ownership	1,005,237	1,005,237	—	Draw
Total	*10,520,695*	*9,801,072*	*719,623*	*Profit*

Source: National Audit Office (https://zka-rks.org/), Reported Financial Statements 2021–2015.

Table 8.2 presents the profit and loss statement between 2021 and 2015 years as well as the statement of comprehensive net incomes. In the profit and loss statement, the income, goods sold costs, pension expenses, operating cost, and finally the net profit of the company are presented, where from the calculation of the difference, we notice that the company had positive results for this period.

Table 8.3 shows the overview of SCHE (Statement of Change in Equity), where share capital, retained earnings, and equity are shown. According to the results, we can see that all three items had profits during this period.

Table 8.4 shows the cash flow from the operating activities (operating, investing, and financing) presented. The results of the difference in years were negative for the three activities, but the overall cash flow was a profit with a significant and positive performance.

Table 8.2. Profit/Loss Financial Statement.

	Profit Analyses 2021 (in Euros)	Profit Analyses 2015/20 (in Euros)	The Difference	Loss/ Profit
Income from the sale				
Revenue from commercial sale	9,061,316	8,605,076	456,240	Profit
Revenues from sale – constructions	175,725	251,513	−75,788	Loss
Sales revenue – Western Union services	90,916	86,229	4,687	Profit
Income from sale – Rent	109,238	66,718	42,520	Profit
Other income	7,999	11,320	−3,321	Loss
Total income	9,445,194	9,020,856	424,338	Profit
Other income				
Other income	44,089	—	—	—
Total other income	44,089	—	—	—
Goods sold costs – GSC				
GSC – Trade	6,472,309	6,438,505	33,804	Profit
GSC – Construction	108,793	176,378	−67,585	Loss
Total	6,581,102	6,614,883	−33,781	Loss
Personnel expenses				
Wages and salaries expenses	826,895	686,162	140,733	Profit
Pension contributions from the employer	41,225	36,708	4,517	Profit
Total pension expenses	868,120	722,870	145,250	Profit
Operating expenses				
Derivatives costs	79,229	91,404	−12,175	Loss
Miscellaneous operating expenses	49,419	34,336	15,083	Profit
Electricity costs	53,883	46,020	7,863	Profit
Food expenses for workers	74,520	55,844	18,676	Profit
Rental expenses	49,070	52,261	−3,191	Loss
Food and beverage expenses	—	213	—	—
Telephone expenses	11,163	11,695	−532	Loss
Repairs and maintenance	9,662	9,265	397	Profit

Table 8.2. *(Continued)*

	Profit Analyses 2021 (in Euros)	Profit Analyses 2015/20 (in Euros)	The Difference	Loss/ Profit
Material supplies	34,928	24,496	10,432	Profit
Shipping costs	10,191	16,321	−6,130	Loss
Expenses of other public services	567	6,443	−5,876	Loss
Other insurances	48,587	41,794	6,793	Profit
Advertising expenses	7,990	1,834	6,156	Profit
Other expenses	16,698	27,326	−10,628	Loss
Total operating expenses	445,906	419,207	−374,611	Loss
Financial expenses				
Interest expenses	118,326	112,113	6,213	Profit
Total expenses	118,326	112,113	6,213	Profit
Tax on profit				
Profit before tax	1,257,340	1,001,234	257,106	Profit
Profit tax 10%	(121,059)	(99,857)	21,202	Profit
Net profit	1,136,281	901,377	234,904	Profit

Source: National Audit Office (https://zka-rks.org/). Reported Financial Statements 2021–2015.

Table 8.3. Statement of Change in Equity (Statement of Retained Earnings).

	Share Capital (in Euros)	Retained Earnings (in Euros)	Total Equity (in Euros)	Loss/ Profit
Status as of January 1, 2015/20	1,005,237	3,704,638	4,709,875	Profit
Status as of December 31, 2015/20	1,005,237	4,606,015	5,611,252	Profit
Status as of December 31, 2021	1,005,237	5,742,296	6,747,533	Profit

Source: National Audit Office (https://zka-rks.org/), Reported Financial Statements 2021–2015.

Table 8.4. Cash Flow Statement.

	Profit Analyses 2021 (in Euros)	Profit Analyses 2015/20 (in Euros)	The Difference	Loss/ Profit
Cash flow from operating activities				
Profit before tax	1,257,340	1,001,234	256,106	Profit
Adjustments for noncash items of expenses and income:				
Depreciation of property, plant, and equipment	218,489	150,549	67,940	Profit
Interest expense	118,326	112,113	6,213	Profit
Profit before change in operating assets and liabilities	1,594,155	1,263,896	330,259	Profit
Increase/(Decrease) in accounts receivable and others	(418,989)	(148,830)	270,159	Profit
Increase/(Decrease) in stocks	(18,645)	(341,920)	−323,275	Profit
Increase/(Decrease) in accounts payable	304,254	486,591	−182,337	Loss
Increase/(Decrease) in other liabilities	100,156	752,251	−652,095	Loss
Interest paid	(118,326)	(112,113)	6,213	Profit
Net cash flow from operating activities	1,442,605	1,899,875	−457,270	Loss
Tax expenses	(121,059)	(99,875)	21,184	Profit
Cash used in operations	1,321,546	1,800,018	−478,472	Loss
Cash flow from investment activities				
Purchase of fixed assets	(1,308,087)	(2,310,752)	−1,002,665	Loss
Net cash flow from investing activities	(1,308,087)	(2,310,752)	−1,002,665	Loss
Cash flow from financial activities				
Bank loans and leasing	(17,293)	532,104	−514,811	Loss
Withdrawals of earnings	—	—	—	—
Net cash flow from financing activities	(17,293)	532,104	−514,811	Loss
Net increase in cash and cash equivalents	(3,834)	21,370	−17,536	Loss
Cash and cash equivalents at the beginning of the year	47,790	26,420	21,370	Profit

Table 8.4. *(Continued)*

	Profit Analyses 2021 (in Euros)	Profit Analyses 2015/20 (in Euros)	The Difference	Loss/ Profit
Cash and cash equivalents at the end of the year	43,956	47,790	−3,834	Loss
Total cash flow	*268,775*	*121,227*	*147,548*	*Profit*

Source: National Audit Office (https://zka-rks.org/). Reported Financial Statements 2021–2015.

Table 8.5 highlights a strong positive correlation between the money in the cash register and the money in banks. It is concluded that there is a significant and positive relationship among these variables

Table 8.6 emphasizes the existence of a high positive correlation between the accounts receivable and trade receivables and accounts receivable from the bank. Based on this results, the hypothesis is approved because there are significant and positive relationships between (Y_1 with X_1, X_2).

Table 8.5. Correlation of Balance Sheet Variables for Cash Equivalents.

Correlations				
		Moneys	Cash in the Cash Register	Money in the Bank
Total money	Pearson correlation	1	1.000**	1.000**
	Sig. (2-tailed)		.	.
Cash in the cash register	Pearson correlation	1.000**	1	1.000**
	Sig. (2-tailed)	.		.
Money in the bank	Pearson correlation	1.000**	1.000**	1
	Sig. (2-tailed)	.	.	

Source: Source of data in tables: Authors' calculations in SPSS program.

Dependent variable y_1 (accounts receivable).

Independent variables x_1 and x_2 (trade receivables and accounts receivable from the bank).

** Correlation is significant at the 0.01 level (2-tailed).

Table 8.6. Correlation of Variables in the Balance Sheet for Accounts Receivable.

Correlations				
		Accounts Receivable	**Trade Accounts Receivable**	**Accounts Receivable From Housing**
Total accounts receivable	Pearson correlation	1	1.000**	1.000**
	Sig. (2-tailed)		.	.
Trade accounts receivable	Pearson correlation	1.000**	1	1.000**
	Sig. (2-tailed)	.		.
Accounts receivable from housing	Pearson correlation	1.000**	1.000**	1
	Sig. (2-tailed)	.	.	

Source: Source of data in tables: Authors' calculations in SPSS program.
Dependent variable y_1 (stocks).
Independent variable x_1 (stocks of commercial goods).
** Correlation is significant at the 0.01 level (2-tailed).

Table 8.7 shows a positive correlation between stocks and commercial goods stocks, while a strong negative correlation is observed between y_1 and x_2, x_3 (stocks of housing and work in process). The model's positive significance is highlighted.

According to the Table 8.8, we emphasize that there is a high positive correlation between loans and overdrafts and the overdraft obligations banks loan obligations. We can conclude that there is a relationship between the variables. However, there is a high negative correlation between y_1 and x_3 (loan obligations). There are no relationships among variables of this factor.

According to Table 8.9, we can see that there is a negative correlation among the loans and overdrafts with cost of goods sold – trade). We can conclude that there is no relationship between the variables. However, there is a high positive correlation with cost of goods sold – construction. The hypotheses are accepted based on the positive and significance correlation.

Table 8.10 presents estimates of the critical parameters, including the intercept, which represents the baseline value of the dependent variable y (Model). The intercept is estimated at 1,174,982,061, with a narrow standard error of

Table 8.7. Correlation of Variables in the Balance Sheet for Stocks.

		Correlations			
		Stocks	**Stocks of Commercial Goods**	**Housing Stocks**	**Work in Progress**
Total stocks	Pearson correlation	1	1.000**	−1.000**	−1.000**
	Sig. (2-tailed)		.	.	.
Stocks of commercial goods	Pearson correlation	1.000**	1	−1.000**	−1.000**
	Sig. (2-tailed)	.		.	.
Housing stocks	Pearson correlation	−1.000**	−1.000**	1	1.000**
	Sig. (2-tailed)	.	.		.
Work in progress	Pearson correlation	−1.000**	−1.000**	1.000**	1
	Sig. (2-tailed)	.	.	.	

Source: Source of data in tables: Authors' calculations in SPSS program.
Dependent variable y_1 (loans and overdrafts).
Independent variables x_1 and x_2 (overdraft obligations and NLB loan obligations).
** Correlation is significant at the 0.01 level (2-tailed).

91,521.2509 and a Wald 95% confidence interval. This indicates a statistically significant baseline level of financial performance. The Wald chi-square test further supports the robustness of this finding ($\chi^2 = 164.823$, df = 1, $p < 0.001$). These results underscore the complex relationship between audited financial statements and business performance measures, offering insights into how financial transparency and management practices influence organizational success and indigenous finance.

5. Conclusions and Future Studies

The paper aimed to illuminate on the performance of the companies through the analysis of the audited financial statements for the two-year period 2019 and 2020. In this analysis, various methods and tools of financial analysis were used to reveal the relationships and influence of financial items on the financial

Table 8.8. Correlation of Balance Sheet Variables for Loans and Overdrafts.

Correlations					
		Loans and Overdrafts	**Overdraft Obligations**	**Obligations to the NLB Loan**	**Liabilities to the PCB Loan**
Loans and overdrafts	Pearson correlation	1	1.000**	1.000**	−1.000**
	Sig. (2-tailed)		.	.	.
Overdraft Obligations	Pearson correlation	1.000**	1	1.000**	−1.000**
	Sig. (2-tailed)	.		.	.
Obligations to the bank loan	Pearson correlation	1.000**	1.000**	1	−1.000**
	Sig. (2-tailed)	.	.		.
Liabilities to the bank loan	Pearson correlation	−1.000**	−1.000**	−1.000**	1
	Sig. (2-tailed)	.	.	.	

Source: Source of data in tables: Authors' calculations in SPSS program.
Dependent variable y_1 (loans and overdrafts).
Independent variable x_1 (cost of goods sold – trade).
** Correlation is significant at the 0.01 level (2-tailed).

performance for firms. The results of this research have highlighted that there are complex and different relationships between elements of financial statements and financial performance. Based on all the financial items related to future indigenous finance, it was identified that the role of cash in bank accounts has a significant, positive impact on the financial performance of companies, particularly in terms of financial analysis and reporting. This suggests that accurate, careful, and detailed management of financial resources is essential for achieving positive results and receiving favorable audit reports. Therefore, these findings underline the importance of a complete analysis of all financial statements in order to understand why there were negative impacts and where these impacts stem from (from the strict rules of the country, from the pandemic, from the lack of innovations, etc.). Therefore, this finding provides an overview of the relationships between the items and factors in each mirror and the effect on the performance, success, and stability of the firm and indigenous finances.

Table 8.9. Correlation of Variables in the Balance Sheet for Cost of Goods Sold.

		Correlations		
		KMSH	**Cost of Goods Sold – Trade**	**Cost of Goods Sold – Construction**
CGS	Pearson correlation	1	−1.000**	1.000**
	Sig. (2-tailed)		.	.
Cost of goods sold – Trade	Pearson correlation	−1.000**	1	−1.000**
	Sig. (2-tailed)	.		.
Cost of goods sold – construction	Pearson correlation	1.000**	−1.000**	1
	Sig. (2-tailed)	.	.	

Source: Source of data in tables: Authors' calculations in SPSS program.
Dependent variable y_1 (loans and overdrafts).
Independent variable x_1 (cost of goods sold – trade).
** Correlation is significant at the 0.01 level (2-tailed).

Based on the findings, this chapter emphasizes the importance of future indigenous finance and provides the following recommendations for companies:

- Increased attention to the control and management of cash and bank accounts, given their strong positive correlation with performance and financial stability.
- Enhancement of accurate inventory management and oversight, as it is positively related to financial performance and position.

5.1 Future Studies

Even though the study identified relationships between variables and factors in each audited financial asset, future studies must have a large number of years, variables, or countries to see a comparative analysis among countries as regards financial indigenous and the effect on the performance of firms.

In summary, this study provided a detailed analysis of how audited financial statements affected company performance, focusing on data from 2019 to 2020. Quantitative findings revealed significant positive correlations, particularly

Table 8.10. Parameter Estimates.

Parameter Estimates							
			95% Wald Confidence Interval		Hypothesis Test		
Parameter	***B***	**Std. Error**	**Lower**	**Upper**	**Wald Chi-Square**	**df**	**Sig.**
(Intercept)	1,174,982.061	91,521.2509	995,603.705	1,354,360.416	164.823	1	0.000
(Scale)	16,752,278,728.652[a]	16,752,278,728.6520	2,359,784,515.741	118,925,622,543.240			

Dependent variable: Model
Model: (Intercept)

a. Maximum likelihood estimate.

Source: Source of data in tables: Authors' calculations in SPSS program.

highlighting the substantial impact of reserves of cash in bank accounts in company results. These results underscored the critical importance of careful management of financial resources for achieving favorable audit results and increasing overall performance. The study's recommendations included improving cash management practices, optimizing inventory control, and improving credit payment and liability policies to strengthen financial stability and operational efficiency. Moreover, these strategies are applied to different economic sectors and industries, suggesting their potential to promote long-term growth and stability in different economic environments.

References

Adeola, O. (2023). Introduction to casebook on indigenous business practices in Africa – Volume II trade, production, and financial services. In O. Adeola (Ed.), *Casebook of indigenous business practices in Africa* (pp. 1–14). Emerald Publishing Limited. https://doi.org/10.1108/978-1-80455-762-420231002

AlMehrzi, A. A., Tipu, S. A., & Sarker, A. E. (2024). Determinants, processes, and impacts of indigenous entrepreneurship: A systematic literature review. *Journal of Enterprising Communities: People and Places in the Global Economy*, *18*(3), 646–681. https://doi.org/10.1108/JEC-10-2022-0155

Boiral, O., Heras-Saizarbitoria, I., & Brotherton, M.-Ch (2019). Corporate sustainability and indigenous community engagement in the extractive industry. *Journal of Cleaner Production*, *235*, 701–711. https //doi.org/10.1016/j.jclepro.2019.06.311

Brown, A. (2018). The financial reporting of the national fisheries corporation of Tuvalu: The case for alternative indigenous reporting mechanisms. *Marine Policy*, *88*, 93–100. https://doi.org/10.1016/j.marpol.2017.11.015

Christine Jonick, E. D. (2017). *Principles of managerial accouting*. University of North Georgia Press Dahlonega.

Colbourne, R. (2017). Indigenous entrepreneurship and hybrid ventures. In A. C. Corbett & J. A. Katz (Eds.), *Hybrid ventures* (*Advances in Entrepreneurship, Firm Emergence and Growth*, Vol. 19, pp. 93–149). Emerald Publishing Limited. https://doi.org/10.1108/S1074-754020170000019004

Denny-Smith, G., & Loosemore, M. (2017). Integrating indigenous enterprises into the Australian construction industry. *Engineering Construction and Architectural Management*, *24*(5), 788–808. https://doi.org/10.1108/ECAM-01-2016-0001

Elliott, B. E. (2009). *Financial Accounting and Reporting*. Pearson.

Escobar, L. H., & Alexandrov, V. (2018). Territorial design optimization for business sales plan. *Journal of Computational and Applied Mathematics*, *340*, 501–507. https://doi.org/10.1016/j.cam.2018.02.010

Falola, A., Mukaila, R., & Abdulhamid, K. O. (2022). Informal finance: Its drivers and contributions to farm investment among rural farmers in Northcentral Nigeria. *Agricultural Finance Review*, *82*(5), 942–959. https://doi.org/10.1108/AFR-08-2021-0116

Hagens, N. J. (2020). Economics for the future—Beyond the superorganism. *Ecological Economics*, *169*, 106520. https://doi.org/10.1016/j.ecolecon.2019.106520

Jha, D. N. (2014). *Introduction to financial accounting*. Himalaya Publishing House.

Jovanova, N. (2013). Accounting. Skopje: Ministry of Education and Science of the Republic of Macedonia.

Kabiru, I., & Dandago, A. S. (2014). *An examination into the quality of audited financial statements of money deposit banks in Nigeria*. HR Mars.

Kuczyńska-Cesarz, A., & Szczerbak, M. (2020). The role of financial audit in assessing company's development ability and increasing investors' protection. *Inżynieria Bezpieczeństwa Obiektów Antropogenicznych*, (2). https://doi.org/10.37105/iboa.61

Le, L. H., & Quang-An, H. (2021). Effects of negative reviews and managerial responses on consumer attitude and subsequent purchase behavior: An experimental design. *Computers in Human Behavior*, *124*, 106912. https://doi.org/10.1016/j.chb.2021.106912

Lulaj, E. (2021a). Accounting, reforms and budget responsibilities in the financial statements. *Accounting and Finance*, *91*(1), 61–69. http://doi.org/10.33146/2307-9878-2021-1(91)-61-69

Lulaj, E. (2021b). Quality and reflecting of financial position: An enterprises model through logistic regression and natural logarithm. *Journal of Economic Development, Environment and People*, *10*(1), 26–50. https://doi.org/10.26458/jedep.v10i1.690

Lulaj, E. (2022). An unstoppable and navigating journey towards development reform in complex financial-economic systems: An interval analysis of government expenses (past, present, future). *Business, Management and Economics Engineering*, *20*(2), 329–357. https://doi.org/10.3846/bmee.2022.17389

Lulaj, E. (2023). A sustainable business profit through customers and its impacts on three key business domains: Technology, innovation, and service (TIS). *Business, Management and Economics Engineering*, *21*(1), 19–47. https://doi.org/10.3846/bmee.2023.18618

Lulaj, E., & Dragusha, B. (2022). Incomes, gaps and well-being: An exploration of direct tax income statements before and during Covid-19 through the comparability interval. *International Journal of Professional Business Review*, *7*(6), e0623. https://doi.org/10.26668/businessreview/2022.v7i6.623

Lulaj, E., Dragusha, B., & Hysa, E. (2023). Investigating accounting factors through audited financial statements in businesses toward a circular economy: Why a sustainable profit through qualified staff and investment in technology? *Administrative Sciences*, *13*(3), 72. https://doi.org/10.3390/admsci13030072

Lulaj, E., Gopalakrishnan, A., & Kehinde Lamidi, K. (2024). Financing, investing in women-led businesses: Understanding strategic profits, entrepreneurial expectations by analysing the factors that determine their company success. *Periodica Polytechnica Social, Management Sciences*. https://doi.org/10.3311/PPso.22532

Lulaj, E., & Iseni, E. (2018). Role of analysis CVP (cost-volume-profit) as important indicator for planning and making decisions in the business environment. *European Journal of Economics and Business Studies*, *4*(2). https://doi.org/10.26417/ejes.v4i2.p104-120

Lulaj, E., Zarin, I., & Rahman, Sh (2022). A novel approach to improving E-government performance from budget challenges in complex financial systems. *Complexity*. https://doi.org/10.1155/2022/2507490

Mahato, J., & Jha, M. K. (2023). Does financial inclusion promote sustainable livelihood development? Mediating effect of microentrepreneurship. *Journal of Financial Economic Policy*, *15*(4/5), 485–499. https://doi.org/10.1108/JFEP-05-2023-0134

Meo, R. (2020). *Accuracy of audit opinion: Factors that influence it*. ResearchGate.

Moro, S. R., Paulo, A., Miguel, C., & Mendes, G. H. S. (2022). Adding sustainable value in product-service systems business models design: A conceptual review towards a framework proposal. *Sustainable Production and Consumption*, *32*, 492–504. https://doi.org/10.1016/j.spc.2022.04.023

Moore, N. M. (2002). From indigenous industry to foreign finance: The changing face of Dublin Docklands. *Land Use Policy*, *19*(4), 325–331. https://doi.org/10.1016/S0264-8377(02)00040-6

Ojera, P. (2018). *Indigenous financial management practices in Africa: A guide for educators and practitioners, indigenous management practices in Africa* (*Advanced Series in Management*, Vol. 20, pp. 71–96). Emerald Publishing Limited. https://doi.org/10.1108/S1877-636120180000020005

Olarewaju, A. D., Adebisi, S. A., & George, O. J. (2018). *Indigenous African financing strategies as a unique engine for entrepreneurship growth, indigenous management practices in Africa* (*Advanced Series in Management*, Vol. 20, pp. 147–166). Emerald Publishing Limited. https://doi.org/10.1108/S1877-636120180000020008

Rayburn, S.-W., Badrinarayanan, V., Anderson, S. T., & Gupta, A. (2021). Continuous techno-training and business-to-business salesperson success: How boosting techno-efficacy enhances sales effort and performance. *Journal of Business Research*, *133*, 66–78. https://doi.org/10.1016/j.jbusres.2021.04.066

Thomas, J., & Mantri, P. (2022). Design for financial sustainability. *Patterns*, *3*, 100585. https://doi.org/10.1016/j.patter.2022.100585

Tong, T., Xu, X., Yan, N., & Xu, J. (2022). Impact of different platform promotions on online sales and conversion rate: The role of business model and product line length. *Decision Support Systems*, *156*. https://doi.org/10.1016/j.dss.2022.113746

Tuovila, A. (2022, May 1). Audit: What it means in finance and accounting, 3 main types. https://www.investopedia.com/terms/a/audit.asp

Uchenna, N., & Akpom, Y. O. (2015). *Determinants of auditor independence: A comparison of the perceptions of auditors and non-auditors in Lagos, Nigeria*. Semantic Scholar.

Uzo, U., Shittu, O., & Meru, A. K. (2018). *Introduction: Indigenous management practices in Africa, indigenous management practices in Africa* (*Advanced Series in Management*, Vol. 20, pp. 1–7). Emerald Publishing Limited. https://doi.org/10.1108/S1877-636120180000020001

Wood, G. J., & Davidson, M. J. (2011). A review of male and female Australian indigenous entrepreneurs: Disadvantaged past – Promising future? *Gender in Management*, *26*(4), 311–326. https://doi.org/10.1108/17542411111144319

Wu, J., Wood, G., Chen, X., Meyer, M., & Liu, Zh (2020). Strategic ambidexterity and innovation in Chinese multinational vs. indigenous firms: The role of managerial capability. *International Business Review*, *29*(6). https://doi.org/10.1016/j.ibusrev.2019.101652

Zheng, B., Hui, W., Amir-Mohammad, G., & Alireza, G. (2022). Impacts of logistics service quality and energy service of business to consumer (B2C) online retailing on customer loyalty in a circular economy. *Sustainable Energy Technologies and Assessments*, *52*, 102333. https://doi.org/10.1016/j.seta.2022.102333

Chapter 9

Revolutionizing the Indigenous Financial System With the Utilization of Artificial Intelligence

Anju Rohilla[a], *Ansh Jindal*[b] *and Priya Jindal*[b]

[a]Department of Business Studies, PIET, Samalkha, India
[b]Chitkara Business School, Chitkara University, Punjab, India

Abstract

Artificial Intelligence (AI) plays a pivot role in the development of the financial sector and promotes the economic development of the country. Indigenous financial sector comprises of private institutions as well as the individuals which are indulged in providing the banking services such as accepting deposits and providing loans. Indigenous financial system (IFS) has transformed with the incorporation of the AI-based technologies which empowers the financial sector to design the tailor-made services and facilitates the digitalization of the services and identifies and prevents the frauds. The exponential rise of the FinTech start-ups demonstrates how the integration of the technology transformed the financial sector. This chapter is dedicated toward identifying the implications of AI in the financial sector. Additionally, it focuses how the use of the AI helps the financial institutions in making sound financial strategies. In order to understand the use of AI in the IFSs and identify its implication on IFSs, the review of the available literature is conducted. The study shows that the employment of AI in financial management operations aids the firm in credit scoring of the customers and firms, designing and providing client-facing chatbots, know your customer, identifying the potential risk of frauds, offering the highly personalized financial services, achieving the time and cost-efficient services, effectively engage the customers and enhances connectivity with customer across the multiple platforms. The study shows that using artificial intelligence in financial management operations and to stay competitive in this digital era.

Indigenous Empowerment through Human-Machine Interactions, 153–165

doi:10.1108/978-1-83608-068-820251009

Keywords: Industry 5.0; Indigenous financial system; financial services; Artificial Intelligence; digitalization; FinTech; banks

1. Introduction

The key factor in the development of financial sector and the promotion of the economic development is the technological revolution. The adoption of computerization, the rise of ATMs, online and mobile banking, e-payments, high frequency trading, and the adoption and use of blockchain technology, artificial intelligence (AI), machine learning (ML), robotic process automation, and quantum computing are all examples of the technological revolution. As per the Cambridge dictionary, the term indigenous means native, and the Indigenous financial institutions (IFIs) stand for the financial institutions which are operating their businesses in the local area. The term "IFIs" refers to the traditional and popular ways that indigenous people around the world manage their finances, including saving and investing. The cultural, social, and economic norms of these indigenous people serve as the foundation of the Indigenous financial system (IFS), which leads to the establishment of the modern banking and financial organizations. It includes both the commercial organizations as well as the private individuals which are engaged in the banking activities such as granting loans and deposits acceptance. This study objective is to investigate the ways in which the IFS is engaging AI as well as how they are influencing and impacted by AI.

AI plays a pivot role in the development of the indigenous financial sector and promotes the economic development of the country. AI-based technology has revolutionized the IFS by enabling the financial industry to provide customized services and streamline the digitalization of the services and detection and prevention of fraudulent activities.

2. IFS and Its Evolution

In the developing economies, IFS plays a crucial role in the generation of employment opportunities (Bhattacharya, 2019) and raising the standard of living. The IFS contributes not only in the growth of the individual but also contribute toward the development of economy through conversion of saving into investment and creation of an ecosystem that supports start-ups and existing business (Aliber, 2015). IFS particularly caters the need of the economically disadvantage societies (Coffie et al., 2021) by offering the bootstrap financing facility to small businesses. The IFS has less stringent conditions such as collateral requirements and provides more affordable payment option in contrast to traditional banking system. The start-ups which don't yet have the credit history to quality the bank loans rely heavily on IFS for funding. Although IFS plays an important role in the lives of those who are without the access of financial resources, the IFI sector has often been seen as "waiting room" or "temporary system" during the transformation of the traditional to the formal economy (Bhattacharya, 2019). The continuous informalization of the IFIs has detrimentally affected the credit standability and further keeps them apart from the discussions on international finance. IFS has historically been linked to rural, low-income, people with low-literacy and urban poor people (Bophela & Khumalo, 2019). Additionally, these groups are frequently neglected by the traditional financial system which forces them to rely on the alternate sources of finance and other financial services. The expansion of the IFS in the urban, educated, and banked populations demonstrated that access to modern financial system does not neglect the importance of the IFIs in spite of the fact that the urban and educated population rely upon the modern financial system more.

IFS in India has been around since the medieval period and persisted until the middle of the 19th century. Historically, IFIs were the backbone of the India's financial system. Everyone from government to individual depends upon the IFIs to get the credit facility. Throughout the Mughal Empire, India had a number of IFSs, such as the Hundi system, Chettiar and Shroff system, and "Nawab's Treasury." Under the Hundi system, the goods can be bought and sold on credit using the documents such as chitti or Hundi. Hundis is a document and an unconditional promise which ensures the payment of certain amount of money that can be changed via legal negotiation. In this system, Hundi brokers assist them in the transfer of the money from one place to another place. In the southern region of India, it is popularly known as the Cooperative Credit Societies. Under the Chit Fund system, pool of specified amount of money is created and disbursed among the members on the request for variety of uses such as investing, launching a new business, expanding a business or for personal use. It is popular in India with the name of committee system.

Regardless their usefulness in providing the financial resources from individuals to entrepreneurs, the popularity of the IFS has declined over the time with the emergence of the international and the commercial banks. The development of the commercial and the cooperative banks in the late 1950s leads toward the intensification of this phenomenon in India. In sub-Saharan Africa, the usage of the IFS has remained prevalent despite a steady growth in their financial inclusion in the mainstream financial system. About 50–95% of the youth is engaged in the usage of the IFS in the West African countries (Kounou et al., 2013). Despite the fact that South Africa has the highest rate of access to mainstream financial services in Africa, i.e., 70% (Lim et al., 2021), 1.5 million people are regularly using the services of IFIs (Biyela et al., 2018). The majority of people have access to banks, but only 33% of them utilize their savings and 12% use their borrowing services, indicating that they are using the IFS, which is an alternative to modern banking (Bophela & Khumalo, 2019). These IFIs provide more affordable services with greater flexibility than the conventional banking system. It is evident from here that merely having the access to the mainstream financial system does not eliminate the need and role of the IFIs in economy. The contemporary functions of IFIs in socioeconomic development are visible regardless of their status, permanent or temporary.

3. IFI's Characteristics and Functions

IFIs possess the various characteristics:

- *Community Based:* IFS is usually based on the indigenous communities and centered on the need and goals of the communities rather than the personal goals.
- *Informal Institutions:* IFIs function independently of the official regulatory structures. IFS depends upon the unofficial neighborhood-based lending circles, rotating savings and credit associations.
- *Social Capital:* IFS usually relies heavily upon the trust and social interactions. Within the communities, transactions are frequently predicted on reciprocity, trust, and mutual assistance.
- *Nonmonetary Exchanges:* IFS may incorporate a variety of deeply ingrained cultural tradition such as giving and taking gifts and bartering.
- *Cultural Values:* The rituals, beliefs, and cultural values of indigenous people influence their financial system. Spirituality, society, and the environment are closely related to economic activity.

IFS works independently from the formal financial system of the country. Examples of IFSs include the use of traditional currencies like cowrie shells or beads by some Africans and indigenous Americans, as well as the worldwide practice of rotating savings and credit associations (Ojera, 2018). This system possesses the unique features such as informal institutions, cultural value-based services and the offers the services which are tailored to meet the requirements and ideals of these communities.

- *Loan Disbursement:* IFS offers the facility of the loan to the community members such as local residents, farmers, and businesses to support them financially in meeting their personal goals such as advancement of the education and business goals.
- *Rendering Services:* Remittance services are provided by the IFIs to make it easier for the money to move/transfer across the various areas. The remittance service facilitates the ease of sending and receiving money to and from the family and friends who resides overseas (Ky et al., 2021).
- *Informal Saving Clubs and Societies:* Indigenous people usually create unofficial saving clubs and societies, in which the members create a pool of money through regularly depositing the money to the communal funds. Subsequently, this money is disbursed among the member of the unofficial club either on the rotating basis or for certain occasions such as holidays, weddings, and funerals.
- *Financing for Inland Trade:* Local banks and merchants can rely on the IFS to help them with funds for transportation, businesses, and meeting other associated cost (Nan et al., 2021). In order to facilitate the seamless exchange of goods and services among the members of the community, they offer more credit and working capital.

4. Literature Review

The fact that the IFS has been around for decades and is still going strong despite efforts to make it more official and civilized shows how strong the IFIs and community activities are (Heeks & Ospina, 2019). Atinaf et al. (2021) states that assessing the resilience of any institutions or system include examining the system before during and after the disruption. One of the biggest disruptions that take place for the IFIs is the colonialism. Living standard and the social economic system were severely disrupted by the colonialism in several countries. The existence of the IFIs before, during, and after the colonialism system is the evidence of the resilience of the IFIs. Another major disruption for the IFIs was technological revolution. The resilience of the IFS has been impacted by the technological revolution. "Robustness, self-organizing and learning" are the three characteristics to evaluate the resilience of IFIs. The degree to which the system can continue to function and exhibit its properties in spite of external shocks and disruption is termed as robustness. The precolonial methods of the IFIs, which included the frequently meeting and giving donation in one or other ways for the betterment of majority, are strikingly similar to the modern ideas, philosophy, operations, and practices of IFIs. The ability of the system to bounce back and reorganize its operations in response to external disruption is measured by its self-organization. After the disruption of the barter system, IFIs has transformed toward the monetary system (Verhoef, 2001). In order to develop new abilities, perspectives, and competences required for experimentation and creativity, the system must be able to generate feedback, and this is how an organization learn and grow (Biyela et al., 2018). The adoption and incorporation of the ICTs is the proof that are IFIs are eager to learn. It enhances the IFIs to interact with the outside system.

5. IFS and Information Communication Technology (ICT)

Particularly, in the unorganized and informal financial sector, ICT has become an indispensable instrument for tacking development issues and concerns. Informal financial sector comprises of the small and microenterprises which is essential for the creation of the jobs and economic expansion for every nation. The emergence of ICT has helped the small businesses to grow their companies and boost their profits (Seetharaman et al., 2019). IFIs are continuously adopting and integrating new technologies, i.e., ICT. It shows that the system can interact with outside knowledge and can use it for betterment of the community as well as the system. ICT also facilitates the reduction in the common problem in the remittances such as unsafe methods of transmitting and receiving funds and storing cash in lock-boxes. Some other common problems faced by the indigenous communities and IFIs were social capital, insurance mechanisms and their accessibility, low interest sources of credit availability to individuals and business. With the advent of ICT and the introduction of the telecommunication and internet facilities, group of traders and small entrepreneurs are more able to communicate more effectively than ever before with their suppliers, customers, and market place (Chatterjee, 2020). The important contributors in the success ICT implementation were the indigenous development, wider program integration, and participatory design (Kariuki & Ofusori, 2017). The ICT has a positive impact in certain circumstances; for instance, Kerala fishermen use their mobile phones for enhancing their market performance (Rangaswamy, 2019), and M Farm in Kenya benefitted for the use of ICT in providing and getting price information to and from farmers (Malanga & Banda, 2021). Despite the fact that ICT brought many benefits, there were many obstacles in the widespread acceptance of the ICTs, particularly in minimizing the digital gap that keeps some out of the global workspace (Tang & Konde, 2020).

6. IFS and Digitalization

An important step toward the greater accessibility, inclusion, and empowerment for the indigenous people throughout the globe is the digitalization of the IFIs. Indigenous people and enterprises are getting access to a greater range of financial goods and services by utilizing the power of digital technologies such as mobile banking, online fund transfer, and AI- driven financial services (Brennen & Kreiss, 2016). It helps them to be more financial resilience and independent in addition to enhance their capacity to manage their money. By fostering cultural awareness in the financial services industry, digitalization also aids in the preservation of indigenous culture and heritage (Chawurura et al., 2021). A more sustainable and equitable financial ecosystem is emerging that enables indigenous people to prosper their needs in the digital age (Jain, 2023). This is due to the collaboration of government agencies, financial institutions and IT companies to develop inclusive digital solutions that are customized to the particular requirement of the indigenous communities and societies. Digitalization transformed the IFS by offering various services. Initially, the mobile banking and internet platforms help to improve access to financial services (Owoseni & Twinomurinzi, 2020) via removing the geographic constraints, enabling indigenous groups to actively participate in the economy and manage their money (Kelikume, 2021). Additionally, digitalization promotes financial inclusion by giving people access to credit, savings, and insurance products that allow them to build wealth and lower risks. Furthermore, it lowers down the risk of exploitation and corruption with IFS by facilitating fast and transparent financial transactions. Digital platforms and apps may also help in the perseverance and revitalize indigenous language and cultures by integrating native symbols, dialects, and language (Van Stam, 2021). Digitalization acts as a significant instrument for boosting indigenous communities' economic growth, empowerment, and cultural perseverance by enhancing access to financial services, encouraging inclusiveness and transparency, and simulating creativity and cultural (Ghosal et al., 2024; Pérez-García, 2021).

7. Incorporation of AI and ML in IFS

The incorporation of the AI and ML into IFS has the potential to swiftly transform the entire system as well as it makes the indigenous people capable to face their financial problems in a very efficient manner. AI-enabled technologies enabled the IFS to enhance their accessibility, transparency, efficiency, and perseverance of their culture (Pérez-García, 2021). AI and ML will have the significant impact over the performance of the IFS. By analyzing the large volume of data, IFIs are able to identify the trends and spot the pattern which helps them to understand the need of the indigenous people in the better way. It will lead toward the creation of the customized financial services and products that are more relevant and accessible to indigenous communities (Tad et al., 2023). For example, AI-enabled chatbots and smartphone applications may offer the customized financial guidance and assistance to them in their native languages and bring down the language barrier and enhance their financial literacy.

Additionally, AI and ML enable technologies contribute to greater financial inclusion through identifying the indigenous communities that are underserved and creating effective ways to contact them. With the use of these technologies, financial institutions got the opportunity to study and evaluate the demographic, geographic, and socioeconomic data and enable them to identify the regions with the poor access to banking services and set-up community-based banking hubs and mobile banking services to cater the needs of the indigenous community. Furthermore, creditworthiness of the persons without possessing traditional credit history can be evaluated with the help of the AI-driven credit scoring models. It enhances their accessibility to get credit facilities for investment into their businesses, education, and for their personal consumption. AI and ML have the potential to enhance the two crucial elements of the IFSs, i.e., efficiency and transparency. Financial institutions may save their administrative overhead and improve operations, resulting into cost saving that can be passed on to clients, by automating regular procedures such as loan processing and account management. Furthermore, the danger of the financial exploitation in IFIs can be reduced with the utilization of the AI-powered fraud detection system to detect the suspicious activities. Additionally, blockchain technology-which is frequently combined with AI & ML capable to develop the transparent, unalterable record of financial transactions as well as enhance the accountability and confidence in the IFIs. Another important aspect of the AI and ML in IFIs is the cultural relevance. AI-empowered technologies enable the development of the financial products and services through analyzing the cultural practices, values, and traditions of indigenous communities that are in line with the cultural norms and preferences of these indigenous communities (Wall et al., 2015). For example: AI-driven recommendation engines have the capabilities to propose investment options that uphold the rights of the indigenous people to their land and the environment, or they may recommend the sustainable development initiatives that promote indigenous community–owned enterprises. Moreover, by integrating the traditional symbols, dialects, and storytelling into financial education material and digital platforms, AI and ML support the perseveration and revival of the indigenous languages and culture.

8. Implications of AI and ML into Indigenous Financial System

The incorporation of the AI into IFIs has significantly contributed toward the autonomy, cultural perseveration and economic prosperity of the indigenous community across the globe. AI-empowered technologies possess the potential and challenges for indigenous communities in managing their funds, obtaining financial resources, and handling the complexities of the contemporary financial systems (Ris et al., 2020). To promote the equitable and sustainable growth of the IFS, it is important to understand both the positive as well as the negative implications of the AI and ML on the IFS.

8.1 Positive Implications of AI and ML on IFS

The incorporation of AI and ML into IFIs brought the various negative implications and challenges that must be addressed.

- *Improved Access to Financial Services:* AI-powered technologies empower the IFIs to more properly evaluate the customer's creditworthiness and which leads toward the higher probability of disbursement of more credit to communities and expanding the opening of saving accounts and availability of insurance products to the IFSs (Malali & Gopalakrishnan, 2020; Rasiwala & Ghosal, 2024). It also helps in bridging the gap between the traditional banking systems and underrated IFIs.
- *Customized Financial Services:* With the considerations for the culture, customs, linguistic preferences, and the value of the indigenous communities, AI-powered financial services will address the unique preferences and requirements of indigenous people (Tyagi et al., 2022). These customized financial services such as personalized investment portfolio, insurance program, and saving strategies make the indigenous people capable to attain their financial objectives.
- *Fraud Prevention and Mitigation of Risk:* Financial institutions may ensure the ethical landing practices and prevent the financial exploitation and fraud with the use of AI-powered risk management solutions (Kalra & Mittal, 2024). Indigenous communities will be shielded from financial loss by utilizing the power of the ML algorithms that are specifically designed to identify fraudulent activity during their financial operations.
- *Efforts toward Community Development:* AI and ML technology will enhance data-driven decision-making in community development projects, allowing indigenous leader to allocate the resources in the efficient way and to prioritize the program that promote long-term economic growth and social well-being (Saluja & Nair, 2024). Predictive analytics may help to identify the various opportunities for the infrastructural development, economic prosperity, and natural resource management.
- *Enhanced Financial Knowledge:* AI-driven educational platforms will provide the specialized literacy training and capacity-building program for the indigenous communities with the objective of fostering the digital literacy abilities and enabling people to make sound financial decisions. Chatbots and virtual assistants will enhance the financial inclusion and literacy through providing tailor-made advices and assistance in their native languages.

- *Preservation of culture:* AI technology will support the digitalization and perseveration of indigenous cultural heritage pertaining to financial and economic activities, protecting traditional knowledge, artifacts, and oral history (Pérez-García, 2021). Blockchain technology will help to ensure indigenous data sovereignty and protect the intellectual rights of indigenous communities.
- *Partnership for Collaborative Innovation:* To address the issues faced by IFIs and to meet the needs of the indigenous population, indigenous organizations work with the government, financial institutions, and tech companies to develop AI and ML solutions (Jindal & Kaur, 2021). The designs of these solutions reflect the cultural ownership as it is designed and executed after considering the suggestions of the indigenous community stakeholders.

8.2 Negative Implications of the AI and ML

- *Digital Divide:* Differences in the digital literacy, technological access, and internet availability can worsen the existing disparities and prevent the indigenous people from accessing the financial services.
- *Loss of Cultural Identity:* AI-driven financial systems could put uniformity and efficiency ahead of traditional values and cultural sensitivity. Consequently, there's a chance that indigenous financial cultural values and customs would be disregarded or ignored, which might result into the loss of cultural identity and legacy.
- *Data Privacy Issues:* While using the AI-driven technologies, indigenous communities could not have the full control over their personal information. This gives rise to worries about security, privacy, and the possible exploitation of personal data by the external parties without proper authorization or supervision.
- *Algorithmic Bias:* Algorithms used in the ML and AI may be biased due to the bias present in data on which it is trained. If preconceived notions and discrimination against Indigenous people are not adequately addressed, it may negatively affect their access to fair and reasonable financial services.

9. Future Perspective Toward the Development of Artificial and ML–Enabled Financial System

In the developing countries, IFS encounters some problems. Many Indigenous entrepreneurs face the various problem which make it difficult for them to access the capital for their businesses such as underdeveloped financial markets, lower per capital income, inadequate market liquidity, poor regulations, and problems associated with the infrastructure development and with services delivery. Further, indigenous people are more likely to be excluded from the mainstream economic activities, especially indigenous women and people with impairments. There is the need to develop such a financial system in the underdeveloped countries which is capable of providing targeted, sensitive, and ethical capital investment in the indigenous business. Along with aiding Indigenous communities' development, it will contribute toward the achievement of the Sustainable Development Goals (SDGs) such as reducing disparities, enhancing decent employment opportunities, and bringing gender equalities. An AI–ML-enabled IFS must be developed to cater the need of the indigenous communities. For developing AI–ML-enabled financial system, global economic policies must be framed to help in the progress of the indigenous communities' development. Furthermore, in order to assist the indigenous enterprises in succeeding and be competitive, they require technical support and training. AI–ML-enabled financial system will assist in developing the Indigenous-led solutions such as concessional financing instrument and risk reduction strategy which encourage the private investment in indigenous firms.

10. Applications of AI and ML Into IFS

AI and ML Application	Indigenous Financial Institutions	Usage in Indigenous Financial System
Credit stack	Navajo nation financial institutions	This application helps the Navajo nations financial institutions to access the credit risk among the individuals and businesses of Navajo and to provide them loan and financial services.
Risk guardian	Massai pastoralist economy	It ensures appropriate lending practices by identifying and reducing the risks connected to lending to Massai pastoralists through the analysis of market trends and historical data.
Fraud shield	Palaw'an indigenous saving and credit cooperatives	It protects against the financial exploitation and fraud by identifying fraudulent activity in financial transactions involving indigenous community in Palaw'an.

Source: Compiled by the author.

11. Conclusion

In the era of the globalization, countries are connected in numerous ways such as trade and technology. But in the certain regions, Indigenous communities experience marginalization and exclusion. Indigenous people have a unique perspective and culture. Indigenous communities employ technology to safeguard their culture and exercise their control over their own lives. This demonstrates that the use of the technology can not only be for financial gain but also for the expression of the individuality, the preservation of the culture, and the choice of lifestyle. ICT allows the indigenous communities to perform financial transactions easily and security without the requirement of the physical infrastructure. With the use of the ICT in IFS, accountability and transparency is improved by recording the transactions in a safe and unchanged ledger. Efficiency, inclusiveness, and cost-effectiveness have all significantly enhanced with the integration of the AI and ML in IFS. The importance of preserving cultural heritage and promoting economic empowerment in the indigenous communities has led to increased focus on supporting and integrating financial services into mainstream financial services. Policies are required for the effective regulations of indigenous financial services due to the digital inequalities, privacy issues, algorithmic bias, and possibility of cultural identity loss. To overcome these obstacles and optimize the advantages of AL and ML, cooperation between the indigenous stakeholders, government agencies, regulators, law makers, and tech firms will be crucial. AI and ML technologies have the abilities to completely transform indigenous communities worldwide by promoting inclusive and sustainable development.

References

Aliber, M. (2015). *The importance of informal finance in promoting decent work among informal operators: A comparative study of Uganda and India*. International Labour Office, Social Finance Programme. ILO. (Social Finance Working Paper, 66).

Atinaf, M., Molla, A., & Anteneh, S. (2021). Towards a resilient information system for agriculture extension information service An exploratory study. arXiv preprint arXiv:2108.09748.

Bhattacharya, R. (2019). ICT solutions for the informal sector in developing economies: What can one expect? *The Electronic Journal of Information Systems in Developing Countries*, *85*(3), e12075.

Biyela, N., Tsibolane, P., & Van Belle, J. P. (2018, August). Domestication of ICTs in community savings and credit associations (stokvels) in the Western Cape, South Africa. In *International development informatics association conference* (pp. 35–47). Springer International Publishing.

Bophela, M. J., & Khumalo, N. (2019). The role of stokvels in South Africa: A case of economic transformation of a municipality. *Problems and Perspectives in Management*, *17*(4), 26.

Brennen, J. S., & Kreiss, D. (2016). Digitalization. In K. B. Jensen, R. T. Craig, J. D. Pooley, & E. W. Rothenbuhler (Eds.), *The international encyclopedia of communication theory and philosophy* (pp. 1–11). Wiley.

Chatterjee, A. (2020). Financial inclusion, information and communication technology diffusion, and economic growth: A panel data analysis. *Information Technology for Development*, *26*(3), 607–635.

Chawurura, T., Chikomo, S., Manhibi, R., van Dijk, J., & van Stam, G. (2021, December). Developing a digital information and consultation platform in

Zimbabwe In International Conference on e-Infrastructure and e-Services for Developing Countries (pp. 80–92). Springer International Publishing,

Coffie, C. P. K., Hongjiang, Z., Mensah, I. A., Kiconco, R., & Simon, A. E. O. (2021). Determinants of FinTech payment services diffusion by SMEs in Sub-Saharan Africa: Evidence from Ghana. *Information Technology for Development*, *27*(3), 539–560.

Ghosal, I., Prasad, B., Ray, N., & Joshi, L. K. (Eds.). (2024). *Sustainable strategic business infrastructure development and contemporary digital practices in Industry 5.0*. CRC Press.

Heeks, R., & Ospina, A. V. (2019). Conceptualising the link between information systems and resilience: A developing country field study. *Information Systems Journal*, *29*(1), 70–96.

Jain, R. (2023). Role of artificial intelligence in banking and finance. *Journal of Management and Science*, *13*(3), 1–4.

Jindal, P., & Kaur, J. (2021, August). Artificial intelligence applications for lending and NPA management. In 2021 Asian Conference on Innovation in Technology (ASIANCON) (pp. 1–6). IEEE.

Kalra, A., & Mittal, R. (2024, March). Explainable AI for improved financial decision support in trading. In 2024 11th International Conference on Reliability, Infocom Technologies and Optimization (Trends and Future Directions)(ICRITO) (pp. 1–6). IEEE.

Kariuki, P., & Ofusori, L. O. (2017). WhatsApp-operated stokvels promoting youth entrepreneurship in Durban, South Africa: Experiences of young entrepreneurs. In Proceedings of the 10th International Conference on Theory and Practice of Electronic Governance (pp. 253–259).

Kelikume, I. (2021). Digital financial inclusion, informal economy and poverty reduction in Africa. *Journal of Enterprising Communities: People and Places in the Global Economy*, *15*(4), 626–640.

Kounou, G., Akpona, C., Ahouantchede, H., Gohoue, R., Belqasmi, F., & Glitho, R. (2013, December). A social network-based architecture for on-line RoSCAs in the developing world. In Proceedings of the 4th Annual Symposium on Computing for Development (pp. 1–2).

Ky, S. S., Rugemintwari, C., & Sauviat, A. (2021). Friends or Foes? Mobile money interaction with formal and informal finance. *Telecommunications Policy*, *45*(1), 102057.

Lim, W. M., Gupta, S., Aggarwal, A., Paul, J., & Sadhna, P. (2021). How do digital natives perceive and react toward online advertising? Implications for SMEs. *Journal of Strategic Marketing*, 1–35.

Malali, A. B., & Gopalakrishnan, S. (2020). Application of artificial intelligence and its powered technologies in the Indian banking and financial industry: An overview. *IOSR Journal of Humanities and Social Science*, *25*(4), 55–60.

Malanga, D. F., & Banda, M. (2021). ICT use and livelihoods of women microenterprises in Malawi. arXiv preprint arXiv:2108.09807.

Nan, W., Zhu, X., & Lynne Markus, M. (2021). What we know and don't know about the socioeconomic impacts of mobile money in Sub-Saharan Africa: A systematic literature review. *The Electronic Journal of Information Systems in Developing Countries*, *87*(2), e12155.

Ojera, P. (2018). Indigenous financial management practices in Africa: A guide for educators and practitioners. In U. Uzo & A. K. Meru (Eds.), *Indigenous management practices in Africa: A guide for educators and practitioners* (pp. 71–96). Emerald Publishing Limited.

Owoseni, A., & Twinomurinzi, H. (2020). Evaluating mobile app usage by service sector micro and small enterprises in Nigeria: An abductive approach. *Information Technology for Development*, *26*(4), 762–772.

Pérez-García, L. (2021). The ICT-Buen Vivir paradox: Using digital tools to defend indigenous cultures. arXiv preprint arXiv:2108.09952.

Rangaswamy, N. (2019). A note on informal economy and ICT. *The Electronic Journal of Information Systems in Developing Countries*, *85*(3), e12083.

Rasiwala, F., & Ghosal, I. (2024). An empirical study on accounting conservatism, integrated reporting in India, and corporate innovation: Evidence from Industry 5.0. In I. Ghosal, B. Prasad, N. Ray, & L. K. Joshi (Eds.), *Sustainable strategic business infrastructure development and contemporary digital practices in Industry 5.0* (pp. 139–159). Apple Academic Press.

Ris, K., Stankovic, Z., & Avramovic, Z. (2020). Implications of implementation of artificial intelligence in the banking business with correlation to the human factor. *Journal of Computer and Communications*, *8*(11), 130.

Saluja, S., & Nair, A. J. (2024). An analysis on frauds affecting the financial security of the Indian banking sector: A systematic literature review. In S. Saluja, V. Nayyar, K. Rojhe, & S. Sharma (Eds.), *Ethical marketing through data governance standards and effective technology* (pp. 201–218).

Seetharaman, P., Cunha, M. A., & Effah, J. (2019). IT for the informal sector in developing countries: A broader perspective. *The Electronic Journal of Information Systems in Developing Countries*, *85*(3), e12093.

Tad, M. S., Mohamed, M. S., Samuel, S. F., & Deepa, M. J. (2023). Artificial intelligence and robotics and their impact on the performance of the workforce in the banking sector. *Revista de Gestão Social e Ambiental*, *17*(6), e03410.

Tang, Y. K., & Konde, V. (2020). Differences in ICT use by entrepreneurial microfirms: Evidence from Zambia. *Information Technology for Development*, *26*(2), 268–291.

Tyagi, S., Jindal, T., Krishna, S. H., Hassen, S. M., Shukla, S. K., & Kaur, C. (2022, December). Comparative analysis of artificial intelligence and its powered technologies applications in the finance sector. In 2022 5th International Conference on Contemporary Computing and Informatics (IC3I) (pp. 260–264). IEEE.

Van Stam, G. (2021). Appropriation, coloniality, and digital technologies. Observations from within an African place. arXiv preprint arXiv:2108.10087.

Verhoef, G. (2001). Informal financial service institutions for survival: African women and stokvels in urban South Africa, 1930–1998. *Enterprise & Society*, *2*(2), 259–296.

Wall, J. D., Stahl, B. C., & Salam, A. F. (2015). Critical discourse analysis as a review methodology: An empirical example. *Communications of the Association for Information Systems*, *37*(1), 11.

Chapter 10

The Role of Education in Indigenous Empowerment

Ruchika Rastogi[a] *and Swati Gupta*[b]

[a]Pranveer Singh Institute of Technology, India
[b]Universal AI University, India

Abstract

The role of education in empowering Indigenous communities is pivotal in fostering cultural resilience, reclaiming identity, and facilitating socioeconomic development. This chapter explores various dimensions of Indigenous education, examining its significance in preserving traditional knowledge, promoting self-determination, and bridging socioeconomic disparities. Drawing upon a review of scholarly literature and case studies, this research elucidates the multifaceted benefits of culturally relevant education in Indigenous empowerment. It highlights the importance of decolonizing educational systems, integrating Indigenous perspectives, and fostering community-led initiatives to ensure meaningful engagement and sustainable progress. Furthermore, this chapter discusses challenges such as inadequate resources, cultural biases, and institutional barriers, emphasizing the need for collaborative efforts between Indigenous communities, governments, and educational institutions to address these issues effectively. By recognizing education as a catalyst for Indigenous empowerment, this study advocates for inclusive and culturally responsive approaches that honor Indigenous worldviews, languages, and ways of knowing, thereby advancing social justice and equity for Indigenous peoples globally.

Keywords: Indigenous community; cultural resilience; socioeconomic development; environmental stewardship; transformative education; culturally responsive education

Indigenous Empowerment through Human-Machine Interactions, 167–177

Published under exclusive licence by Emerald Publishing Limited
doi:10.1108/978-1-83608-068-820251010

1. Introduction

Literacy and access to education are powerful tools for Indigenous individuals and communities. When resilience peoples have access to education and literacy, they become more resistant in the face of adversity. These skills are vital in navigating legal frameworks and can empower individuals to challenge discriminatory practices. It allows them to advocate for the recognition and protection of their land, culture, and overall human rights.

Empowerment through education is one of the major strategies used worldwide to increase social cohesion, which in turn can be used as tools to reduce poverty (Tsey et al., 2009). When offered in a particular workplace or community settings, empowerment programs have been shown to improve workforce retention, job satisfaction, performance, service delivery, and concern for others (Fulford & Enz, 1995). In community settings, they have been shown to improve individual self-worth, resilience, problem-solving abilities, health and interpersonal communication, with a subsequent reduction in interpersonal violence (Haswell et al., 2010).

The common delivery method of empowerment programs involves a group setting as this allows participants to explore topics such as "beliefs and attitudes," "conflict resolution," "crises" and "relationships," etc., and to compare their views and behaviors to others. Through their participatory nature, empowerment programs can empower individuals to take responsibility for their everyday lives (Jain et al., 2024; Maton, 2008; Tsey & Every, 2000; Tsey et al., 2010; Wallerstein, 2006; Whiteside et al., 2006; Zimmerman et al., 1992) and positively influence individuals to make better choices and modify their behaviors (Wallerstein, 1992).

2. Literature Review

Conventional teacher-centered education is a traditional approach to classroom instruction where the teacher takes on a central role in delivering information and directing the learning process. This method focuses on transmitting a specific set of facts, competencies, and established knowledge to students, with the aim of preparing them for future social and economic success. In this model, the teacher typically leads lectures, directs class activities, and assesses student progress through tests and assignments. The emphasis is often on memorization, rote learning, and adherence to established curriculum standards. While this approach has been widely used in education systems around the world, there is increasing recognition of the need for more student-centered and interactive approaches that foster critical thinking, creativity, and problem-solving skills. Developed to equip students with the necessary skills and knowledge to succeed in an industrialized society, in this type of education process, teachers played an essential role in transmission of knowledge (Boudreault et al., 2013). Their duties were to instill upon students' minds, by habit or persistent instructions, "a limited body of facts and principles, mainly through rote learning" (Hall-Dennis Report, 1968) whereby successful students could "replicate what they were taught" (C21, 2015).

Classroom rules were developed by teachers rather than co-constructed with students. Learning expectations were fixed for all students rather than structured to reflect and/or accommodate individual differences in students' learning abilities (Garrett, 2008).

In traditional teacher-centered education, teachers typically assume the primary leadership role in the classroom, with limited involvement from members of the larger community where students reside. This approach often isolates learning within the classroom walls, without leveraging the wealth of knowledge and resources available in the community to enrich students' educational experiences.

In such a setting, community members, including local experts, professionals, and cultural leaders, are not actively invited to participate in the educational process. As a result, students may miss out on valuable opportunities to connect their learning to real-world contexts, engage with diverse perspectives, and develop a deeper understanding of their community and its resources.

However, there is a growing recognition of the importance of community involvement in education. By inviting members of the community into schools, educators can tap into a wide range of expertise and experiences, providing students with authentic learning opportunities that extend beyond the confines of the classroom. This collaborative approach not only enhances learning outcomes but also strengthens ties between schools and their communities, fostering a sense of belonging and mutual support.

Compared to conventional teacher-centered education systems that focused on the transmission of facts and procedural knowledge, the transformative view of education is a social process that positions teachers as facilitators and students as co-constructors of their learning (C21, 2015). 21st-century learning represents a "shift in emphasis from the instruction of facts to a model which focuses on competencies such as critical thinking, character, creativity, innovation, as well as digital and computer literacy" (Boudreault et al., 2013, p. 3).

> While schooling tends to condition people to live within a given structure, to cooperate with authority, and to think only within the box, innovation arises out of creativity and compassion, allowing us to shift toward more natural and ecological models. (Cameron, 2006, p. 1)

This need for education reform has been linked to an unprecedented advancement in technological innovation and international political development (Ananiadou & Claro, 2009; Boudreault et al., 2013; C21, 2012; Prasad & Ghosal, 2022; Schleicher, 2015).

The current movement within the education system is focused on equipping learners with the skills and knowledge necessary to thrive in an information-rich era and rapidly changing global economy. This entails preparing students to effectively navigate and make sense of the vast amount of information available to them, as well as to adapt to the evolving demands of the modern workforce.

In this context, there is an emphasis on developing critical thinking, problem-solving, communication, collaboration, and digital literacy skills. These competencies are considered essential for students to analyze complex information, think creatively, communicate effectively, work collaboratively, and leverage technology to solve problems and innovate.

Furthermore, there is a recognition of the importance of fostering a growth mindset and lifelong learning habits among students. As technology and industries continue to evolve, the ability to continuously acquire new knowledge and skills will be crucial for individuals to remain competitive and successful in their careers.

Overall, the goal of the current education movement is to empower learners to thrive in a dynamic and interconnected world by providing them with the necessary tools, skills, and knowledge to navigate and succeed in the information age and global economy.

3. Research Methodology

Employing a normative descriptive research methodology, this study examines the intersection of education systems and the right to education concerning the right to self-determination within Indigenous communities. Through a descriptive lens, the research elucidates the historical experiences of an Indigenous group within an assimilative educational framework and its implications on the denial of self-determination. Furthermore, employing a normative research approach, the study utilizes secondary sources to identify gaps in current educational systems and proposes strategies for addressing them. Drawing insights from a specific case study, the findings are extrapolated to inform broader contexts. Despite potential temporal disparities between the case studies, efforts have been made to select relevant cases that offer unique perspectives on the role of education as an empowerment tool.

4. 21st-Century Education Competencies

According to the Organization for Economic Co-operation and Development, "A competency is more than just knowledge or skills. It involves the ability to meet complex demands, by drawing on and mobilizing psychosocial resources (including skills and attitudes) in a particular context" (Rychen & Salganik as cited in Ananiadou & Claro, 2009, p. 4). In terms of the 21st-century discourse (Table 10.1), this definition has been extended to include the requirement for growth and development not just in the cognitive and psychomotor knowledge domains but also in the affective knowledge domain which is associated with social skill development, ethical reasoning, attitudes, and feelings.

5. Creativity and Innovation

In a 21st-century education system, creativity and innovation are defined as the pursuit of new ideas, concepts, or products that address real-world needs, as well as the realization of these ideas to make meaningful contributions to specific

Table 10.1. Learning and Leading Competencies for 21st Century (Salas Pilco, 2013).

Competency	Definition
Creativity and innovation	Fresh concepts and daring opportunities
Critical thinking	Novel perspectives and resilient resolutions
Collaboration partnerships that work	Effective collaborations
Communication making sense and expanding perspectives	Communication making sense and expanding perspectives
Character	Character
Culture and ethical citizenship	Sharing what we value
Computer and digital technology	Transforming how we learn and lead

fields. Creativity and innovation, characterized by being innovative and resourceful, hold significant importance in the lives of Indigenous peoples, particularly in northern regions where collaboration, flexibility, and perseverance are essential for survival.

Creativity and innovation extend beyond academic contexts to encompass everyday life, including the challenges of survival in extreme climates. Indigenous communities often rely on these qualities to navigate and adapt to their environments, whether in harvesting activities, hunting expeditions, or other aspects of their traditional lifestyles. In the realm of education, creativity and innovation are crucial for accommodating students' unique needs, such as allowing flexibility for participation in cultural practices like harvesting and hunting, without compromising their educational opportunities. Thus, creativity and innovation play a vital role in both the cultural resilience and educational success of Indigenous communities.

6. Critical Thinking

Critical thinking entails the capacity to design and manage projects, solve problems, and make effective decisions using a variety of tools and resources (Fullan, 2013; Kapoor & Ghosal, 2022). This skill involves acquiring, processing, interpreting, rationalizing, and critically analyzing large volumes of often conflicting information to make informed decisions and take timely action (C21, 2012). An example of critical thinking in leadership is demonstrated through problem-solving. For instance, in recent years, fires have tragically destroyed schools in Arctic regions. For instance, the Peter Pitseolak School in Cape Dorset was destroyed by a fire in 2015, resulting in an urgent need for space and resources to accommodate 150 students and 22 staff. In such situations, critical thinking skills are essential for swiftly assessing the situation, identifying available resources, and devising effective solutions to address the immediate needs of the affected community.

7. Collaboration

The essence of collaboration lies in the capacity to engage positively and respectfully with others to generate fresh ideas and produce outcomes collectively. This entails collaborating with individuals from diverse cultural backgrounds, working in teams to pursue common objectives. For instance, leadership fosters skill development through observation, collaborative problem-solving, and decision-making via discussions. Talking circles exemplify this collaborative approach in leadership, enabling participants to contribute to goal achievement and problem resolution in inclusive ways that honor everyone's voices and perspectives.

8. Communication

In the realm of 21st-century communication, effectiveness extends beyond verbal and written forms to encompass various facets of interaction, including positive interventions within the classroom. For instance, communication skills are instrumental in developing relationships and fostering teamwork through activities such as team building exercises. Creating safe and nurturing school environments that embrace newcomers and celebrate diversity is essential for fostering a sense of belonging, openness, and acceptance among students and staff alike. These efforts contribute to cultivating a positive atmosphere characterized by goodwill, openness, and inclusivity.

9. Character

In the context of 21st-century education, the development of character encompasses essential qualities necessary for personal effectiveness in navigating a complex world. These qualities include grit, tenacity, perseverance, resilience, reliability, and honesty (Fullan & Scott, 2014). In schools, character skills are often instilled through dedicated education programs that progress from nurturing self-care to fostering care for others and eventually extending to serving and contributing to the wider community.

Character education emphasizes the importance of caring for individuals, as well as serving and supporting family and community. An example of how leadership can promote character development is by implementing breakfast and snack programs in schools to address hunger-related barriers to learning. By ensuring that students' basic needs are met, such initiatives support their ability to focus, engage, and thrive academically, ultimately contributing to their personal growth and character development.

10. Cultural and Ethical Citizenship

The development of cultural and ethical citizenship competency is multifaceted and dynamic. It entails appreciating diversity at various societal levels – local, national, and global – and involves the ability to critically analyze historical and contemporary contexts to inform future actions (C21, 2012). Learners are required to grasp key concepts such as democracy, social justice, and human rights, alongside cultivating the disposition and skills necessary for effective civic engagement.

An integral aspect of cultural and ethical citizenship is thoughtful consideration for the environment. Demonstrating respect and care for the land, animals, and the environment, as well as for fellow human beings, exemplifies this citizenship. Examples of such practices include engaging in cultural awareness activities, involving elders in traditional knowledge sharing within schools, and participating in community environmental initiatives. These actions not only foster environmental stewardship but also promote a sense of responsibility toward the well-being of both natural and human communities.

11. Computer and Digital Technology

Because 21st-century learning and leadership is immersed in technology and digital literacy, "the capacity to use computers and digital resources to access information and create knowledge, solutions, products and services" (C21, 2012) is key. An example of how to support digital learning through computer and digital technology is development of skills through gathering, understanding, and using information to improve community life.

> Inuit leaders and educators have guided the development of locally owned and operated broadband networks, equipment, and the associated education applications in their communities. With these developments, people living in remote areas are innovating and creating choices in the delivery of new training programs and services. With digital tools and networks… Parents and children are now able to remain in their communities to complete their education in familiar and safe spaces. (Tomczyk & Potyrała, 2021)

12. Extending the Conversation With an Indigenous Knowledge Perspective

The integration of Indigenous knowledge perspectives into the Canadian educational system has been a long-standing topic of discussion among education leaders. However, recent developments, such as the release of the Truth and Reconciliation Commission of Canada's final report, have heightened awareness among the general Canadian public regarding the necessity of this inclusion. As stated in the report, there is a recognition that in order to build a better future, Canadians must reflect on and learn from the past (Truth and ReconciliationCommission of Canada, 2015, p. 8).

In the context of 21st-century education, there is a growing emphasis on extending this conversation beyond mere acknowledgment to practical integration. This involves considering Indigenous societal principles as foundational for learning and educational leadership. By incorporating Indigenous perspectives into educational frameworks, institutions can foster a more holistic understanding of history, culture, and societal dynamics. This integration not only promotes cultural diversity but also enriches the learning experiences of all students, contributing to a more inclusive and equitable educational environment.

13. Indigenous Knowledge and Principles of Learning

For individuals of First Nations, Métis, and Inuit ancestry, Indigenous knowledge is regarded as inherently true. It embodies a dynamic understanding of the relationship between people and their natural environment, incorporating traditional wisdom regarding flora and fauna, cultural beliefs, and history to enrich their lives (Abotaleb & Dutta, 2024; Semali & Kincheloe, 1999, p. 3). The First Peoples Principles of Learning, developed by the First Nations Education Steering Committee and utilized in British Columbia schools, serve as a foundational resource for understanding the learning needs of Indigenous students.

In the realm of leadership, it is crucial for leaders to not only acknowledge these principles but also to integrate them into their vision. By connecting Indigenous knowledge with leadership practices, leaders can develop a vision that respects the past and validates the experiences of First Peoples. This approach fosters reconciliation and promotes a more inclusive and culturally responsive leadership process that benefits all individuals involved.

14. Findings and Conclusion

Dynamic Nature of Cultural and Ethical Citizenship: Competency development in cultural and ethical citizenship is multifaceted and continuously evolving. It involves appreciating diversity at various levels while critically analyzing historical and present contexts to inform future actions. Learners are expected to grasp key concepts such as democracy, social justice, and human rights, alongside developing the disposition and skills for effective civic engagement.

Incorporation of Environmental Awareness: Cultural and ethical citizenship includes a significant component of environmental stewardship. Respect and care for the environment, animals, and community members are vital aspects. Activities like cultural awareness initiatives, involvement of elders for traditional knowledge sharing, and planning environmental projects demonstrate this aspect of citizenship.

Importance of Computer and Digital Technology: In the contemporary landscape, proficiency in digital literacy is indispensable for effective learning and leadership. The ability to leverage computers and digital resources to access information, generate knowledge, and contribute to community improvement is highlighted. Initiatives such as locally owned and operated broadband networks in Indigenous communities exemplify how technology can empower remote populations and enhance educational opportunities.

Integration of Indigenous Knowledge: The recognition of the importance of integrating Indigenous knowledge perspectives into educational frameworks is growing. The final report of the Truth and Reconciliation Commission of Canada stressed the significance of learning from historical experiences to construct a more promising future. Indigenous knowledge is esteemed as a dynamic source of wisdom deeply intertwined with the natural environment and cultural heritage. Principles like the First Peoples Principles of Learning offer valuable insights into the educational requirements of Indigenous students and serve as a cornerstone for educational leadership that honors and acknowledges Indigenous experiences.

Leadership Vision and Indigenous Principles: Educational leadership should not only acknowledge Indigenous principles of learning but also integrate them into the leadership vision. This involves aligning leadership practices with Indigenous knowledge systems, fostering reconciliation, and honoring the diverse cultural perspectives within educational contexts.

In conclusion, fostering cultural and ethical citizenship, embracing digital technology, and integrating Indigenous knowledge are essential components of contemporary educational leadership. By understanding and incorporating these elements, educational leaders can create inclusive, equitable, and culturally responsive learning environments that empower learners and contribute to societal well-being.

The role of education in indigenous empowerment is multifaceted and crucial for the socioeconomic and cultural development of indigenous communities worldwide. In conclusion, education plays a pivotal role in empowering indigenous communities by preserving culture, reducing disparities, promoting sustainable development, advancing indigenous rights, building capacity, fostering intercultural dialog, and adopting a holistic approach to learning. Efforts to improve access to quality education for indigenous peoples are essential for achieving social justice, cultural revitalization, and sustainable development.

References

Abotaleb, M., & Dutta, P. K. (2024). Optimizing bidirectional long short-term memory networks for univariate time series forecasting: A comprehensive guide. In *Hybrid information systems: Non-linear optimization strategies with artificial intelligence* (p. 443). De Gruyter.

Boudreault, F. A., Haga, J., Paylor, B., Sabourin, A., Thomas, S., & van der Linden, C. (2013). *Future tense: Adapting Canadian education systems for the 21st century*. Action Canada. http://actioncanada.ca/wp-content/uploads/2021/06/TF2-Report_Future-Tense_EN.pdf

C21 Canada (Canadians for 21st Century Learning and Innovation). (2012). Shifting minds 1.0: A 21st century vision of public education for Canada. http://www.c21canada.org/wp-content/uploads/2012/11/Shifting-Minds-Revised.pdf

C21 Canada (Canadians for 21st Century Learning and Innovation). (2015). Shifting minds 3.0: Redefining the learning landscape in Canada. http://www.c21canada.org/wp-content/uploads/2015/05/C21-ShiftingMinds-3.pdf

Fulford, M. D., & Enz, C. A. (1995). The impact of empowerment on service employees. *Journal of Managerial Issues*, *7*(2), 161–175.

Fullan, M., & Scott, G. (2014). *New pedagogies for deep learning whitepaper: Education PLUS*. Collaborative Impact SPC.

Garrett, T. (2008). Student-centered and teacher-centered classroom management: A case study of three elementary teachers. *Journal of Classroom Interaction*, *43*(1), 34–47.

Hall, E. M., & Dennis, L. A. (1968). *Living and learning: The report of the provincial committee on aims and objectives of education in the schools of Ontario*. Ontario Department of Education. http://www.connexions.org/CxLibrary/Docs/CX5636-HallDennis.htm

Haswell, M. R., Kavanagh, D., Tsey, K., Reilly, L., Cadet-James, Y., Laliberte, A., Wilson, A., & Doran, C. (2010). Psychometric validation of the growth and empowerment measure (GEM) applied with indigenous Australians. *Australian and New Zealand Journal of Psychiatry*, *44*(9), 791–799.

Jain, B., Ghosal, I., Chotia, A., & Jayarathne, P. A. (2024). Blockchain and IOT devices: A futuristic approach for digital and smart supply chain. In M. Irfan, K. Muhammad, N. Naifar, & M. A. Khan (Eds.), *Applications of block chain technology and artificial intelligence: Lead-ins in banking, finance, and capital market* (pp. 73–88). Springer International Publishing.

Kapoor, R., & Ghosal, I. (2022). Will artificial intelligence compliment or supplement human workforce in organizations? A shift to a collaborative human–machine environment. *International Journal on Recent Trends in Business and Tourism (IJRTBT)*, *6*(4), 19–28.

Maton, K. I. (2008). Empowering community settings: Agents of individual development, community betterment, and positive social change. *American Journal of Community Psychology*, *41*(1–2), 4–21.

Prasad, B., & Ghosal, I. (2022). Forecasting buying intention through artificial neural network: An algorithmic solution on direct-to-consumer brands. *FIIB Business Review*, *11*(4), 405–421.

Salas Pilco, S. Z. (2013). Evolution of the framework for 21st century competencies. *Knowledge Management & E-Learning: An International Journal*, *5*(1), 10–24.

Semali, L. M., & Kincheloe, J. L. (1999). What is indigenous knowledge and why should we study it? In L. M Semali & J. L. Kincheloe (Eds.), *What is indigenous knowledge? Voices from the academy*. Falmer Press.

Tomczyk, Ł., & Potyrała, K. (2016). Parents' knowledge and skills about the risks of the digital world. *South African Journal of Education*, *41*(1).

Truth and Reconciliation Commission of Canada. (2015). *Canada's residential schools: The final report of the truth and reconciliation commission of Canada* (Vol. 1). McGill-Queen's Press-MQUP. http://www.trc.ca/websites/trcinstitution/File/2015/Honouring_the_Truth_Reconciling_ for_the_Future_July_23_2015.pdf

Tsey, K., & Every, A. (2000). Evaluating aboriginal empowerment programs: The case of Family WellBeing. *Australian and New Zealand Journal of Public Health, 24*(5), 509–514.

Tsey, K., Harvey, D., Gibson, T., & Pearson, L. (2009). The role of empowerment in setting a foundation for social and emotional wellbeing. *Australian e-Journal for the Advancement of Mental Health, 8*(1), 1–10.

Tsey, K., Whiteside, M., Haswell-Elkins, M., Bainbridge, R., Cadet-James, Y., & Wilson, A. (2010). Empowerment and indigenous Australian health: A synthesis of findings from family wellbeing formative research. *Health & Social Care in the Community, 18*(2), 169–179.

Wallerstein, N. (1992). Powerlessness, empowerment, and health: Implications for health promotion programs. *American Journal of Health Promotion, 6*(3), 197–205.

Wallerstein, N. (2006). *What is the evidence on effectiveness of empowerment to improve health?* World Health Organization.

Whiteside, M., Tsey, K., McCalman, J., Cadet-James, Y., & Wilson, A. (2006). Empowerment as a framework for Indigenous workforce development and organisational change. *Australian Social Work, 59*, 422–434.

Zimmerman, M. A., Israel, B. A., Schulz, A., & Checkoway, B. (1992). Further explorations in empowerment theory: An empirical analysis of psychological empowerment. *American Journal of Community Psychology, 20*, 707–727.

Chapter 11

Leveraging the Internet of Things (IoT) for Financial Empowerment in Indigenous Communities

Chinhita Sanyal[a], *Pradipta Banerjee*[b] *and Indrajit Ghosal*[a]

[a]Brainware University, India
[b]Sidho Kanho Birsha University, India

Abstract

This research investigates the potential of leveraging the Internet of Things (IoT) to foster financial empowerment within Indigenous communities. This study aims to explore the challenges and opportunities of implementing IoT solutions in Indigenous communities through a literature review and case studies. By addressing factors like infrastructure, culture, and regulations, it seeks to develop tailored strategies for leveraging IoT to enhance financial inclusion. This study's implication is in line with the growing scholarship in the employment of technology for the social and economic uplift of needy groups. The study will have some limitations that can influence the validity of the findings, i.e., some aspects could be sampling bias where it is hard to reach some of the Indigenous communities or groups, hence limiting the sample and the possibility of introducing biases. Constraints in the study are derived from the limited access to technology, the conflicts of regulations, shortage of time and resources, and ethical issues, which also make the study's limitations magnified, hence the need for a cautious approach in the interpretation of results. The study highlights the trends and challenges related to Indigenous people's financial capacity and requirements, providing valuable insight for policymakers, financial organizations, and other communities in formulating policies concerning Indigenous populations.

Keywords: Indigenous communities; Internet of Things (IoT); financial empowerment; regulatory framework; socioeconomic dynamics; technology adoption; policy recommendations

Indigenous Empowerment through Human-Machine Interactions, 179–195

Published under exclusive licence by Emerald Publishing Limited
doi:10.1108/978-1-83608-068-820251011

1. Introduction

Internet of Things (IoT): Refers to scenarios where network connectivity and computing capabilities extend to objects, sensors, and everyday items not typically considered computers, allowing these devices to generate, exchange, and consume data with minimal human intervention (Atzori et al., 2010; Rose et al., 2015). While IoT offers transformative potential, challenges such as data privacy and security vulnerabilities persist. These devices, equipped with an array of sensors, actuators, and connectivity features, span from household appliances and wearable gadgets to industrial machinery and smart city infrastructure (Al-Fuqaha et al., 2015). The IoT ecosystem facilitates seamless automation, real-time monitoring, and remote control across various domains, thereby enhancing operational efficiency, productivity, and user convenience (Botta et al., 2016). From this insight, it is clear that IoT has great abilities to bring change within communities in terms of connectivity, efficiency, and even aesthetical value. IoT connects everyday objects and systems to the internet with computing capabilities and makes them capable of creating, exchanging, and acting on data independently. IoT can effectively be used in smart cities in community development. By leveraging sensors and data analytics, IoT enables improved management of essential services such as transportation systems, energy distribution, and waste disposal. The consequence of this is optimization in resource use, lower operation expenses as well as the provision of quality services. However, while IoT holds transformative potential, it also presents challenges such as data privacy concerns, security vulnerabilities, and interoperability issues, underscoring the need for robust governance frameworks and cybersecurity measures (Zhang & Wang, 2019). Numerous indigenous communities suffer from digital divide issues affecting their social, cultural, and economic well-being (Hasan, Bao, & Miah, 2022). In the finance sector, IoT can streamline operations, automate processes, and enrich customer experiences (Chowdhury et al., 2018). For instance, IoT-enabled devices can gather real-time data on consumer behavior, spending patterns, and transaction histories, empowering financial institutions to provide personalized services and targeted marketing campaigns (Zhang & Wang, 2019). In community development, IoT technologies play a pivotal role in addressing urban challenges and enhancing residents' quality of life (Hanna & Noh, 2017). Through smart city initiatives, IoT sensors and data analytics monitor and manage critical infrastructure, such as transportation networks and energy grids, leading to optimized resource allocation and improved service delivery (Bibri & Krogstie, 2017). Financial empowerment is the process through which individuals and communities gain control over their financial situation, knowledge, and decision-making abilities, leading to the achievement of financial goals and improved well-being (Collins & O'Rourke, 2010). It encompasses various elements, including financial literacy, access to resources, and the development of skills necessary for effective financial management (Huston, 2010). By promoting financial inclusion and resilience, financial empowerment contributes to broader social and economic development objectives, fostering economic stability and social equity within communities (Duflo & Banerjee, 2019). Indigenous communities often face barriers to financial inclusion, including limited access to mainstream financial services, historical injustices, and systemic discrimination. Financial empowerment initiatives tailored to Indigenous contexts can empower community members by providing them with the knowledge, skills, and resources necessary to manage their finances effectively and make informed decisions. By fostering financial literacy and capability, these initiatives enable Indigenous individuals to navigate complex financial systems, advocate for their rights, and participate more fully in economic opportunities (Braunstein & Welch, 2002). Overall, financial empowerment plays a pivotal role in empowering indigenous communities to overcome economic challenges, reclaim their autonomy, and build a prosperous future for generations to come.

2. Review of Literature

2.1 Indigenous Communities

Within the field of Indigenous Compass, Voinov et al. (2024) focuses on bridging the digital divide in Latin American Indigenous communities through AI and wireless networks. In as much as the opportunities that come with IoT, the paper assesses the chances the new technology presents in, maintaining indigenous customs, handling resources and social issues. But at the same time, it is also essential to question the IoT's privacy and data ownership issues as well as risks associated with culture appropriation when designing IoT solutions. In a similar vein, Johnson and Henderson (2021) point out the need to decolonize IoT conversations to include Indigenous people in decision-making processes that affect them and respect their way of knowledge, wisdom, and rights. Together, these works which illustrate the social relations of IoT technologies with the development of community and occupying Indigenous territories speak to the potentiality of a critical IoT.

2.2 Internet of Things

In the domain of the financial sector, the works of Dorri et al. (2016, 2017) propose a hierarchical blockchain-based lightweight architecture to enhance IoT security and privacy while minimizing computational and bandwidth overhead. The information provided features a more detailed field of their work that concerns smart payments, asset tracking, and risk management, among others to explain how IoT is impacting the finance industries and shifting the existing financial practices. Based on this, AT Kearney (2018) presents a guide that includes the IoT discussion concerning the financial sector to outline the possibilities of change.

Moving to the effectiveness of IoT in applying the community development concept, Pritiprada et al. (2023) provide a comprehensive analysis of the literature. From the conclusions of their work, in detail, their analysis identifies the versatility of the IoT to solve different issues, from smart farming to health and environmental management to foster sustainable community development.

2.3 Financial Empowerment

In the field of finance, discuss the IoT implementation regarding banking companies to consider its influence on business strategies and risks. Li and Wang identified other fields of application including smart branches, tracking of assets, and fraud analysis. On the other hand, Dedeoglu et al. (2019) are concerned with the interaction between blockchain and IoT in the finance industry. Their survey paper also covers areas such as supply chain finance involving the financing of the supply chain, smart contracts involving automated business transactions, and decentralized identity involving identity management.

2.4 Cultural Considerations

Moving to community development, Mukherjee et al. (2022) based on a systematic literature review investigate the state-of-art and future directions of IoT for sustainable development in various domains such as agriculture, energy, health care, and smart cities. Through these articles, they also look at the consequences, issues, as well as prospects of sustainability through IoT.

2.5 Regulatory Framework

In the same vein, Sharma et al. (2024) consider the prospects of IoT technologies for improving the quality of life of people, in particular, those living in rural areas by ensuring necessary service delivery including, for instance, health care, education, and support in the sphere of agriculture. Muhammad et al.'s review also discusses in detail how it's possible to meet the objectives of rural development projects through the use of IoT solutions.

2.6 Socioeconomic Dynamics

Despite the vast literature concerning the opportunities of IoT in financial services and its impact on community development, it is possible to detect the lack of research dedicated to the application of IoT in the financial enfranchisement of Indigenous populations.

2.7 Technology Adoption

One possible void is the absence of more and especially qualitative studies on the possibilities of using IoT technologies to promote financial inclusion and capability in Indigenous populations. This involves understanding how IoT can enhance the delivery of financial services, increase people's awareness of it, and contribute to economic growth strategies as well as respecting Indigenous people's institutions and worldview regarding money.

2.8 Policy Recommendations

Additionally, despite the presence of the literature on Indigenous data sovereignty and governance, there are no studies that focus on data privacy, security, and ownership concerning the financial data generated by IoTs within Indigenous communities. This includes exploring mechanisms for ensuring community control over data and addressing concerns about data exploitation and external surveillance. Overall, there is a gap in the literature regarding the specific intersection of IoT and financial empowerment for Indigenous communities, highlighting the need for further research and interdisciplinary collaboration in this area.

Summary of Literature Review.

Domain	Authors	Focus	Key Contributions
Financial Sector	Dorri et al. (2016)	Addressing IoT security and privacy challenges using a blockchain-based lightweight architecture	Proposing a hierarchical blockchain architecture that enhances IoT security and privacy while minimizing computational and bandwidth overhead.
Financial Sector	AT Kearney (2018)	IoT transformation in finance	Examines how IoT is transforming customer experience, operational efficiency, and risk management. Provides examples like smart ATMs and focused banking services.
Community Development	Pritiprada et al. (2023)	IoT and sustainable community development	Analyzes IoT's versatility in addressing issues like smart farming, health, and environmental management for sustainable community development.
Indigenous Studies	Voinov et al. (2024)	Bridging the digital divide in Latin American Indigenous communities through	The paper explores how AI and wireless networks can enhance Internet access for indigenous

(Continued)

(Continued)

Domain	Authors	Focus	Key Contributions
		AI and wireless networks	communities in Latin America, addressing challenges and opportunities.
Indigenous Studies	Johnson and Henderson (2021)	Decolonizing IoT for Indigenous communities	Highlights the need to include Indigenous people in IoT decision-making processes and respect their knowledge, wisdom, and rights.
Financial Sector	Mohaghar et al. (2021)	IoT in banking	Discusses the influence of IoT on banking strategies and risks, covering areas like smart branches, asset tracking, and fraud analysis.
Financial Sector	Dedeoglu et al. (2019)	Blockchain and IoT in finance	Explores the interaction between blockchain and IoT in finance, focusing on supply chain finance, smart contracts, and decentralized identity.
Community Development	Mukherjee et al. (2022)	IoT for sustainable development	Investigates the state-of-art and future directions of IoT for sustainable development in agriculture, energy, health care, and smart cities.
Community Development	Sharma et al. (2024)	IoT for rural quality of life	Explores IoT's prospects for improving the quality of life in rural areas through health care, education, and agricultural support.

(Continued)

Domain	Authors	Focus	Key Contributions
Indigenous Studies	Kukutai and Taylor (2016)	Indigenous data sovereignty	Discusses Indigenous Peoples' data sovereignty, control, and accessibility. Explores policies and practices supporting Indigenous data management rights.
Indigenous Studies	Rainie et al. (2019)	Indigenous Data Sovereignty Agenda	Summarizes Indigenous people's rights in data management and presents principles, practices, and policies in support of Indigenous data sovereignty self-determination frames.
Financial Sector	Dorri et al. (2017)	Developing a lightweight blockchain-based architecture for IoT to enhance security and privacy while minimizing computational overhead	The paper proposes a hybrid blockchain model that optimizes energy consumption for IoT devices by using a private immutable ledger and a distributed public blockchain, significantly reducing processing and bandwidth overhead.
Community Development	Pal et al. (2021)	The study examines how privacy concerns hinder the adoption of IoT-based smart homes, using a resistive modeling approach grounded in Innovation Resistance Theory (IRT) and Multidimensional Development Theory (MDT)	The paper identifies multiple antecedents of privacy concerns – individual, technological, and environmental – and empirically validates their impact on smart-home adoption resistance, providing theoretical and practical insights.

Source: Authors' own works.

3. Gaps in the Literature

3.1 Lack of Case Studies or Empirical Research

Absence of Specific Examples: To the best of the author's knowledge, the literature is devoid of rich case studies or empirical studies dedicated to describing how IoT technologies can be applied to improve Indigenous financial literacy in particular. This gap implies that the existing literature provides little information on the actual implementation of the various frameworks and their effectiveness.

Practical Implementations: Studies should be conducted on real-life applications of IoT where the focus must target how such technologies may be used to enhance access to the financial sector, enhance the understanding of financial services, and provide economic endeavors that sustain the indigenous population's economic systems and systems of worth.

3.2 Cultural, Social, and Environmental Considerations: Cultural, Social, and Environmental Considerations

Cultural Appropriateness: Some key considerations discussed are awareness and inclusion of the indigenous people's consideration, values, and management systems in attaining sustainable IoT solutions.

Community Involvement: More empirical investigations would be conducted to investigate ways through which ecosystem implementation and IoT can engage indigenous communities in planning and implementing ecosystem improvement and innovation.

4. Methods

4.1 Synthetic Framework of IoT, Financial Enabling, and Indigenous Science

IoT technologies along with strategic financial empowerment and Indigenous studies make up the effective framework to increase the level of financial empowerment within Indigenous groups. This framework starts with an identification of the key constructs from each sphere. IoT refers to the connection of devices into a network that incorporates a mechanism of gathering data, sharing it, and getting it done.

The relationships and interactions are presented in the framework as important aspects of the domains. The IoT can notably advance the levels of financial inclusion by providing services such as remote banking and mobile money in regions with poor coverage. Regarding the indigenous people, IoT can assist in the retention of languages, customs, and other aspects of indigenous cultures as well as help sustain indigenous peoples' management of resources that do not threaten or harm the indigenous peoples and their environment.

The last of the framework calls for the implementation of the selected IoT project by way of participatory design, whereby the targeted indigenous community is involved in the design of the IoT solution because the solutions that are designed with their input are most likely to effectively meet their needs and be culturally sensitive. Sustainability and scalability are important parts of the implemented strategy. This presented interrelated theoretical system wants to offer the general direction of the research and practice related to IoTs, financial enfranchisement, and Indigenous research that would foster the holistic, culturally sensitive, and efficient improvement of the financial enfranchisement of Indigenous people.

Empowerment theory is very relevant in an endeavor to explain how people and/or communities can assert influence on their own lives and on their surroundings. This theory focuses on the expansion of abilities of individuals and assembling to make decisions and the conversion of those decisions into wanted actions and results. From the perspective of financial emancipation, it is the improvement of individuals' financial education, their ability for financial services and economic opportunities. Empowerment theory suggests that when people and communities need to be empowered, certainly one must provide them with resources, still, one must ensure that they can appropriate them. This also covers the aspect of decision-making, which is significant when it comes to the integration of IoT solutions in Indigenous communities.

Another valuable theoretical framework to consider when it comes to the implementation of IoT technologies regarding Indigenous people refers to the theories of community development. These theories are based on the activities and actions leading to the enhancement of the capability of improving the welfare of the people in the community and making the enhancement sustainable. The process concept that lies at the heart of community development is participatory development which entails involving the populace in developmental processes.

TAM and diffusion of innovation (DOI) theories give a useful background as to how technologies percolate through users and are integrated into their everyday practices. According to the work of Davis (1989) on the TAM, perceived usefulness and perceived ease of use have been identified as important determinants of an individual's decision to use a certain technology. This model stresses that whenever developing IoT, solutions must be easy to understand and show exactly how they will be useful for the user.

In summary, integrating empowerment theory, community development theories, and technology adoption frameworks provides a robust theoretical foundation for developing and implementing IoT solutions aimed at financial empowerment in Indigenous communities. Together, these perspectives ensure that IoT interventions are not only effective and sustainable but also culturally respectful and empowering for Indigenous communities.

5. Findings and Discussions

5.1 Case Study: Smart Agriculture in Indigenous Communities in Australia

Lack of access to reading and writing put blind people at a serious disadvantage in nineteenth-century society. Text was one of the primary methods through which people engaged with culture, communicated with others, and accessed information; without a well-developed reading system that did not rely on sight, blind people were excluded from social participation. While disabled people, in general, suffered from discrimination, blindness was widely viewed as the worst disability, and it was commonly believed that blind people were incapable of pursuing a profession or improving themselves through culture. This demonstrates the importance of reading and writing to social status at the time: without access to text, it was considered impossible to fully participate in society. Blind people were excluded from the sighted world but also entirely dependent on sighted people for information and education.

5.2 Case Study: IoT-Enabled Financial Services in Rural India

5.2.1 Overview

At the start of 2019, a FinTech start-up began an IoT-based finance project in rural India with indigenous communities. Digital banking services such as mobile payments, microloans, and savings accounts were all enabled via IoT devices.

5.2.2 Successes

Enhanced Financial Inclusion: Access to banking services was extended via IoT-based products and some regions now had access, which previously did not have a proper reach in the formal system.

Increased Financial Literacy: Some of the educational programs implemented alongside have led to increased financial literacy among recipients (Johnson & Sherraden, 2007; Lusardi & Mitchell, 2014).

Economic Development: The availability of the microloan enabled community members to open up small businesses which in turn helped in economic development.

Challenges: Infrastructure Limitations: Poor internet connectivity in rural areas reduced performance of IoT devices.

Trust Issues: People used cash to handle most of their transactions, so establishing trust in digital financial services was a challenge.

Technical Support: The need for ongoing technical support and maintenance of IoT devices was significant.

Insights: This case highlighted the critical role of reliable infrastructure and the importance of trust-building measures when introducing digital financial services.

5.3 Case Study: IoT and Natural Resource Management in Canada

An IoT Initiative for Natural Resource Management in Indigenous Territories, Canada: Summary – This outlines a project to use IoT sensors for monitoring and management of fisheries with an intention to monitor water quality and fish populations.

Successes: Sustainable Resource Management – Real-time data from the fishery caught with IoT sensors helped keep fishing within sustainable levels.

Economic Impact: Better management practices resulted in an increased sustainability of fish stocks, which helped the economic activities related to fishing villages.

Preservation of Culture: The project combined Nonguiali's traditional fishing knowledge with technology in order to ensure the application of the IoT infrastructure is in line with the community's cultural values.

5.3.1 Problems

The marginalized and small-scale development (MSD) community needed training that was more intense because the general speculation and practical uses of IoT technology had not been widely practiced.

Funding: There must be an attempt to properly fund and perpetuate the IoT infrastructure's lifecycle in due course. Data Ownership: It was crucial to make sure that data collected through IoT devices were left under the control of the community.

5.3.2 Lessons

This case displayed how sustainable economic activities can be supported by IoT technologies, while at the same time keeping cultural practices alive.

5.3.3 Assessment of Impacts

IoT-Enabled Financial Empowerment Initiatives on Indigenous Communities.

Indigenous communities will witness a tremendous change following such initiatives as they are based on internet-connected devices (IoTs) within three scopes namely social, economic, and cultural, respectively.

5.3.4 Economic Impact

For example, IoT devices could help increase economic efficiency by providing access to financial services and promoting sustainable economic development. Such systems can enhance the economic inclusion of Indigenous peoples by enabling digital banking or microcredit services that support mobile payments (Dorri et al., 2016, 2017). Besides, community members would be empowered in their decision-making on investment and resource use as they become financially literate through the implementation of educational programs accompanying IoTs. In addition, IoT-based practices of agriculture and natural resources management can increase productivity, and resource optimization, and create new economies within native economies (Pritiprada et al., 2023).

5.3.5 Social Impact

IoT initiatives could transform communities socially by facilitating development and empowerment. Having better ways to access finance will alleviate dependence on informal exploitative finance facilities leading to self-reliance and resilience (AT Kearney, 2018).

5.3.6 Cultural Impact

In terms of culture, IoT technology adoption should take into account native values, knowledge, and governance structures. For successful implementation to happen, collaboration and co-design with the locals are needed to make sure that IoT solutions suit the applicable cultural demands as well as practices (Johnson & Henderson, 2021). Consequently, there is a possibility that these projects will play their part toward reinforcing cultural identity and sovereignty through the preservation and incorporation of indigenous cultural practices in IoT-enabled financial services as well as resource management. Nevertheless, addressing issues such as upholding data sovereignty or bridging digital divides amid initial resistance to technology uptake is crucial in avoiding the erosion of culture while allowing for community control over data and decision-making processes (Dedeoglu et al., 2019).

5.4 Policy Implications for Promoting Financial Empowerment Among Indigenous Communities Through IoT

For governments, nongovernmental organizations (NGOs), and stakeholders interested in the promotion of financial empowerment among indigenous communities through IoT, the following are some key policy implications to consider for equitable and culturally sensitive development.

5.4.1 Digital Infrastructure

In indigenous regions, policy initiatives must focus on developing a sound digital infrastructure by extending broadband connectivity, increasing mobile network coverage, and facilitating affordable internet services. Furthermore, there is a need to educate community members about IoT-based financial services so that they can be able to utilize them properly (Pritiprada et al., 2023).

5.4.2 Regulatory Frameworks

Data sovereignty rights protection for Indigenous communities in IoT deployments is one of the important roles played by government regulations. In this context, policies should tackle issues like data ownership control and consent while ensuring that indigenous people maintain their autonomy over data (Johnson & Henderson, 2021). Regulations must align with community needs and values by including Indigenous leaders in their formulation through collaborative governance models (Rainie et al., 2019).

5.4.3 Cultural Sensitivity

Cultural appropriateness is a vital aspect that needs to be followed when introducing a new technology. It is thus policy-relevant that decision-making systems should promote regions' participation in biochemical substances while staying with and valuing indigenous knowledge and institutions (Dedeoglu et al., 2019). Engaging people in the design of solutions for IoT guarantees that the developed technology embodies the culture and demands of the community; hence, acceptance and sustainability are guaranteed.

5.4.4 Key Findings

- Empowerment through Technology Adoption: The paper educating the theoretical aspect of IoT also calls for it to help indigenous people's communities economically by enabling them to have access to financial services while also improving operational gains (Pritiprada et al., 2023). By integrating smart payment systems and digital banking which are part of IoT, specific communities can be able to overcome financial exclusion and barriers to economic recovery.
- Cultural and Social Considerations: An understanding of culture and recognition of the people in our society and their needs are very crucial things to note when it comes to the use of IoT. In the paper, the co-design with Indigenous knowledge systems, as well as knowledge governance (Johnson & Henderson, 2021), is highlighted. Besides, the acceptance and sustainability are also increased, and the cultural and historical values are conserved together with the community connections.

Policy and Regulatory Frameworks: Again, a good policy environment plays a critical role in ensuring data sovereignty and indigenous people's privacy in these communities. Since the IoT system involves collecting and processing large amounts of data, regulatory frameworks must establish a mechanism to achieve community control of such data and how such networks reflect the community's culture and goals.

- Thus, this paper aims to offer a literature review of information on how IoT contributes toward financial enfranchisement and other community development among Indigenous people.
- Perceiving research opportunities, and thus identifying gaps in existing literature, this study focuses on how dispositive IoT could be used to empower indigenous cultures financially.

- Other recommendations include integration of cultural competence in IoT adoption, policy interactions, and collaborative structures in the governance of the technology and adoption of IoT policies.

For stakeholders, including governments, NGOs, and technology developers:

- Policy Development: Subsequently, based on the concepts identified within the theoretical framework, stakeholders can formulate policies that would help in building more infrastructure in the Indigenous people's areas and increase the level of information literacy among these communities.
- Community Engagement: Community engagement approaches are recommended to enhance understanding of IoT projects for the development of the local society, thus embracing the principles of trust and sustainability (Dedeoglu et al., 2019).
- Capacity Building: Further development is crucial for IoT growth and to overcome IoT issues like technology and society acceptance.

This theoretical foundation demonstrates how IoT can be utilized to promote financial inclusion among Indigenous people. Therefore, additional investigations and interprofessional cooperation are essential to address IoT's opportunities and challenges and guarantee that new technologies support Indigenous populations' rights and improve their quality of life.

5.4.5 Research Avenues

Impact Assessment and Longitudinal Studies: Performing integrated impact analysis and performing long-term analysis of Indigenous individuals' IoT-based financial services to identify their lasting consequences. This includes measuring economic outcomes, social impacts, and cultural preservation over time.

- Cultural and Ethical Considerations: Going deeper into the aspects of culture/ethic issues on the IoT adoption and how these technologies could reflect the indigenous knowledge system and its governance.
- Policy and Regulatory Frameworks: Moving forward the knowledge on policy enablers for data ownership sovereignty, privacy, and ownership at the community level of IoT-generated data. It encompasses supporting a policy environment that provides fair opportunities to command and benefit from IoT while protecting Indigenous people's rights.
- Capacity Building and Digital Literacy: Designing methodologies for the acquisition of the Indigenous peoples' awareness of and skills to use IoT technologies for, inter alia, personal finance and enterprise development for sustainability.

5.4.6 Emerging Technologies and Methodologies

- Blockchain and Decentralized Finance (DeFi): Examining the implementation of blockchain and DeFi solutions for improving the efficiency and security of indigenous populations' P2P lending.
- IoT in Environmental and Resource Management: Using IoT in the efficient sustainable utilization of resources such as natural resources essential to Indigenous people's lifestyle. A case is whether such applications could be deployed in sectors such as agroforestry, agriculture, or even in the protection of the environment.
- AI and Predictive Analytics: Merging AI and big data with IoT in financial planning, risk assessment, and economic development plans for Indigenous people.

5.4.7 Practical Implementation

- Community-Driven Innovation: Promoting Bottom-up innovation and culture of participation in the implementation of IoT solutions and systems. This makes sure that technologies introduced in the society fit the culture of that place, address the needs of the people, and the people are able to lead their own development.
- Public–Private Partnerships: About developing partnerships and coalitions between the national ministries, international NGOs, private companies, and indigenous associations to adopt and finance IoT projects. This involves the translation of effective pilot applications into broader diffusion, as well as coordination on the utilization of available expertise and assets.

6. Conclusion

Therefore, it can be concluded that the development of the mentioned approach for the further use of financial IoT for the indigenous population's financial enfranchisement depends heavily on advanced interdisciplinary research, the emergence of new technologies, and, most importantly, actively involved native communities' development. By leveraging these areas, the stakeholders can unlock the future of indigenous people through development that is socially, economically, technologically, and culturally sustainable. Further cooperation and proper investment into these fields will be critical to contribute to IoT's potential regarding Indigenous peoples' engagement worldwide.

References

Al-Fuqaha, A., Guizani, M., Mohammadi, M., Aledhari, M., & Ayyash, M. (2015). Internet of Things: A survey on enabling technologies, protocols, and applications. *IEEE Communications Surveys & Tutorials*, *17*(4), 2347–2376.

AT Kearney. (2018). IoT in financial services: A primer.https://www.atkearney.com

Atzori, L., Iera, A., & Morabito, G. (2010). The Internet of Things: A survey. *Computer Networks*, *54*(15), 2787–2805.

Bibri, S. E., & Krogstie, J. (2017). Smart sustainable cities of the future: An extensive interdisciplinary literature review. *Sustainable Cities and Society*, *31*, 183–212.

Botta, A., De Donato, W., Persico, V., & Pescapé, A. (2016). Integration of cloud computing and Internet of Things: A survey. *Future Generation Computer Systems*, *56*, 684–700.

Braunstein, S., & Welch, C. (2002). Financial literacy: An overview of practice, research, and policy. *Federal Reserve Bulletin*, *88*, 445–457. https://doi.org/10.17016/bulletin.2002.88-11

Chowdhury, M., Uddin, M., Dey, N., Zomaya, A., & Hassan, M. (2018). The role of Internet of Things in smart banking. *Journal of Network and Computer Applications*, *103*, 189–202.

Collins, J. M., & O'Rourke, C. M. (2010). Financial education and counseling—Still holding promise. *Journal of Consumer Affairs*, *44*(3), 483–498.

Davis, F. D. (1989). Perceived usefulness, perceived ease of use, and user acceptance of information technology. *MIS Quarterly*, *13*(3), 319–340. https://doi.org/10.2307/249008

Dedeoglu, V., Jurdak, R., & Dorri, A. (2019). Blockchain technologies for IoT. In *Advanced applications of blockchain technology*. Springer Nature. [Chapter 3]. https://doi.org/10.1007/978-981-13-8775-3_3

Dorri, A., Kanhere, S. S., & Jurdak, R. (2016). Blockchain in Internet of Things: Challenges and solutions. *arXiv*. https://doi.org/10.48550/arXiv.1608.05187

Dorri, A., Kanhere, S. S., & Jurdak, R. (2017, April). Towards an optimized blockchain for IoT. In *Proceedings of the second international conference on Internet-of-Things design and implementation* (pp. 173–178). IEEE.

Duflo, E., & Banerjee, A. (2019). *Good economics for hard times*. Public Affairs.

Hanna, R., & Noh, Y. (2017). Smart cities: A case study in Seoul. *Cities*, *63*, 62–73.

Hasan, N., Bao, Y., & Miah, S. J. (2022). Exploring the impact of ICT usage among indigenous people and their quality of life: Operationalizing Sen's capability approach. *Information Technology for Development*, *28*(2), 230–250. https://doi.org/10.1080/02681102.2021.1951150

Huston, S. J. (2010). Measuring financial literacy. *Journal of Consumer Affairs*, *44*(2), 296–316.

Johnson, J., & Henderson, J. (2021). Reclaiming the Indigenous voice in IoT: Perspectives and challenges. *Journal of Indigenous Studies*, *42*(2), 87–101.

Johnson, L., & Sherraden, M. (2007). From financial literacy to financial capability among youth. *Journal of Sociology and Social Welfare*, *34*(3), 119–145.

Kukutai, T., & Taylor, J. (Eds.). (2016). *Indigenous data sovereignty: Toward an agenda* (Vol. 38). ANU Press. http://www.jstor.org/stable/j.ctt1q1crgf

Lusardi, A., & Mitchell, O. S. (2014). The economic importance of financial literacy: Theory and evidence. *American Economic Journal: Journal of Economic Literature*, *52*(1), 5–44.

Mohaghar, A., Sadeghi Moghadam, M. R., Ghourchi Beigi, R., & Ghasemi, R. (2021). IoT-based services in the banking industry using a business continuity management approach. *Journal of Information Technology Management*, *13*(4), 16–38.

Mukherjee, A. A., Singh, R. K., Mishra, R., & Bag, S. (2022). Application of blockchain technology for sustainability development in agricultural supply chain: Justification framework. *Operations Management Research*, *15*(1), 46–61.

Pal, D., Zhang, X., & Siyal, S. (2021). Prohibitive factors to the acceptance of Internet of Things (IoT) technology in society: A smart-home context using a resistive modelling approach. *Technology in Society*, *66*, 101683.

Pritiprada, P., Satpathy, I., Patnaik, B. C. M., Patnaik, A., & Khang, A. (2023). Role of the Internet of Things (IoT) in enhancing the effectiveness of the self-help groups (SHGs) in smart city. In A. Khang, S. K. Gupta, S. Rani, & D. A. Karras (Eds.), *Smart cities* (pp. 183–203). CRC Press.

Rainie, S. C., Kukutai, T., Walter, M., Figueroa-Rodrígues, O. L., Walker, J., & Axelsson, P. (2019). Indigenous data sovereignty. In T. Davies, S. Walker, M. Rubinstein, & F. Perini (Eds.), *State of open data: Histories and horizons* (pp. 300–319). African Minds. https://doi.org/10.5281/zenodo.2677801

Rose, K., Eldridge, S., & Chapin, L. (2015). The internet of things: An overview. *The Internet Society (ISOC)*, *80*(15), 1–53.

Sharma, R. S., Loucif, S., Kshetri, N., & Voas, J. (2024, November). Global initiatives on "safer" and more "responsible" artificial intelligence. *Computer*, *57*(11), 131–137. https://doi.org/10.1109/MC.2024.3447488

Voinov, I. A., Chung, J., Bordel, B., & Alcarria, R. (2024, May). Artificial intelligence and wireless networks in indigenous villages: Potentials and challenges to bridge the digital divide. In *International conference on management, tourism and technologies* (pp. 119–128). Springer Nature.

Zhang, Y., & Wang, S. (2019). Security and privacy in the Internet of Things: A survey. *Pervasive and Mobile Computing*, *52*, 1–26.

Chapter 12

Influence of Indigenous Marketing Practices on Consumer Behavior and Brand Loyalty in Retail

Vijay Prakash Gupta[a], *Shebin Sharief*[b], *Vishal Kumar Tyagi*[c] *and Rajesh Manchanda*[d]

[a]GLA University, India
[b]DC School of Management and Technology, India
[c]Institute of Technology & Science, India
[d]Omkarananda Institute of Management and Technology, India

Abstract

This chapter examines the relationship of indigenous marketing practices and its influence on consumers' buying behaviors and the level of brand loyalty among the retail clients. As for indigenous marketing, it differs from the traditional marketing that is mainly based on advertisement; instead, it blends cultural factors into each stage of the marketing, which makes the buyers and users feel touched emotionally by brands. Thus, by combining symbols and stories along with sustainable materials into the clothing, this approach is culturally sensitive and appeals to young consumers. This chapter also covers the historical background and the characteristics of indigenous marketing stressing the roles of the emotional and cultural relations in the attachment to the brands. It also examines how today's technologies can complement indigenous marketing trends, illustrating traditional methods with updated technologies. Hence, this chapter underscores cultural sensitivity in marketing communication in relation to the key stakeholder audience and brand loyalty given the growing consideration for culture in consumerism.

Keywords: Indigenous marketing; consumer behavior; brand loyalty; retail sector; cultural heritage; social influence

---

Indigenous Empowerment through Human-Machine Interactions, 197–214

Published under exclusive licence by Emerald Publishing Limited
doi:10.1108/978-1-83608-068-820251012

1. Introduction

Indigenous marketing strategies play a significant role in influencing consumer behavior in retail. By using culturally meaningful storytelling, brands can form stronger emotional connections and create a sense of belonging among their customers. Incorporating traditional symbols and stories not only enhances a brand's identity but also pays tribute to the heritage of indigenous communities, making a personal and cultural impact on consumers (Dholakia et al., 2018).

These marketing techniques also emphasize sustainability and ethical sourcing, addressing the rising consumer demand for environmentally friendly and socially responsible products. This commitment to sustainability boosts brand loyalty, as consumers are more likely to support brands that uphold ethical practices and environmental care (Cherian & Jacob, 2012).

Understanding the influence of indigenous marketing on consumer behavior and brand loyalty provides valuable insights for the modern retail industry. By blending cultural heritage with contemporary commerce, these strategies help brands build deeper and more meaningful connections with their customers (Kolar & Zabkar, 2010). In a market where personal relationships with consumers are essential for growth, indigenous marketing stands out due to its cultural roots and focus on community values and ethics (Maghnati et al., 2024).

In today's retail environment, creating personal and meaningful connections with customers is crucial for growth. Indigenous marketing practices set themselves apart by leveraging cultural heritage, community values, and ethical considerations (Brown et al., 2019). These methods offer a unique way to engage with customers, rooted in the traditional practices of indigenous communities.

1.1 About an Indigenous Marketing Practices

1.1.1 Community-Centric Approach

Indigenous marketing frequently centered on the community. Unlike mainstream marketing, which usually targets individual consumers, Indigenous approaches prioritize the collective well-being and requirements of the community (Dyck et al., 2023).

1.1.2 Oral Tradition and Storytelling

Indigenous cultures rely heavily on storytelling, which is an important part of their marketing strategies. Stories are used to explain the history, significance, and distinguishing characteristics of things (Goody, 2006). This oral tradition aids in the development of a strong emotional connection and trust between the vendor and the customer, as stories are perceived as a more real and engaging mode of communication.

1.1.3 Relationship-Building

Indigenous marketing relies heavily on relationship-building and retention. Transactions are viewed as more than just business dealings; they are part of ongoing partnerships. Trust and respect are essential, and many businesses survive on repeat clients and word-of-mouth referrals from the community. The focus is on long-term relationships rather than one-time purchases.

1.1.4 Holistic and Sustainable Practices

Indigenous marketing emphasizes sustainability and environmental stewardship. Products are frequently manufactured utilizing traditional processes that are ecologically beneficial and sustainable. Marketing these items entails emphasizing their environmental benefits and the ethical principles used in their manufacture. This comprehensive strategy appeals to consumers that respect sustainability (Sharma & Iyer, 2012).

1.1.5 Cultural Authenticity

Maintaining cultural authenticity is critical. Indigenous marketing strategies guarantee that products accurately reflect the community's culture and traditions. This authenticity is a unique selling factor, as consumers are increasingly looking for things with cultural relevance and a story to tell. Maintaining credibility and respect requires ensuring that cultural heritage is respected and appropriately represented.

1.1.6 Using Local Resources and Knowledge

Indigenous marketers incorporate local resources and traditional knowledge into their activities. This involves employing locally sourced materials and traditional manufacturing techniques to create products. This helps to assist local economies while also preserving traditional crafts and skills. Marketing techniques frequently stress this local experience and the distinct traits it adds to the product.

1.1.7 Participatory and Inclusive Methods

Participation and inclusivity are crucial in Indigenous marketing. Community members frequently have a say in marketing strategy and tactics. This inclusive strategy guarantees that marketing efforts are consistent with the beliefs and requirements of the community. It also instills a sense of ownership and pride among the community members.

1.2 Historical Context: Evolution of Indigenous Marketing

The social, cultural, and economic advancements of indigenous tribes in diverse locations are closely linked to the creation of indigenous marketing strategies, making for a complex story. The journey illustrates how trade and business have changed, but it also shows how resilient and adaptive indigenous peoples have been in maintaining their culture despite shifting market conditions.

1.2.1 Precolonial Era (Before the 15th Century)

Complex trade networks existed among indigenous population of tribal people prior to the arrival of the Europeans. Some of the products that were exchanged in this trade include food stuffs, furs, and artefacts; the highly developed Mississippian culture, which flourished between 800 and 1600 CE, had complex trading systems as well as market centers such as Cahokia. The trans-Saharan trade routes in Africa became a necessity for linking sub-Saharan Africa with North Africa and allowing the interchange of commodities such as salt and gold as early as 500 CE. By about 2000 BCE, Austronesian-speaking populations in Southeast Asia had developed vast maritime trading networks across the Indian and Pacific Oceans, exchanging ceramics, textiles, and spices.

1.2.2 Colonial Period (15th to 19th Century)

A by-product of colonization was that indigenous trade networks were both asserted to and supplanted by new economic organizations, and indigenous commodities were commodified by Europeans. In Americas, the fur trade became one of the most important trade fields where Native Americans exchanged fur for European merchandize. In Africa, the Atlantic slave trade – where the slaves were bought from the Africans and trade of goods exchanged from the 16th to 19th century – altered the trade relationships and entangled Africans in the European trading system.

1.2.3 Postcolonial and Modern Period (20th Century to Present)

Native American marketing in the early to mid-20th century changed as countries became independent and indigenous people started looking for economic independence to feed their people; this has been modified by information and communication technology and globalization systems. The Navajo, Hopi, and Zuni Native American natives, for instance, were able to integrate traditional skills of arts, silver, and weaving and put them in contemporary approaches that are more of a commerce. It is also urgent to mention the Santa Fe Indian Market, which has been existing since 1922 and helped indigenous art gain recognition on the international level.

In Australia, the aboriginal and Torres Strait islander people popularized their arts and tourism. There were other newly founded organizations such as Arnhem Land Progress Aboriginal Corporation (ALPA) which promoted indigenous business ownership. The appreciation of the art went global, and artists such as Emily Kame Kngwarreye started being famous internationally.

After the European colonial rule in Africa, the webs of trade that were existing in African continent were given a new birth. The Maasai of East Africa used artisans to sell their products, especially to the tourist, while they were being innocent of being employed by anybody, thus earning them a source of income and embracing the culture of their forefathers. Some of the benefits of fair trade organizations include providing fair wages and market access for indigenous producers.

1.2.4 Current Period: Digital Technology and Indigenous Marketing Practices

Digital technology has been integrated into Native American marketing even more and more fully in the 21st century. Local business owners have found new means of engaging the global market through the internet and the social media. Some market regulations have been eliminated, and with the help of internet sales, artisans are now able to sell their products directly to customers all over the world with the help of such platforms as Etsy, Indigenous-owned web businesses, and others.

In particular, the Indigenous Tourism Association of Canada (ITAC) has applied the use of digital media while trying to popularize indigenous tourism and attract visitors to indigenous-owned facilities and attractions. Likewise, Māori business people in New Zealand have taken up e-commerce campaigns to market traditional products including arts, health, and cultural tours.

2. Consumer Behavior in Indigenous Marketing of Retail Sector

A sophisticated approach is necessary to comprehend consumer behavior in indigenous retail marketplaces since these markets are shaped by a distinct combination of social, cultural, and economic elements.

Some key factors that influence consumer behavior in these contexts are as follows:

2.1 Traditions and Cultural Heritage

Consumer behavior is significantly influenced by cultural heritage. Indigenous groups frequently have deeply ingrained customs that affect the things they choose to buy (Kolar & Zabkar, 2010).

Example: *In the Navajo Nation in the United States, purchasing decisions for traditional crafts such as rugs and pottery are often influenced by recommendations from respected community elders. These endorsements can significantly boost the sales of specific artisans.*

2.2 Social Influence and Community Ties

Social structures and relationships have a significant influence on consumer behavior in indigenous cultures. Leaders in the family, clan, or community frequently have an impact on decisions.

Example: Recommendations from esteemed community elders are frequently the deciding factor when it comes to buying traditional products like ceramics and rugs in the Navajo Nation in the United States. The sales of particular craftsmen can be greatly increased by these recommendations.

2.3 Economic Factors

Consumer behavior is closely influenced by economic factors, such as income levels and resource accessibility. The economic dynamics of indigenous communities may differ from those of urban areas.

Example: In rural India, seasonal agriculture is a significant source of income for tribal communities like the Bhils and Gonds. Their purchasing power is impacted by the agricultural cycle, which modifies when and how they buy goods including machinery, seeds, and everyday necessities.

2.4 Geographical Access and Infrastructure

Accessibility to markets is one way that the physical location and infrastructure development of indigenous territories can impact consumer behavior.

Example: In the Amazon Rainforest, indigenous communities like the Kayapo may have limited access to mainstream retail markets due to remote locations. This limitation leads to a higher reliance on local markets and barter systems, where goods are traded within the community.

2.5 Education and Awareness

Consumer decisions are influenced by the degree of education and awareness regarding various goods and services. Products that fit with indigenous people' traditional knowledge and traditions may be given priority.

Example: In Aboriginal communities in Australia, there is a growing emphasis on "bush tucker" – native Australian foods. Education about the health benefits and cultural significance of these foods has led to increased consumer interest and demand for these products in local markets.

2.6 Trust and Authenticity

In indigenous retail markets, consumer confidence in the genuineness of products is essential. Products that are manufactured with traditional methods and are viewed as authentic are valued by consumers.

Example: Authentic Māori art and carvings are highly valued by the Māori population in New Zealand. Authentic handcrafted items by Māori artists are more valuable and sought after than mass-produced replicas.

2.7 Government Policies and Support

Government programs and policies that assist indigenous companies and promote traditional goods have the power to change consumer behavior.

Example: The indigenous Zapotec people of Oaxaca produce and sell hand-woven textiles, and the Mexican government has encouraged *this activity, which has improved the products' marketability and visibility. Customers have been enticed to purchase these traditional goods both domestically and abroad through subsidies and marketing campaigns.*

2.8 Environmental Concerns

In indigenous markets, environmental sustainability is becoming a more significant component in determining customer behavior. Customers may favor eco-friendly products as they become more conscious of how their purchases affect the environment.

Example: In Canada's Inuit communities, *people choose things manufactured from sustainable sources. For example, traditional sealskin apparel is preferred not just for its cultural value but also for its sustainability when compared to synthetic substitutes.*

2.9 Socioeconomic Factors

Various socioeconomic differences can restrict cost affordability and thus affect the supply and demand levels. Eradication of credit and loans hinders investments that are made toward marketing and inventories.

3. Consumer Perception and Loyalty in Indigenous Retail Marketing

Both perception and consumer loyalty retain some influences on the consumption pattern of retail brands and, therefore, any indigenous population's marketing approaches (Kataria et al., 2023). Through the analysis of the drivers that cause consumers to make the particular buying decision and consumers' brand loyalty in Indigenous retailers, one gets a sense of the strategies that are necessary to attract and build long-term customer loyalty (Das, 2014).

- *Cultural Influences:* Here, it can be ascertained that based on the designated culture, consumer behavior of Indigenous people of Canada and Australia does not vary significantly. They are the norms and beliefs, the past practice, and traditions that define the tendencies and actions including the tendencies to consumption. For instance, the products that relate to conventional uses or the products that are inherently expanded or sourced may enjoy a higher preference.

- *Community and Social Networks:* This means that the indigenous consumers depend on recommendations on the products to be bought or the services to be consumed from word of mouth or their circle of acquaintances. These suggestions from the perceptual influences such as word of mouth and recommendations from the community may have quite an impact in such circumstances (Mikalef et al., 2013).
- *Economic Factors:* For this reason, economic factors such as income levels and access to credit are key components that influence buying power and related decisions. Other features that may run in the Indigenous population include concerns with the price and therefore the worth for the amount paid.
- *Trust and Relationship-Building:* An example of Indigenous marketing practices found that trust is an essential factor in Indigenous retail marketing. Customers also prefer companies that are ethical and display willingness to be truthful with their clients and show respect for Indigenous people and their cultures.

4. How Indigenous Marketing Influences Consumer Behavior

4.1 The Role of Indigenous Marketing in Shaping Consumer Behavior

4.1.1 Understanding Consumer Behavior

Thus, to understand the impact of marketing, the consumer concept should be addressed firstly. This term is one which concerns the psychological actions that are realized by the customer when buying goods. Age, gender, education, ethnicity, and other factors that may determine consumer behavior determine perception to and reaction toward marketing stimuli.

4.1.2 Role of Marketing

Marketing is essentially the intermediary between businesses and customer needs for the purpose of delivering information to the consumers to influence them to purchase products or services. Through these tactics, firms aim to influence consumers' buying behavior, encouraging them to spend their money on the firm's products rather than on competing products offering similar value.

4.1.3 Emotional Influence in Marketing

Emotional appeal is a powerful marketing strategy, helping to establish long-lasting relationships with consumers. By triggering emotions like joy, excitement, or nostalgia, emotional marketing creates a strong connection that endures beyond the initial brand interaction.

4.1.4 The Power of Advertising

Advertisements shape consumer behavior by influencing desires and thoughts. Through captivating visuals and storytelling, brands build emotional connections, creating positive associations and gaining a competitive edge in consumers' minds.

4.1.5 Impact of Social Media

In the digital age, social media has transformed communication and buyer behavior. It shapes social identity, reinforces perceived norms, and strengthens emotional connections through visual appeal, familiar content, and timely interaction.

4.2 Impact of Indigenous Marketing on Brand Loyalty and Customer Relationships

- Indigenous marketing enhances customer loyalty since the promotion and communication requires usage of the native culture including but not limited to traditions, values, and languages.
- Relevance of culture fosters the creation of emotional connection with the consumers, hence guaranteeing customer flow and recommendations.
- There is a high probability that consumers will go for brands which respect and recognize their race/ethnicity.
- Support of community events and programs is not only beneficial for the company's image and the brand but for the society as well.
- Companies that employ localized communications strategies and locally relevant products contribute toward a better customer experience, hence customer satisfaction and loyalty.
- Ethnic marketing ensures that brands are easily noticed and gets to have long-term contractual relations with the necessary ethnic target markets.

4.3 Impact of Brand Loyalty on the Indigenous Marketing

- *Cultural Relevance:* Indigenous marketing involves the use of the cultural aspects of a particular culture such as the language and even traditions in the communications about the brand.
  This tends to have a positive working with consumers who feel connected to the representation, therefore relating with the brand.
- *Authenticity and Trust:* Some of the brands that can embrace the indigenous culture will be more credible than the others, positively influencing the interaction.
  Cultural realism and performances that are free from prejudices create an outstanding image of a brand among the audience.

- *Emotional Connection:* Indigenous marketing is usually associated with the use of stories in describing a subject which creates responsibility.
  Cultural interactions which lead to the creation of an emotional bond with the main plot have a long-lasting impact and make consumers loyal to brand.
- *Community Engagement:* Brands that participate in community activities as well as support charitable organizations show the society how much they care for it.
  Thus, the more the community members are engaged, it increases the exposure of the brand and the loyalty toward the same.
- *Personalized Customer Experience:* Speaking of the customers, adapting goods and services to the specific culture helps achieve better and more satisfying results.
  They always get more than one thing, and what they get is relevant to their cultural background which makes them keep coming back for more.
- *Word-of-Mouth Advocacy:* The society culture is significant since satisfying customers who appreciate culturally sensitive products are more likely to act as brand ambassadors for a particular brand.
  Customers' recommendation of the brand within the community improves the brand image, thus increasing repeat patronage.
- *Long-Term Relationships:* Engaging in indigenous marketing contributes to establishing mutual trust and respect as well as creating deeper understanding of the clients' needs and enhancing long-lasting relationships.

5. Strategies and Techniques for Retailers in Indigenous Marketing

5.1 Think Local with Search Engine Optimization

Local keywords (city, neighborhood, landmarks) have to be included in search engine optimization (SEO) and ads to increase its conspicuousness. It is good to search for local events and interact with other related influencers to help gain relevance and backlinks.

5.2 Try Geo-Targeting

For geo-targeted advertising through Google AdWords and Facebook AdWords, utilize GPS data. Engage in cross-promotions by partnering with other local organizations to offer joint deals and advertise each other's services.

5.3 Get Involved in Community Business Organizations

Try to become a member of local organizations concerned with trade and commerce and more so attend their meetings in order to increase popularity among the traders and business people and the ability to participate in most of the community activities that are of much importance.

5.4 Give Back to Your Community

Special campaigns in local causes and charities should be supported; promotions made through social media and the stores should always ensure they have face time with the people.

6. Strategies for Cultural Engagement in Indigenous Marketing

- *Mobile Applications for Cultural Engagement:* Develop apps that promote indigenous customs and creations, such as virtual tours of cultural institutions and information about artists. Example: "Our Story" app featuring interactive storytelling.
- *Voice Assistants as a Preserve of Languages:* Teach indigenous languages to voice assistants like Alexa or Google Assistant, enabling users to learn key phrases and immerse themselves in the culture.
- *Virtual Reality for Experience:* Use VR technology to transport users to indigenous environments, allowing them to experience rituals and celebrations, such as a virtual tour of the Maori haka dance.
- *Machine Learning for Real Engagement:* Utilize AI to analyze customer data and interests related to indigenous products, allowing for targeted marketing. Example: AI recommendations for indigenous crafts based on consumer interest.

Example: Social Media Campaigning for Indigenous Peoples:

Launch campaigns on platforms like Instagram and TikTok to raise awareness of indigenous narratives and sustainable practices, such as promoting Indigenous Day to celebrate diverse cultures.

Utilize AI to analyze customer data and interests related to indigenous products, allowing for targeted marketing. Example: AI recommendations for indigenous crafts based on consumer interest.

Social Media Campaigning for Indigenous Peoples: Launch campaigns on platforms like Instagram and TikTok to raise awareness of indigenous narratives and sustainable practices, such as promoting Indigenous Day to celebrate diverse cultures.

6.1 Brief Case Studies of Leveraging Technology to Enhance Indigenous Marketing

6.1.1 Apple Inc. and Indigenous Cultural Preservation

Apple Inc. has leveraged its technological prowess to support indigenous cultural preservation through initiatives like the "Apple Maps Connect" project. This project allows indigenous communities to add and update points of interest that are significant to their culture, such as sacred sites, heritage landmarks, and cultural centers. By using Apple's advanced mapping technology, indigenous communities can share their heritage with a global audience, promoting cultural awareness and tourism (Cleveland & Laroche, 2007).

Example: The Yolŋu community in Australia utilized the tools in Apple Maps to tag the area as culturally sensitive so that any changes would not only occur to the place but also for the outsiders to learn the meaning of those sites. Thus, this initiative not only increases the representation of cultures but also contributes to the development of sustainable tourism.

6.1.2 Reliance Jio and Indigenous Peoples' Digital Engagement

The innovation of RIL Jio which offered financially subsidized 4G has made several deprived rural and indigenous populations in India connected with the internet. This digital inclusion has enabled these communities make transactions in e-commerce, learning resources as well as the general markets.

Example: Jio internet facilities in the regions of north-eastern India have brought indigenous handmade art to e-commerce *opportunities through apps like MyJio. This has implied new sources of income and made it possible for cultures to export their artifacts and other products.*

6.1.3 Selling Among Amazon and Indigenous Product Marketplaces

Another way in which Amazon has impacted indigenous peoples is through making special, open market place for indigenous items where the indigenous craftsmen have an opportunity to sell their products to people from different parts of the world. From the foregoing therefore, it can be argued that since indigenous sellers are able to sell their products on Amazon and consumers are also able to access those products, this means that geographical barriers are eliminated due to the strong logistics and e-commerce structures of Amazon.

Example: Amazon Karigar from India similarly speaks for the indigenous artisans, where they get a platform on selling the handmade embroidered fabrics or accessories such as tribal jewelry. It also assists in the growth of economic processes within these communities as well as in the preservation of the local folklore and artisanship.

6.1.4 OYO and Indigenous Tourism Promotion

Companies such as OYO and Indigenous Tourism Promotion play the role of promoting indigenous tourism. OYO has used its technological platform to help support indigenous tourism using hospitality assets directly with the help of the locals. Their app is quite unique because it helps the travelers to book places to stay in that have environmental friendly accommodation such as the eco lodges and indigenous homesteads.

Example: In Kerala, India, OYO provides tourists with vacations in the native tribe's conventional bamboo houses, with tours *of Western Ghats, giving the travelers cultural experience and contributing to native economy.*

6.1.5 Tesla and Indigenous People Organizations

Another good example of Tesla adopting sustainable practices is that the company has engaged indigenous people in ways to offer sustainable electricity solutions by applying the firm's proficiency in renewable energy devices. Such partnerships play a role in decreasing the usage of conventional sources of energy and encouraging environmentalism.

Example: Tesla applied solar panels and batteries into homes of the Haisla Nation in Canada to supply a sustainable energy for the communities. This undertaking not only serves the cause of energy self-sufficiency but also consonant with the *indigenous people's stewardship of the environment.*

7. Conclusion: Summary of Key Findings

In this chapter, the author has theoretically examined the long-term impact of indigenous forms of marketing on consumers' attitude and behaviors in the retail industry.

Our key findings can be summarized as follows:

7.1 Cultural Connection and Emotional Bonding

Techniques in indigenous marketing appeal to people's emotions since they utilize storytelling and symbols typical to the culture of Aboriginal people. These approaches contribute to brand identity while at the same time appealing to consumers' self-identity and their cultural values.

7.2 Sustainability and Ethical Practices

Part of indigenous marketing is its focus on sustainability and ethical supply chain management related to goods. In this respect, uniting with the growing consumer concern about the environmental and social impact of their purchases, the companies that adopt such policies increase their customers' confidence and, as a result, become more loyal.

7.3 Unique Consumer Behavior Factors

The process of consuming goods and services in indigenous retail markets depends on a unique combination of the social, cultural, and economic factors. Cultural influence, peer pressure, the economic situation, geographical location, level of knowledge, credibility, government laws on the usage of products, and environmental issues all have their parts to play on the decision to purchase a product.

7.4 Brand Loyalty and Customer Relationships

Indigenizing marketing thus affects brand equity in two ways since consumers are culturally aligned, organismic, emotionally attached, socially inclusive, and receive customized treatment. These aspects lock consumers emotionally with brands and thus give assurance of long term business and referrals.

7.5 Leveraging Technology for Enhancement

It is disclosed that advancement in technology provides new opportunities for developing indigenous marketing. Mobile applications, voice assistance, virtual reality, machine learning, and social media campaigns can be used to solve the problem of relations with indigenous peoples, to preserve languages and cultural values. Examples such as Apple, Reliance Jio, Amazon, OYO, and Tesla show that technology is the enabler that can bring about improvements and uplift indigenous communities' economic state.

7.6 Managerial Implications for Retailers and Marketers

7.6.1 Enhanced Brand Connection Through Cultural Relevance

Retailers can strengthen their brand connection with consumers by integrating indigenous cultural elements into their marketing strategies. Utilizing traditional symbols, stories, and practices in branding and advertising fosters deeper emotional connections and a sense of belonging. These strategies honor the cultural heritage of indigenous communities, differentiate the brand in a crowded market, and potentially increase customer loyalty and advocacy.

7.6.2 Sustainability and Ethical Sourcing

Emphasizing sustainability and ethical sourcing in indigenous marketing meets the growing consumer demand for eco-friendly and socially responsible products. By prioritizing sustainable practices and transparent supply chains, retailers can build trust and loyalty among consumers who base their purchasing decisions on a brand's ethical stance. This commitment enhances the brand's reputation and attracts a broader customer base that values ethical consumption.

7.6.3 Community Engagement and Support

Active engagement with indigenous communities fosters mutually beneficial relationships. Retailers should collaborate with local artisans to create authentic products that reflect indigenous culture. Supporting these communities through fair trade, infrastructure investments, and educational initiatives preserves cultural heritage and enhances the brand's image as socially responsible. This community-focused approach can drive positive word of mouth and strengthen long-term customer relationships.

7.6.4 Leveraging Technology for Cultural Preservation and Marketing

Using modern technology to preserve and promote indigenous cultures creates unique marketing opportunities. Mobile apps, social media campaigns, virtual reality experiences, and machine learning can engage and educate consumers. For example, virtual tours of indigenous sites and interactive storytelling on social media can offer immersive cultural experiences, boosting consumer interest and engagement.

7.6.5 Localized Marketing Strategies

Implementing localized marketing tactics like geo-targeting and local SEO can effectively reach indigenous groups. Retailers should tailor their efforts to the cultural and social contexts of these areas, making communications more relevant and enhancing their effectiveness in driving consumer behavior and brand loyalty.

8. Future Scope for Indigenous Marketing

Here are some key areas of future development for Indigenous Marketing:

- *Integration With Advanced Technologies:* It is stated that the concept of integration with advanced technologies is not new, as it has been implemented in organizations for quite some time. The application of technology appears applicable and useful in the concept of indigenous marketing. New technologies such as mobile applications and virtual reality increase the understanding of culture while machine learning provides better experiences. Furthermore, it ensures the authenticity and origin of indigenous products, thereby improving their quality.
- *Expansion of Global Reach:* One of the more distinct advantages is that with the onset of the digital sphere, location plays no role on where the indigenous products can be sold. E-commerce and key consumers help increase market access through access to prompt consumer's requirements to access artisans' creations from any location. The programs like Amazon Karigar give a glimpse of how the indigenous products can get better visibility or sell more.
- *Increased Understanding of Consumers:* In future marketing strategies, there will be some focus placed on trying to increase consumers' awareness of the cultural importance of indigenous products. Further implementation of storytelling across modes of media can lead to improved brand affect and an increased level of perceived value, providing differentiation as well as improved brand loyalty.
- *Mainstream Brand Engagement and Partnership:* There is little doubt that linkages between indigenous arts and production or mainstream designers can pioneer new product designs with an indigenous tenor that improves the exposure of indigenous populations. Positive experiences here could go a long way in making other brands to adopt the indigenous aspect in their products.

9. Recommendations and Suggestions for the Start-Ups Entrepreneur of Indigenous Marketers

The following are the recommendations and suggestions for the start-ups entrepreneurs going into Indigenous Marketing:

- *Embrace Cultural Authenticity:* It is vital that you also make your brand very relevant with the cultures and practices of the community. It is advisable to employ real symbolism and live anecdotes from elderly people in order to increase the credibility of a story.
- *Focus on Sustainability and Ethics:* Act responsibly in social context and be as open as possible in managing and coordinating operations. Source materials for products responsibly and the companies briefly illustrate these actions in marketing to consumers.
- *Leverage Technology for Wider Reach:* Use social media as well as e-commerce platforms to market indigenous products and link artisans, for instance, Amazon Karigar.
- *Community Engagement and Empowerment:* Treat locals as stakeholders. This is most efficiently achieved through making employees participate in the process of employment, partnership, and sharing of profits.
- *Tailor Marketing Strategies to Local Nuances:* Understand cultural nuances and consumer buying behavior for the store, so that marketers can come out with culturally appealing strategies.
- *Use Storytelling to Connect Emotionally:* Build an emotional connection around your product through the sharing of motivating and inspiring stories to make consumers loyal.
- *Invest in Education and Awareness:* Help consumers understand the social, historical, or ecological importance of your products through workshops and cooperation with brands.

10. Significance and Potential Impact of the Chapter

10.1 Significance

This chapter exclusively discusses the indigenous approach to the concept of marketing in regards to consumer behavior and effects on brand loyalty in retail business. On the other hand, indigenous marketing is a little more artistic that makes use of culture and myth to sway the emotion of the consumer unlike the usual systematic mass marketing. Any such approach not only responds to the requirement of the culture and history of the nation but also adapts to the new trends in the marketplace like organic, earthy, and the "buy local" culture in favor of the product. Therefore, this chapter expands current insights on the impact that such multicultural techniques have on consumer retention and behaviors, namely as valuable contributions to the global search for marketing interventions that culturally improve consumer orientation.

10.2 Potential Impact

- *Enhanced Brand Loyalty:* When the practices of indigenous marketing are combined with cultural values, it is possible to create powerful emotions that will brand consumers for a long time.
- *Sustainable Practices Adoption:* Promoting sustainability and appropriate sourcing keeps the companies on the right side of the law, improving their reputation as well as that of anyone who buys from them.
- *Cultural Sensitivity in Marketing:* The failure was an eventual result of resistance of indigenous cultures to aggressive marketing strategies that were insensitive to their culture. During marketing processes, it is advisable to respect the recipients' indigenous cultures in order to develop mutual trust.
- *Integration of Technology and Tradition:* One can combine the conventional and technological appeals with the cultural sanctions and arrive at creative marketing appeals that have applicability in the contemporary world.
- *Educational Resource:* This chapter can be useful as a marketing tool for marketers, for students learning or studying indigenous marketing systems, and for researchers investigating cultural variations in course of marketing.

References

Brown, D., Foroudi, P., & Hafeez, K. (2019). Marketing management capability: The construct and its dimensions: An examination of managers' and entrepreneurs' perceptions in a retail setting. *Qualitative Market Research: An International Journal*, *22*(5), 609–637.

Cherian, J., & Jacob, J. (2012). Green marketing: A study of consumers' attitude towards environment friendly products. *Asian Social Science*, *8*(12), 117.

Cleveland, M., & Laroche, M. (2007). Acculturaton to the global consumer culture: Scale development and research paradigm. *Journal of Business Research*, *60*(3), 249–259.

Das, G. (2014). Linkages of retailer awareness, retailer association, retailer perceived quality and retailer loyalty with purchase intention: A study of Indian food retail brands. *Journal of Retailing and Consumer Services*, *21*(3), 284–292.

Dholakia, R. R., Dholakia, N., & Chattopadhyay, A. (2018). Indigenous marketing practices and theories in emerging economies: Consumer behavior and retail transformations in India. *Journal of Business Research*, *86*, 406–415.

Dyck, B., Manchanda, R. V., Vagianos, S., & Bernardin, M. (2023). Sustainable marketing: An exploratory study of a sustain-centric, versus profit-centric, approach. *Business and Society Review*, *128*(2), 195–216.

Goody, J. (2006). From oral to written: An anthropological breakthrough in storytelling. *The Novel*, *1*, 3–36.

Kataria, P., Gupta, V. P., Kumar, S., & Gupta, R. (2023). Effect of collaborative consumption on sustainable homestay development: An empirical study. *Consumer Behavior in Tourism and Hospitality*, *18*(4), 515–531.

Kolar, T., & Zabkar, V. (2010). A consumer-based model of authenticity: An oxymoron or the foundation of cultural heritage marketing? *Tourism Management*, *31*(5), 652–664.

Maghnati, F., Rousta, A., & Asayesh, F. (2024). Presenting an indigenous model of brand loyalty in the retail industry. *International Journal of Innovation Management and Organizational Behavior (IJIMOB)*, *4*(2), 118–125.

Mikalef, P., Giannakos, M., & Pateli, A. (2013). Shopping and word-of-mouth intentions on social media. *Journal of Theoretical and Applied Electronic Commerce Research*, *8*(1), 17–34.

Sharma, A., & Iyer, G. R. (2012). Resource-constrained product development: Implications for green marketing and green supply chains. *Industrial Marketing Management*, *41*(4), 599–608.

Chapter 13

Techno-Entrepreneurship: Catalyst for Profitability and Efficiency in the Handicraft Industry

Mrinal Kanti Das[a], *Soumya Mukherjee*[b] *and Amina Omrane*[c]

[a]Kanchrapara College, India
[b]Techno India (Hooghly Campus), India
[c]IHEC Carthage, Tunisia

Abstract

This study examines the transformative role of techno-entrepreneurship in the handicraft industry, focusing on its potential to reduce operating costs, enhance profit margins, and improve efficiency and productivity. By leveraging technological innovations, the handicraft sector has the opportunity to drive revenue growth and achieve sustainable success in an evolving marketplace. Using data from 390 participants, the study employs partial least squares structural equation modeling (PLS-SEM) to assess how techno-entrepreneurship influences cost reduction and profit enhancement. The findings confirm the strategic importance of techno-entrepreneurship in revolutionizing the industry, identifying productivity and efficiency as key mediators that connect entrepreneurial innovation to organizational success. These mediating factors play a crucial role in initiatives aimed at cutting costs and boosting profit margins. The research offers new insights for industry stakeholders, policymakers, and entrepreneurs, emphasizing the need to embrace technological advancements and entrepreneurial ingenuity to navigate the complexities of the handicraft sector. The study highlights the connections between innovation, productivity, efficiency, and sustainable growth, providing a road map for businesses looking to succeed in this rapidly changing landscape.

Keywords: Technology; entrepreneurship; productivity and efficiency; cost reduction and profit improvement; handloom industry

Indigenous Empowerment through Human-Machine Interactions, 215–235

doi:10.1108/978-1-83608-068-820251013

1. Introduction

1.1 Concept of Entrepreneurship in the Light of Handicraft Industry

Although business management research primarily centers on entrepreneurship, its influence extends across various disciplines, encompassing science, the arts, engineering, and even handicrafts (Kirzner, 2009). The overwhelming number of academic books and articles devoted to entrepreneurship highlights the legitimacy of this field of study (Apostolopoulos et al., 2021). Creating value and subsequent positive societal impact is central to most conceptualizations of entrepreneurship (Jones et al., 2020). However, entrepreneurship can also have negative consequences, depending on one's social standing (Ratten, 2019). Although entrepreneurship has attracted much attention in fields like technology and tourism, much research has not been done on it in more established industries like handicrafts (Jones et al., 2020; Ratten, 2021). In order to contextualize entrepreneurial phenomena and encourage more scholarly research, researchers must explore uncharted areas of entrepreneurship, including its implications for the sustainability and revitalization of the handicraft industry (Kuhn & Galloway, 2015; Ratten, 2014; Ratten, 2016).

Shane and Venkataraman (2000, p. 218) presented one of the most well-known definitions of entrepreneurship, characterizing it as "how, by whom and with what effects opportunities to create future goods and services are discovered, evaluated and exploited." Although it can be expressed in various ways (Lounsbury & Glynn, 2001), this definition centers on opportunity recognition. Examples include emphasizing new markets or technological advancements (Phan et al., 2009). The market environment, people, and human capital all play a role in their capacity to seize opportunities (Meek et al., 2010). Current definitions highlight the process component of entrepreneurship and provide insights into how creativity and innovation in business endeavors change over time (Nambisan, 2017). For example, *entrepreneurship* is defined by Elia et al. (2020, p. 3) as the "process of identifying potential business opportunities and exploiting them through the creation of new resources or the recombination of existing ones to develop and commercialize new products and services." This article defines *entrepreneurship* as identifying business-related opportunities by utilizing new, existing, or repurposed resources in an inventive and creative way, focusing on how it can be applied to the handicraft industry. This definition is given from both the opportunity-seeking and process perspectives.

Entrepreneurial endeavors are greatly impacted by the regulatory environment (Rashid & Ratten, 2021). Institutional frameworks and macro-cultural contexts shape the capacity for entrepreneurship. By utilizing their human resources, entrepreneurs adjust to changing market conditions (Ratten, 2020a), negotiating a terrain shaped by a combination of public and private organizations (Ratten, 2020b). Entrepreneurs are often characterized by personal qualities like resilience and determination (Ratten, 2021), and their pursuit of success is driven by proactive behavior and a need for accomplishment (Ratten & Jones, 2021).

Since the COVID-19 pandemic, entrepreneurship has become increasingly important to society (Solomon & Mathias, 2020; Zhou, 2004). This has led to a resurgence of interest in entrepreneurship's place in traditional industries like handicrafts. Still, given the contextual subtleties surrounding the term "entrepreneurship" (Ratten et al., 2016), there are disagreements about its exact definition (Ratten & Usmanij, 2021). Some see entrepreneurship as a mindset or a behavioral trait, while others associate it with innovation in a business setting (Solomon & Mathias, 2020). Because of these various interpretations, the literature on entrepreneurship today does not thoroughly examine its definitions, difficulties, and prospects in relation to traditional industries like handicrafts.

1.2 The Paradigm Shift of Entrepreneurship Concept

The integration of technology into entrepreneurship marks a paradigm shift in modern business, with technological innovation driving efficiency, productivity, and profitability. This study investigates the role of techno-entrepreneurship in the handicraft industry, focusing on how it enhances profit margins, reduces costs, and boosts productivity. Drawing on foundational works such as Dorf and Byers' *Technology Ventures* (2005) and Prodan's *A Model of Technological Entrepreneurship* (2007), the research emphasizes how technology opens new opportunities for growth. Brem and Borchardt's (2014) insights on managing entrepreneurial activities in tech-intensive environments underscore the potential for cost reductions through strategic use of technology, while Örnek and Danyal's (2015) study highlights the shift from traditional to techno-entrepreneurship, especially in developing nations. Frau and Moi's (2019) exploration of digital technologies and local ecosystems further contributes by showing how these innovations support local growth in sectors like handicrafts. This paper provides a novel analysis of the interplay between techno-entrepreneurship, efficiency, productivity, cost containment, and profit margins, offering a theoretical framework where productivity and efficiency mediate these outcomes. The study aims to deliver actionable insights for handicraft industry stakeholders, policymakers, and entrepreneurs, highlighting techno-entrepreneurship as a strategic driver of organizational success and sustainable growth.

2. Theoretical Framework

2.1 Harnessing Techno-Entrepreneurship for Enhanced Productivity and Efficiency

The differences in productivity between economies have long piqued the interest of economists, leading to in-depth research into the underlying factors. Although a great deal of research has been done on elements like financial markets, human capital, and technology diffusion (Barro & Sala-i-Martin, 1997; Lucas, 1988; Moll, 2014), the importance of entrepreneurship and the condition of the entrepreneurial ecosystem are receiving more and more attention. Academics such as Acemoglu and Zilibotti (2001), Caselli and Coleman (2006), and Griffith et al.

(2004) emphasize the significance of comprehending the relationship between entrepreneurship and growth in total factor productivity (TFP).

As part of the larger investigation into the relationship between entrepreneurship and economic growth, our study takes a thorough approach to look at how entrepreneurship affects productivity (Acs et al., 2018; Atkinson & Lind, 2018; Baumol & Strom, 2007; Lafuente et al., 2016; Prieger et al., 2016). An argument crops up concerning national entrepreneurship system quality, highlighting a greater impact on TFP than differences in technology and factor endowments.

The idea of the "national system of entrepreneurship," as put forth by Acs et al. (2018), is fundamental for the analysis. This framework recognizes the diversity of entrepreneurship across national boundaries and considers the institutional context that influences entrepreneurial decisions. The goal is to give readers a better understanding of how, by incorporating contextual factors, entrepreneurship boosts national productivity and efficiency (Acs et al., 2018).

Additionally, the study differentiates between various types of entrepreneurship and how they affect TFP. Using Kirzner's (1973) and Schumpeter's (1934) writings, this research distinguishes between Schumpeterian entrepreneurship, linked to disruptive innovation and creative destruction, and Kirznerian entrepreneurship, defined by resource mobilization and opportunity exploitation. The study distinguishes these forms' relative contributions to productivity growth by examining how the country's entrepreneurial system fosters them (Aghion & Howitt, 1992; Schumpeter, 1934).

To put it briefly, our study aims to clarify the complex connection between productivity, entrepreneurship, and efficiency. The goal is to identify the mechanisms that underlie productivity disparities between countries over time by examining how the national system of entrepreneurship shapes various forms of entrepreneurship and their effects on TFP. Policymakers and other stakeholders in the handicraft industry can benefit greatly from this understanding, which can aid in creating plans to boost output and promote economic expansion.

H1. A positive correlation exists between techno-entrepreneurship and productivity and efficiency.

2.2 The Effects of Techno-Entrepreneurship on Cost Reduction and Profit Margin

The key to a successful business is profitability, which is the capacity of an enterprise to produce excess revenue from its activities. Profitability is attained when total revenue exceeds total expenses during a given reporting period (AccountingTools). As Lingesiya (2012) pointed out, the Cottage and Small Scale Industries (CSSI) sector is essential to promoting economic growth. Numerous advantages, such as the creation of jobs, the mobilization of domestic savings, the reduction of poverty, and the advancement of regional development, are facilitated by this industry.

Sometimes, looking at financial structures in the context of business operations is necessary. Financial ratios that provide important information about an organization's financial health include receivables turnover, debt management, and inventory optimization (BDC). A guiding concept that embodies the values of stewardship and resilience in a variety of spheres of life is sustainability. Sustainable development, as defined by the World Commission on Environment and Development (1987), is addressing current needs without compromising the capacity of future generations to address their own. El Ebrashi (2013) highlights the sustainability paradigm as a fundamental component of social entrepreneurship. This strategy includes an all-encompassing business plan emphasizing creating social impact as a vital component of organizational success. While tackling societal issues, social entrepreneurs seek to advance inclusive growth and sustainable development.

Essentially, the interaction among sustainability, profitability, and entrepreneurship highlights the importance of matching corporate goals with more general social objectives. In order to effectively navigate the ever-changing business landscape and promote long-term success and societal well-being, entrepreneurs operating in the handicraft sector must integrate the principles of sustainability and profitability.

H2. Techno-entrepreneurship can yield positive effects on cost reduction and profit margin

2.3 Productivity and Efficiency Impact on Cost Reduction and Profit Margin Enhancement

Techno-entrepreneurship is the keystone at the intersection of "Reduce Cost & Improved Profit Margin" and "Productivity & Efficiency" in the handicraft industry, creating a mutually beneficial partnership essential for long-term business viability. Techno-entrepreneurship emphasizes increasing efficiency and productivity by skillfully utilizing cutting-edge technologies (Dorf & Byers, 2005). Entrepreneurs in the handicraft industry use these technologies to optimize resource allocation, streamline operational workflows, and promote a continuous improvement culture. They improve output quality and cut costs related to manual labor simultaneously by automating processes and decreasing waste.

In the handicraft industry, the goal of increased productivity results in real cost savings in several operational areas. Techno-entrepreneurs use technology to find inefficiencies and put them into practice to reduce waste, lower production costs, and maximize resource use (Prodan, 2007). Furthermore, technological developments frequently make economies of scale possible, which lowers costs per unit and improves cost-effectiveness.

Reductions in costs and increased productivity directly translate into higher profit margins. The difference between revenue and expenses grows as operational efficiencies rise and costs fall, which boosts profitability for handicraft businesses (Brem & Borchardt, 2014). Techno-entrepreneurs take advantage of this increased margin by reinvesting profits into cutting-edge technological advancements or extending their market reach, strengthening their competitive advantage and long-term growth prospects. Techno-entrepreneurship essentially stimulates a positive feedback loop that drives higher output, lower costs, and higher profit margins in the handicraft industry. Using technology, entrepreneurs can optimize organizational efficiency, reduce expenses, and increase profitability, establishing a strong basis for sustained prosperity and adaptability in ever-changing business environments.

H3. Organizations that prioritize productivity and efficiency initiatives Cost Reduction and Profit Margin Enhancement

2.4 Mediating Role of Productivity and Efficiency Over Techno-Entrepreneur and Reduce Cost and Improved Profit Margin

The research highlights how important productivity and efficiency are as intermediaries between technological entrepreneurship and cost savings, resulting in higher profit margins for the handicraft industry. According to Dorf and Byers (2005), techno-entrepreneurs skilled at utilizing cutting-edge technologies promote increased productivity and efficiency. By deftly integrating these technologies into their operations, they maximize resource utilization and process efficiency, creating opportunities for cost savings in various business operations contexts (Prodan, 2007). These developments result in real advantages for the handicraft industry, like lower production costs, less waste, and more efficient workflows. For example, automation and digital tools can improve production process efficiency and yield higher output levels with less resource consumption. Handicraft entrepreneurs can also reduce operating costs by implementing cutting-edge supply chain optimization and inventory management methods.

Reducing production overhead costs that follow a spike in productivity and efficiency helps handicraft businesses' bottom lines. This phenomenon is consistent with research by Brem and Borchardt (2014), who show that increased productivity results in substantial cost savings and higher profit margins. When these cost-cutting strategies are combined, the gap between revenue and expenses widens noticeably, increasing the handicraft businesses' overall profitability. Simply put, the deliberate incorporation of technological entrepreneurship into the handicraft industry promotes increases in output and efficiency and establishes the framework for significant cost savings and profit margin expansion. Handicraft business owners can successfully navigate the changing industry landscape while maximizing profitability and sustainability by embracing innovation and utilizing technology-driven solutions (Frau & Moi, 2019).

H4. Techno-entrepreneurship and cost reduction and profit margin are mediated by productivity and efficiency, which boosts organization to excel.

Based on this groundwork, our proposed conceptual model (Fig. 13.1) is:

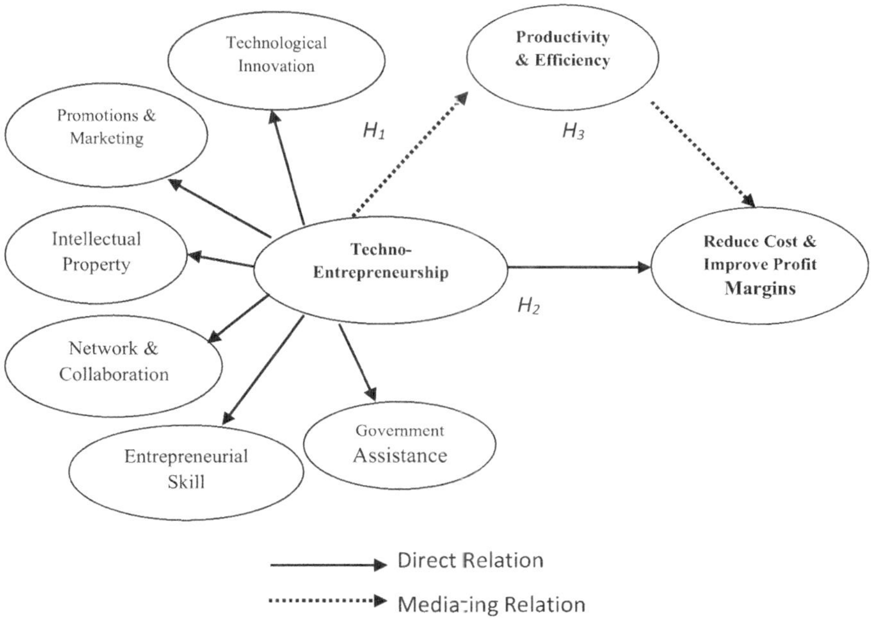

Fig. 13.1. Proposed Conceptual Model. *Source:* Authors' own creation.

3. Methodology

This study uses an exploratory, cross-sectional design to examine techno-entrepreneurship's impact on profits and cost reduction in India's handicraft sector. Data were collected from 390 respondents via a semi-structured questionnaire with seven-point Likert scales, employing purposive sampling to ensure diverse industry representation. A pilot study was conducted to refine the questionnaire, focusing on 10 key items related to techno-entrepreneurship's impact on costs and profitability. A G* Power analysis indicated 137 samples were sufficient for 5% statistical significance (Faul et al., 2007, 2009). The sample showed gender diversity (45.84% women, 54.16% men) and included respondents from various ages and occupations, ensuring diverse perspectives on techno-entrepreneurship in the handicraft sector. This study employed Structural Equation Modeling (SEM) using Smart PLS 3.3.2 to analyze the relationships between techno -entrepreneurship, productivity, efficiency, cost reduction, and profit margin in the handicraft industry. SEM effectively handles complex interactions among variables (Hair et al., 2017, 2019, 2020; Nitzl et al., 2016; Richter et al., 2016; Rigdon et al., 2017; Ringle et al., 2013, 2014). PLS is ideal for our exploratory study due to its ability to handle small samples, non-normal data, and complex models (Hair et al., 2017, 2019, 2020; Nitzl et al., 2016; Richter et al., 2016; Rigdon et al., 2017; Ringle et al., 2013, 2014; Yadav et al., 2023). By combining SEM with PLS, the study provided valuable insights for policymakers and stakeholders on fostering sustainable growth in the handicraft sector.

To capture the complexities of our research model, a comprehensive questionnaire was designed with both general and specific questions. General inquiries covered demographic variables such as age, gender, family income, and employment, providing a broad view of respondents' experiences outside the handicraft sector. Specific questions assessed key elements of the research model, including government support, technology innovation, marketing, intellectual property, networking, entrepreneurial proficiency, productivity, efficiency, cost reduction, and profit margins. These questions were adapted from existing measurement scales, ensuring relevance to the handicraft sector. For instance, seven items addressed productivity and efficiency, while four focused on cost reduction and profit margins. Minor adjustments were made to ensure clarity and accuracy for the target respondents, enabling a detailed examination of factors influencing the handicraft industry's growth and success.

4. Results

4.1 Measurement Model Assessment

With a strong methodology, the study focused on assuring internal reliability and validating convergent validity through confirmatory techniques in partial least squares structural equation modeling (PLS-SEM). The complexities of techno-entrepreneurship were explored through the deft application of Composite Analysis (CCA) and reflective formative assessment. A rigorous two-phase assessment approach was used to carefully examine the constructs, with first-order constructs being examined in detail before moving on to assessing second-order constructs. The same care was taken with the validation metrics that were used. Composite reliability, Dijkstra and Henseler's rho, and Cronbach's alpha were used to evaluate internal reliability, and they all exceeded the generally accepted cutoff point of 0.70, confirming the measurement's robustness. Thus, there was a dire need to carefully evaluate convergent validity using the Average Variance Extracted (AVE) values, which consistently exceeded 0.50, to support the robustness of our findings. This robustness guarantees the validity of the measurements by indicating the coherence and consistency of the examined constructs.

In addition, the research shows impressive composite reliability values that are higher than the predetermined cutoff point of 0.8 but still fall below the 0.9 sophistication threshold. This careful balance, which guarantees a high degree of internal consistency without inflating reliability estimates artificially, reflects our rigorous methodology and strengthens the credibility of the findings. Table 13.1 provides an unambiguous and transparent explanation of the methodological rigor, including a detailed account of internal reliability and convergent validity validation. This thorough synopsis gives stakeholders assurance regarding the strength of the research methodology. Furthermore, Table 13.2 provides critical

Table 13.1. Construct Reliability and Validity.

Constructs	Cronbach's Alpha	Composite Reliability (rho_a)	Composite Reliability (rho_c)	Average Variance Extracted (AVE)
Entrepreneurial skill	0.940	0.944	0.957	0.848
Government assistance	0.950	0.953	0.968	0.909
Intellectual property	0.917	0.923	0.937	0.750
Network and collaboration	0.877	0.877	0.908	0.625
Productivity and efficiency	0.892	0.918	0.908	0.585
Promotions and marketing	0.936	0.939	0.954	0.839
Reduce cost and improve profit margin	0.934	0.935	0.953	0.835
Technological innovation	0.9150	0.919	0.937	0.714

Source: Authors' calculations.

information about the uniqueness of every construct studied, allowing for a more in-depth and perceptive examination of our findings. By carefully defining the distinctiveness of every variable, this study ensured a thorough analysis of the research results, which improved their comprehensibility and complexity.

Overall, the validity and integrity of the research findings are protected by the meticulous procedural approach, which strengthens their reliability and robustness. This methodological rigor improves the coherence and clarity of the analytical framework used, as well as the credibility of our study.

4.2 Quality Criterion for Reflective Model Assessments and Composite Model

The Heterotrait-Monotrait ratio of correlations (HTMT) was an essential instrument in this study for evaluating the uniqueness of the investigated constructs. All HTMT values must stay below 1, according to the HTMT inference approach developed by Henseler et al. (2015) and Gold et al. (2001). Table 13.3 lists the HTMT values and shows that they all comply with this crucial requirement. This observation supports the validity and integrity of the analytical framework used in this study by offering strong evidence of the uniqueness and autonomy of each construct.

In this study, it was crucial to utilize bootstrapping, a robust re-sampling technique, to evaluate the structural model's predictive power and determine our hypotheses' significance. To guarantee the accuracy of the analysis, there was a need to carefully consider formative measurement models and resolve any

Table 13.2. Discriminant Validity Using Fornell and Larcker.

Constructs	Entrepreneurial Skill	Government Assistance	Intellectual Property	Network and Collaboration	Productivity and Efficiency	Promotions and Marketing	Reduce Cost and Improve Profit Margin	Techno-Entrepreneur_Global	Technological Innovation
Entrepreneurial skill	0.921								
Government assistance	0.392	0.953							
Intellectual property	0.374	0.677	0.866						
Network and collaboration	0.411	0.373	0.341	0.791					
Productivity and efficiency	0.680	0.513	0.408	0.552	0.765				
Promotions and marketing	0.315	0.364	0.309	0.429	0.462	0.916			
Reduce cost and improve profit margin	0.736	0.373	0.350	0.336	0.655	0.258	0.914		
Techno-entrepreneur_global	0.567	0.393	0.367	0.382	0.515	0.286	0.563	1.000	
Technological innovation	0.606	0.438	0.428	0.425	0.609	0.330	0.573	0.637	0.845

Source: Authors' calculations.

Table 13.3. Discriminant Validity Using HTMT.

Constructs	Entrepreneurial Skill	Government Assistance	Intellectual Property	Network and Collaboration	Productivity and Efficiency	Promotions and Marketing	Reduce Cost and Improve Profit Margin	Techno-Entrepreneur_Global	Technological Innovation
Entrepreneurial skill									
Government assistance	0.414								
Intellectual property	0.398	0.725							
Network and collaboration	0.451	0.405	0.376						
Productivity and efficiency	0.638	0.548	0.432	0.629					
Promotions and marketing	0.335	0.384	0.331	0.466	0.512				
Reduce cost and improve profit margin	0.783	0.395	0.375	0.367	0.624	0.273			
Techno-entrepreneur_global	0.582	0.402	0.38	0.404	0.488	0.295	0.582		
Technological innovation	0.653	0.469	0.465	0.472	0.607	0.359	0.62	0.662	

Source: Authors' calculations.

possible collinearity issues using Variance Inflation Factors (VIF), confirming that there were none. By bootstrapping the PLS algorithm, we examined path coefficients and searched for values in the ideal range of −1 to +1. Specifically, at the 1% significance level, this study explores that "Productivity & Efficiency" and the endogenous construct "Reduce Cost and Improve Profit Margin" formed a second-order composite, further supported by outer weights statistically different from zero.

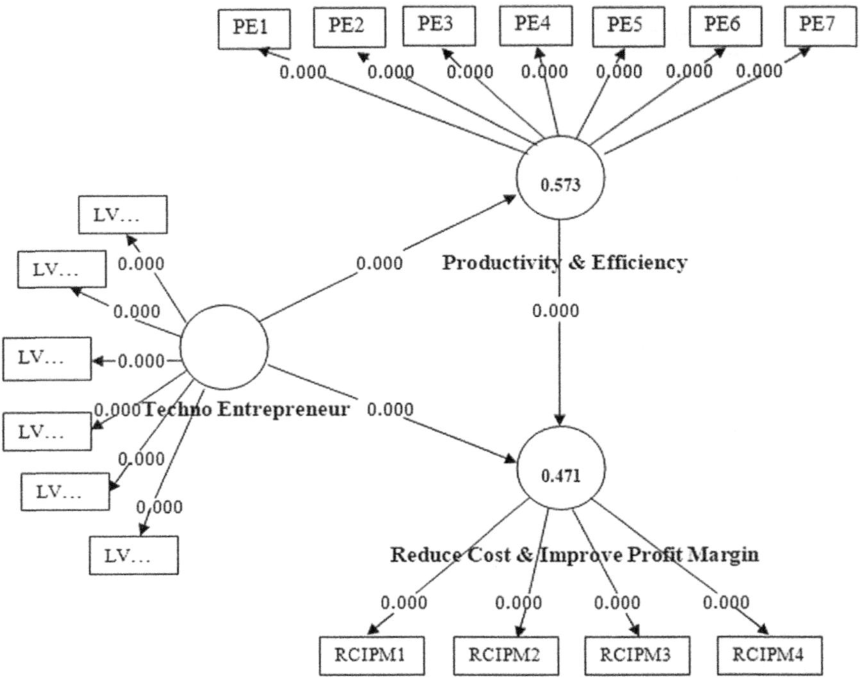

Fig. 13.2. Structural Model Assessments With Control Variables.
Source: Output of PLS-SEM.

Table 13.4 illustrates the techno-entrepreneur's role in fostering "Productivity and Efficiency" and "Reduce Cost and Improve Profit Margin." Crucially, the relationship is complex, with "Productivity and Efficiency" being recognized as a mediating factor in the relationship between "Techno-Entrepreneur" and "Reduce Cost and Improve Profit Margin." This mediation highlights the complex dynamics in the techno-entrepreneurial space, where increased efficiency and productivity function as conduits for maximizing profit and reducing costs.

Table 13.4. Structural Model Assessment.

Hypothesis	Path Relationship	Std. Beta	Sample Mean (M)	t Values	*p* Values	CI2.5%	CI97.5%	Supported/Not Supported
H1	Productivity and efficiency → Reduce cost and improved profit margin	0.309	0.307	5.545	0	0.197	0.417	Supported
H2	Techno-entrepreneur → Productivity and efficiency	0.757	0.759	41.678	0	0.721	0.792	Supported
H3	Techno-entrepreneur → Reduce cost and improved profit margin	0.422	0.425	7.703	0	0.315	0.532	Supported

Source: Authors' calculations.

This thorough examination shows that every hypothesis in the research is strongly supported. The results (Fig. 13.2) highlight the direct impact of techno-entrepreneurship on cost-cutting and productivity and the indirect path that productivity and efficiency take to increase profit margins. The study's hypotheses are thus supported, providing insightful information about the complex ways in which techno-entrepreneurial activities affect organizational performance metrics.

4.3 Mediation Effect

To analyze the direct and indirect impacts of techno-entrepreneurship, this research used rigorous methodologies to understand the complex dynamics within the handicraft industry. As part of the methodology, a thorough examination of the data was applied for advanced techniques to guarantee the validity and dependability of our conclusions.

Firstly, this study carefully bootstrapped the path coefficients, a well-known method for its resilience in parameter estimation even with small sample sizes. As a result, this research was able to obtain *t*-values and bias-corrected confidence intervals, which gave a strong basis for examining how techno-entrepreneurship directly affects "Reducing Costs and Improving Profit Margins." The findings were clear-cut, showing that techno-entrepreneurship has a major and substantial impact on these important organizational outcomes in the handicraft industry.

However, the investigation did not end there. The study explored the domain of mediation effects after realizing the intricate relationship between techno-entrepreneurship, productivity, efficiency, and organizational performance. It became apparent that "Productivity and Efficiency" played a pivotal role in mediating between techno-entrepreneurship and the intended results of profit margin improvement and cost reduction. Based on the recommendations provided by Hayes and Scharkow (2013), this research conducted a simulation study using bootstrapping to evaluate these indirect effects thoroughly. It is thus necessary to calculate the product of indirect effect coefficients using this method, which strongly confirms the mediating function that efficiency and productivity play. Strong findings supported the existence of a partially complementary mediating effect, which improved the comprehension of the complex dynamics at work within the handicraft industry.

This analysis offers an in-depth exploration of the techno-entrepreneurial landscape in the handicraft industry. By detailing both direct and indirect channels (Table 13.5) through which techno-entrepreneurship influences organizational performance, the study provides valuable insights for stakeholders to guide strategic decisions and foster long-term growth and prosperity in the sector.

4.4 Importance Performance Map Analysis

To achieve a more pragmatic and complicated interpretation of the study results, this study utilized an advanced analytical tool called the Importance Performance

Table 13.5. Direct, Indirect, and Total Effects.

Relationship	**Estimate**	***p* Value**
Techno-entrepreneur → Productivity and efficiency → Reduce cost and improved profit margin	0.234	0
Techno-entrepreneur → Reduce cost and improved profit margin	0.112	0

Source: Authors' calculations.

Matrix, also called the Impact Performance Map. This methodological decision, motivated by the observations of Ringle and Sarstedt (2016), allowed us to explore in greater detail the complex relationships between productivity and efficiency and techno-entrepreneurship and how these relationships affect the main objective of reducing cost and improving profit margin enhancement.

With this strategy, this research determined the relative importance of techno-entrepreneurship, productivity, and efficiency and their combined effects on the crucial reduction cost and improvement profit margin metric. This study domain's complex interplay of factors is fully reflected in the performance score of 46.566 obtained and painstakingly displayed in Table 13.6.

The practical implications of the findings go beyond the simple statistical associations through in-depth analysis. Creating an Impact Performance Map, which acts as a compass for stakeholders in the handicraft sector, is essential to our investigation. This map points them toward strategic decisions that capitalize on the combined power of techno-entrepreneurial activities and increased productivity to advance organizational success.

By taking a broad approach, the analysis provides decision-makers with useful insights and enhances the comprehension of the underlying dynamics. Equipped with this understanding, interested parties can make well-informed decisions that direct their initiatives toward profitable and long-term expansion.

Simply put, the Impact Performance Map (Table 13.6) is a road map that shows the way to success in the handicraft sector. It draws attention to the critical role that efficiency and techno-entrepreneurship play in boosting organizational performance and provides helpful advice for navigating the industry's complicated terrain. Stakeholders can steer toward prosperity and resilience by adopting the insights obtained from our analysis, thereby guaranteeing a more promising future for the handicraft industry.

Table 13.6. Importance-Performance Map (Construct-wise Unstandardized Effects).

Constructs	**Importance**	**Performance**
Productivity and efficiency	0 309	48.955
Techno-entrepreneur	0.656	52.553
Mean	0.482	50.754

Source: Authors' calculations.

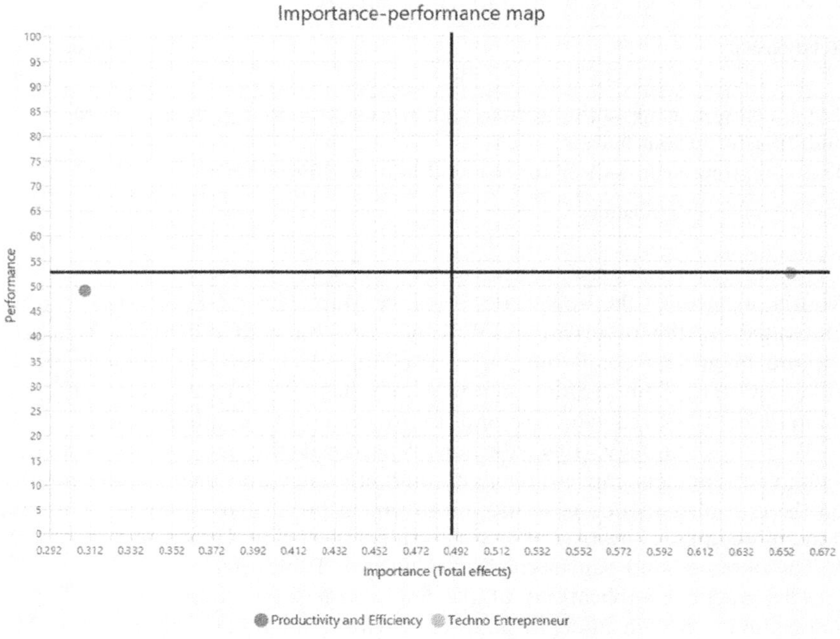

Fig. 13.3. Adjusted Importance Performance Matrix for Reduce Cost and Improve Profit Margin. *Source:* Output of PLS-SEM.

Fig. 13.3 demonstrates clearly how "productivity and efficiency" and "techno-entrepreneur" affect the crucial result of "Reduce Cost and Improve Profit Margin." A careful examination shows that "Reduce Cost and Improve Profit Margin" rises to 47.222 with a one-unit increase in techno-entrepreneur and an overall effect of 0.656. Similarly, "Reduce Cost and Improve Profit Margin" is improved to 46.956 by a corresponding one unit increase in productivity and efficiency, for a total effect of 0.309.

This incisive illustration highlights the crucial roles that productivity, efficiency, and techno-entrepreneurship play in advancing the main goals of cost containment and profit maximization. The observable rise in "Reduce Cost and Improve Profit Margin" that follows small steps forward in techno-entrepreneurship, productivity, and efficiency highlights the mutually beneficial relationship between these elements and the performance of organizations.

The results highlight the necessity of techno-entrepreneurial endeavors and productivity-boosting measures to attain the targeted results of cost containment and profit maximization. In order to propel organizational success in a dynamic and competitive landscape, fostering a conducive environment that nurtures innovation, technological advancement, and operational efficiency is imperative. This holistic perspective supports this need.

5. Conclusion

This study is a guiding thread in the complex handicraft industry, weaving detailed insights into the symbiotic relationship between productivity and techno-entrepreneurship and their significant impact on the sector's economic fabric. This study explores the underlying dynamics using an exacting approach based on PLS-SEM. It sheds light on routes to organizational success in this dynamic industry. The results, painstakingly extracted from in-depth measurement model evaluations, provide crucial compass points for negotiating the intricate terrain of the handloom sector.

The dynamic world of handicrafts is found amid the lively atmosphere of artisan workshops, where skilled hands painstakingly shape raw materials into works of art. We explore this domain to identify the fundamental underpinnings that sustain its robustness and vitality. The investigation centers on techno-entrepreneurship, a powerful force propelling development and change in the handicraft industry. The results highlight techno-entrepreneurship's important role in fostering innovation and progress. It serves as a catalyst, stoking the creative spark and opening doors for business owners and artisans.

One of the most important conclusions drawn from our study is the complex relationship between efficiency, productivity, and techno-entrepreneurship. These components create a symbiotic relationship that promotes the sector's growth and prosperity. Techno-entrepreneurship positively impacts productivity and efficiency, making them invaluable partners in the quest for operational excellence. Furthermore, the study reveals the complex function that productivity and efficiency play as intermediaries of techno-entrepreneurship. They act as channels for the seeds of innovation to sprout, enhancing the business environment and encouraging a creative and efficient culture. As they blend, they create an intricate web of creativity and effectiveness that opens the door to optimizing the handicraft industry's profit and cost.

6. Future Directions of the Study

At the brink of the handicraft industry's potential, the lure of uncharted territory invites us to embark on an exploration voyage. Future research projects in the handicraft industry must explore the depths of temporal dynamics and contextual nuances to reveal a deeper understanding of its complexities. Studies that trace the development of techno-entrepreneurial endeavors in the context of changing market conditions have the potential to yield priceless knowledge. Researchers can learn about the resilience and adaptability of the industry by seeing how these initiatives flourish in the face of shifting circumstances.

Moreover, exploring moderating variables like legislative structures and cultural influences will enhance our understanding of the complex relationships determining the industry's course. Crucial insights into the opportunities and challenges faced by artisans and entrepreneurs can be gained by comprehending how regulatory policies and cultural norms affect entrepreneurial endeavors. By adopting a comprehensive viewpoint that considers micro and macro-level dynamics, forthcoming research can elucidate avenues toward sustainable development and prosperity for the handicraft sector. Research along these lines can help create a dynamic and resilient handicraft industry that adapts to change and thrives by illuminating opportunities for innovation, adaptation, and cooperation.

7. Limitations of the Study

While our study provides valuable insights into the handicraft sector, it is important to acknowledge its limitations to avoid overgeneralization. Although our results provide insightful information about the workings of the handicraft industry, care must be used when applying them outside of this field. Additionally, the cross-sectional nature of our data may obscure causal relationships and temporal dynamics within the handicraft industry. Despite using a rigorous methodology, unforeseen variables and biases may still affect the results. However, the research offers a foundation for understanding how techno-entrepreneurship, productivity, and efficiency can drive transformation in the handicraft sector. Embracing these elements is crucial for fostering resilience and prosperity in an ever-changing industry.

References

Acemoglu, D., & Zilibotti, F. (2001). Productivity differences. *The Quarterly Journal of Economics*, *116*(2), 563–606.

Acs, Z. J., Estrin, S., Mickiewicz, T., & Szerb, L. (2018). Entrepreneurship, institutional economics, and economic growth: An ecosystem perspective. *Small Business Economics*, *51*(2), 501–514.

Aghion, P., & Howitt, P. (1992). A model of growth through creative destruction. *Econometrica*, *60*(2), 323–351.

Apostolopoulos, N., Al-Dajani, H., & Holt, R. (2021). Entrepreneurial orientation, entrepreneurial climate and firm performance: The mediating role of corporate entrepreneurship. *International Entrepreneurship and Management Journal*, *17*(1), 137–163.

Atkinson, R. D., & Lind, M. (2018). *Big is beautiful: Debunking the myth of small business*. The MIT Press.

Barro, R. J. (1991). Economic growth in a cross section of countries. *The Quarterly Journal of Economics*, *106*(2), 407–443.

Barro, R. J., & Sala-i-Martin, X. (1997). Technological diffusion, convergence, and growth. *Journal of Economic Growth*, *2*(1), 1–26. Springer.

Baumol, W. J., & Strom, R. J. (2007). Entrepreneurship and economic growth. *Strategic Entrepreneurship Journal*, *1*, 233–237.

Brem, A., & Borchardt, J. (2014). Technology entrepreneurship, innovation, and intrapreneurship-managing entrepreneurial activities in technology-intensive environments. In F. Thérin (Ed.), *Handbook of research on techno-entrepreneurship*. Edward Elgar Publishing.

Caselli, F., & Coleman II, W. J. (2006). The world technology frontier. *American Economic Review*, *96*(3), 499–522.

Dorf, R. C., & Byers, T. H. (2005). *Technology ventures*. McGraw Hill.

El Ebrashi, R. (2013). Social entrepreneurship theory and sustainable social impact. *Social Responsibility Journal*, *9*(2), 188–209. http://doi.org/10.1108/SRJ-07-2011-0013

Faul, F., Erdfelder, E., Buchner, A., & Lang, A.-G. (2009). Statistical power analyses using G*Power 3.1: Tests for correlation and regression analyses. *Behavior Research Methods*, *41*(4), 1149–1160. https://doi.org/10.3758/BRM.41.4.1149

Faul, F., Erdfelder, E., Lang, A.-G., & Buchner, A. (2007). G*Power 3: A flexible statistical power analysis program for the social, behavioral, and biomedical sciences. *Behavior Research Methods*, *39*(2), 175–191. https://doi.org/10.3758/BF03193146

Frau, M., & Moi, L. (2019). Digital technologies, techno-entrepreneurship and regional ecosystems: The case of the net value. In F. Thérin, F. P. Appio, & H. Yoon (Eds.), *Handbook of research on techno-entrepreneurship* (3rd ed.). Edward Elgar.

Gold, A. H., Malhotra, A., & Segars, A. H. (2001). Knowledge management: An organizational capabilities perspective. *Journal of Management Information Systems*, *18*(1), 185–214.

Griffith, R., Redding, S., & Van Reenen, J. (2004). Mapping the two faces of R&D: Productivity growth in a panel of OECD industries. *The Review of Economics and Statistics*, *86*(4), 883–895.

Hair, J. F., Hult, G. T. M., Ringle, C. M., & Sarstedt, M. (2017). *A primer on partial least squares structural equation modeling (PLS-SEM)* (2nd ed.). Sage.

Hair, J. F., Risher, J. J., Sarstedt, M., & Ringle, C. M. (2019). When to use and how to report the results of PLS-SEM. *European Business Review*, *31*(1), 2–24. https://doi.org/10.1108/EBR-11-2018-0203

Hair, J. F., Sarstedt, M., Ringle, C. M., & Gudergan, S. P. (2020). *Advanced issues in partial least squares structural equation modeling (PLS-SEM)*. Sage.

Hayes, A. F., & Scharkow, M. (2013). The relative trustworthiness of inferential tests of the indirect effect in statistical mediation analysis: Does method really matter? *Psychological Science*, *24*(10), 1918–1927. https://doi.org/10.1177/0956797613480187

Henseler, J., Ringle, C. M., & Sarstedt, M. (2015). A new criterion for assessing discriminant validity in variance-based structural equation modeling. *Journal of the Academy of Marketing Science*, *43*(1), 115–135.

Jones, P., Comfort, D., Hillier, D., & East, D. (2020). Developing entrepreneurship education in the arts. *Industry and Higher Education*, *34*(3), 209–220.

Kuhn, K. M., & Galloway, T. L. (2015). With a little help from my competitors: Peer networking among artisan entrepreneurs. *Entrepreneurship Theory and Practice*, *39*(3), 571–600. https://doi.org/10.1111/etap.12053

Kirzner, I. M. (1973). *Competition and entrepreneurship*. University of Chicago Press.

Kirzner, I. M. (2009). The alert and creative entrepreneur: A clarification. *Small Business Economics*, *32*(2), 145–152.

Lafuente, E., Szerb, L., & Acs, Z. J. (2016). Country level efficiency and national systems of entrepreneurship: A data envelopment analysis approach. *Journal of Technology Transfer*, *41*(6), 1260–1283.

Lingesiya, Y. (2012). Identifying factors to indicate the business performance of small scale industries: Evidence from Sri Lanka. *Global Journal of Management and Business Research*, *12*(21), 75–82.

Lounsbury, M., & Glynn, M. A. (2001). Cultural entrepreneurship: Stories, legitimacy, and the acquisition of resources. *Strategic Management Journal*, *22*(6–7), 545–564.

Lucas, R. E. (1988). On the mechanics of economic development. *Journal of Monetary Economics*, *22*(1), 3–42. https://doi.org/10.1016/0304-3932(88)90168-7

Meek, W. R., Pacheco, D. F., & York, J. G. (2010). The impact of social norms on entrepreneurial action: Evidence from the environmental entrepreneurship context. *Journal of Business Venturing*, *25*(5), 493–509.

Moll, B. (2014). Productivity losses from financial frictions: Can self-financing undo capital misallocation? *American Economic Review*, *104*(10), 3186–3221.

Nambisan, S. (2017). Digital entrepreneurship: Toward a digital technology perspective of entrepreneurship. *Entrepreneurship Theory and Practice*, *41*(6), 1029–1055.

Nitzl, C., Roldan, J. L., & Cepeda, G. (2016). Mediation analysis in partial least squares path modeling: Helping researchers discuss more sophisticated models. *Industrial Management & Data Systems*, *116*(9), 1849–1864. https://doi.org/10.1108/IMDS-07-2015-0302

Örnek, A. S., & Danyal, Y. (2015). Increased importance of entrepreneurship from entrepreneurship to techno-entrepreneurship (startup): Provided supports and conveniences to techno-entrepreneurs in Turkey. *Procedia-Social and Behavioral Sciences*, *195*, 1146–1155. https://doi.org/10.1016/j.sbspro.2015.06.164

Phan, P. H., Wright, M., Ucbasaran, D., & Tan, W. L. (2009). Corporate entrepreneurship: Current research and future directions. *Journal of Business Venturing*, *24*(3), 197–205.

Prieger, J. E., Bampoky, C., Blanco, L. R., & Liu, A. (2016). Economic growth and the optimal level of entrepreneurship. *World Development*, *82*, 95–109.

Prodan, I. (2007). A model of technological entrepreneurship. In F. Thérin (Ed.), *Handbook of research on techno-entrepreneurship* (pp. 26–50). Edward Elgar.

Rashid, Z., & Ratten, V. (2021). Regulatory focus and entrepreneurship: A review of recent research. *International Journal of Entrepreneurial Behavior & Research*, *27*(1), 19–42.

Ratten, V. (2014). Entrepreneurial potential and potential entrepreneurs in business and sport organizations. *International Entrepreneurship and Management Journal*, *10*(4), 729–738.

Ratten, V. (2016). Entrepreneurial universities: Emerging models in the new social and economic landscape. *Journal of Enterprising Communities: People and Places in the Global Economy*, *10*(4), 421–430.

Ratten, V. (2019). Entrepreneurship and sport. *International Journal of Entrepreneurial Behavior & Research*, *25*(3), 421–434.

Ratten, V. (2020a). Entrepreneurial processes and the internationalization of firms in dynamic economies. *International Journal of Entrepreneurial Behavior & Research*, *26*(6), 1161–1177.

Ratten, V. (2020b). The role of networking in business and sport: What do they have in common? *International Journal of Entrepreneurial Behavior & Research*, *26*(6), 1193–1206.

Ratten, V. (2021). Entrepreneurial creativity and innovation in the sports industry. *Journal of Business Research*, *134*, 620–623.

Ratten, V., & Jones, P. (2021). Exploring creativity and innovation in entrepreneurship education: Insights from industry. *Industry and Higher Education*, *35*(1), 2–10.

Ratten, V., & Usmanij, P. (2021). The impact of entrepreneurship education on entrepreneurial intention: A Malaysian perspective. *Education + Training*, *63*(4), 477–491.

Richter, N. F., Sinkovics, R. R., Ringle, C. M., & Schlägel, C. (2016). A critical look at the use of SEM in international business research. *International Marketing Review*, *33*(3), 376–404. https://doi.org/10.1108/IMR-09-2014-0304

Rigdon, E. E., Sarstedt, M., & Ringle, C. M. (2017). On comparing results from CB-SEM and PLS-SEM: Five perspectives and five recommendations. *Marketing ZFP*, *39*(3), 4–16.

Ringle, C. M., & Sarstedt, M. (2016). Gain more insight from your PLS-SEM results: The importance-performance map analysis. *Industrial Management & Data Systems*, *116*(9), 1865–1886.

Ringle, C. M., Sarstedt, M., & Schlittgen, R. (2014). Genetic algorithm segmentation in partial least squares structural equation modeling. *OR Spectrum*, *36*(1), 251–276.

Ringle, C. M., Sarstedt, M., Schlittgen, R., & Taylor, C. R. (2013). PLS path modeling and evolutionary segmentation. *Journal of Business Research*, *66*, 1318–1324.

Schumpeter, J. A. (1934). *The theory of economic development*. Harvard University Press.

Shane, S., & Venkataraman, S. (2000). The promise of entrepreneurship as a field of research. *Academy of Management Review*, *25*(1), 217–226.

Solomon, G. T., & Mathias, B. D. (2020). "This is not your grandmother's business school": Provocations for the future of entrepreneurship education. *Journal of Small Business Management*, *58*(1), 1–21.

World Commission on Environment and Development. (1987). *Our common future* (p. 27). Oxford University Press.

Yadav, U. S., Tripathi, R., Tripathi, M. A., Ghosal, I., Kumar, A., Mandal, M., & Singh, A. (2023). Digital and innovative entrepreneurship in the Indian handicraft sector after the COVID-19 pandemic: Challenges and opportunities. *Journal of Innovation and Entrepreneurship*, *12*(1), 69.

Zhou, L. (2004). Entrepreneurship education in China: Issues and challenges. *Education + Training*, *46*(8/9), 483–493.

Chapter 14

Integrating AI and ML Technologies Into Indigenous Financial Systems: Applications of AI and ML in Indigenous Financial Management

Rajiv Iyer and Vedprakash C. Maralapalle

Amity University, India

Abstract

This chapter of the book is inspired by the specific aim to investigate the possibilities that Artificial Intelligence (AI) and machine learning (ML) generate on indigenous financial systems particularly in addressing financial decision-making, risk management, and customer relations. Through the implementation of AI/ML, such microbanks that exist to cater for indigenous citizens can enhance credit rating, fraud identification, and service delivery so as to enhance on financial services that are in consonance with the indigenous people's needs. Concerning the advantages of the discussed AI and ML integration, this chapter focuses on the following insights: accurate and effective financial decisions as a result of big data analyses, risk management with the help of monitoring and detection of unusual patterns, and enhancing customer relations with smart founded bots and recommendations. This chapter also analyzes various potential challenges some of which are the availability and quality of data, the requirement for expertise in the technological field, and lastly, the culture and structure of the organization that might resist the implementation of such technologies. In this chapter, examples of successful AI and ML implementation in indigenous financial systems across the globe are explained, and this section offers the indigenous financial institutions specific ideas, lessons learned from other places, and experiences that will prove useful as they continue to advance their digital transformations. The last part of this chapter provides the concluding remarks in the given discussion on possible future directions and opportunities of AI and ML integration for development, such as enabling finance

Indigenous Empowerment through Human-Machine Interactions, 237–251

doi:10.1108/978-1-83608-068-820251014

inclusion and supporting indigenous peoples and organizations and sustainable economic initiatives.

Keywords: Indigenous financial systems; financial inclusion; financial management; financial decision-making; risk management; machine learning (ML); Artificial Intelligence (AI)

1. Introduction to Artificial Intelligence (AI) and Machine Learning (ML) in Indigenous Financial Systems

The current efforts to implement AI and ML technologies into the indigenous financial systems imply a potential to improve the existing financial management strategies in the indigenous population. Through the adoption of these technologies, indigenous players in the financial services sector will be able to enhance credit risk assessment, authentication, and business effectiveness, thus positively impacting the manner in which indigenous people's financial needs will be met in the future (Herrmann, 2023). AI and ML in this case can be of assistance to the indigenous financial systems by supporting operational efficiency in terms of workflows and processes, supporting the analysis of current markets, their changes, the customers, and their attitudes toward financial products (Goodell et al., 2021). They can also help in managing assets, evaluating risks, and giving advice in investment as well as in fraud (Moutinho & Smith, 2000). Thus, thanks to the trainability and the access to enormous data, AI and ML algorithms are steadily narrowing the gap and getting the world closer to a fully automated financial system (Polychroniou & Giannikos, 2009). However, the integration of AI and ML into the indigenous financial systems is not without some of these challenges. Some of the challenges that institutions dealing with these technologies encounter include unrealistic expectations, lack of understanding on how the technology functions, and the need to work with experienced partners that will help you to develop the best models (Vorobyev & Krivitskaya, 2022). Moreover, excessive data and the requirement to secure data further, which has to solve many regulatory requirements, are crucial issues (Canhoto, 2021). However, the prospects of incorporating AI and ML organization into indigenous financial structures offer huge benefits. These technologies can enhance customer experience and boost clients' repeat rate due to the efficient organization and reduced human errors (Nagurney, 2015). In addition, AI and ML can be used to improve risk management since, through the analysis of historical data, they are able to determine new risks and characteristic patterns regarding fraudulent behavior (Singh & Best, 2019).

Therefore, it is apparent that assimilation of AI and especially of ML into indigenous financial environments will play a significant role in enhancing decision-making concerning financial issues, handling risk factors, and managing customers. As these technologies remain adopted by indigenous financial institutions, they must be in a position to develop better and sustainable services that include the indigenous people (Cao et al., 2022).

1.1 Ethical Considerations for Integrating AI and ML Technologies Into Financial Systems

The integration of AI into the finance industry raises several important ethical considerations that must be carefully addressed:

- Transparency and Explainability: Another issue closely related to ethics is the opaqueness and the inability to explain the decisions made by these AI in finance. AI systems play an important role in financial institutions advising their clients, and therefore, those institutions bear the responsibility of ensuring that recommendations as well as actions taken by an AI system are understandable to the clients. The schemes learned by AI models may be very large and complex which makes it difficult to interpret or even explain why a specific output was produced, and this is where trust is at risk and where it could lead to adverse results (Farbmacher et al., 2022).
- Bias and Fairness: AI models, just as other AI networks, possess properties due to which they can strengthen and, in some cases, even amplify biases inherent in the training dataset. This could lead to prejudices in credit or lending decisions, credit discrimination, or prejudice investment decisions that work negatively to some people or groups. Banking and lending companies must also check for fairness in the AI solutions implemented to avoid discriminating clients (Azzutti, 2022).

2. Enhancing Financial Decision-Making Through AI and ML

The implementation of existing AI and ML technologies into indigenous financial systems will consequently improve the decision-making options for financial solutions. The following are some of the advantages that come with the said advanced technologies to enable the indigenous financial institutions take better decisions:

In financial decision-making, and in general, AI and ML are very useful in the ability to analyze vast amounts of data. Due to their capability of using complex mathematical formulas, these technologies are capable of recognizing certain patterns, trends, and insights that may be hard for analysts to discover. This way, indigenous financial institutions are in a better position to forecast the current market trend, consumers' behavior, and even any risks likely to occur (Manahov, 2015).

Furthermore, AI and ML can free the process from many existing biases and mistakes of people which are usual in traditional financial decisions. These technologies can be used to measure credit rating, identify fraud, and rate investment possibilities in a more wet and exact manner, that results in decision-making that is fair and based on facts. AI and ML can also be useful to indigenous financial institutions in matters to do with personalization. Through the employment of these technologies, clients are able to receive personal financial solutions, product preferences, or risk control that suits their specific needs. Such levels of customization may improve customer satisfaction and loyalty and, consequently, their financial status within the indigenous populations.

In addition, the use of AI and ML can help the indigenous financial institutions to make faster decisions on relevant opportunities as a result of integration. The set-up inherent in the real-time data analysis and decision-making that are evident in the technology could assist in keeping these indigenous financial institutions relevant in the changing financial environment. Therefore, the amalgamation of AI and ML technologies in indigenous systems of financial decision-making may signify efficiency. Thus, indigenous financial institutions can use business intelligence and analytics to make far more sound, fair, and relevant decisions that can positively impact their clients and the overall community's advancement in the long run (Prix et al., 2007). Fig. 14.1 shows how AI capabilities apply financial decision-making:

3. Improving Risk Management in Indigenous Financial Institutions

These institutions are relevant in ensuring effective and affordable financial services to the people and a way of boosting the economy of the indigenous people. However, these institutions are not very effective when it comes to the management of risks as they encounter several challenges as pointed out by different scholars. Hence, the adoption of AI and ML in risk management frameworks in indigenous financial institutions enables credible identification, evaluation, and management of different types of risks that will ultimately improve the financial institutions' performance. The use of AI and ML in risk management brings the following benefits, one of which is the ability to analyze large volumes of data from various sources including previous financial records, the market trends, and consumer behaviors. These technologies can find structures of relationships and

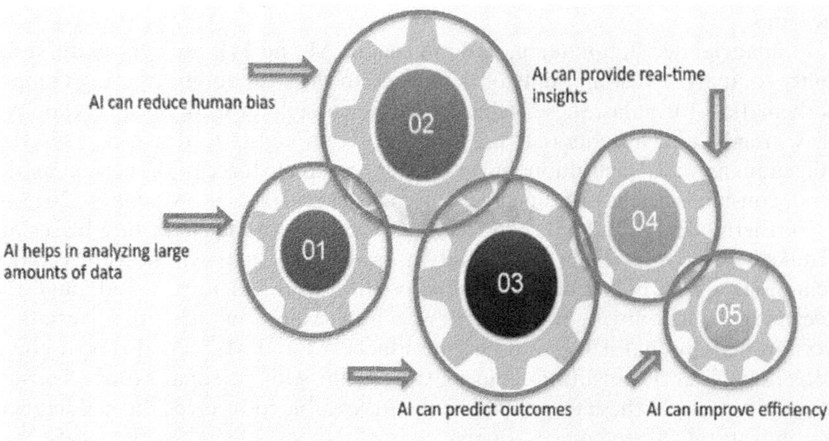

Fig. 14.1. Using AI in Decision-Making.

know when something is wrong, such as fraud or credit defaults. Due to its capability to learn from new data, AI and ML algorithms can be used to navigate through the existing and innovative schemes in the native markets and counter threats, hence providing indigenous financial institutions with a competitive edge. Further, the nature of innovations in the financial organizations of indigenous communities means that enhanced levels of AI and ML can support risk valuation. These technologies regarding data collection and analysis can help to shorten time and capitals for risk assessment and let institutions pay more attention to constructing and implementing the right strategies of risk mitigations. Also, the use of AI in risk assessment can increase the efficiency, covering a greater number of factors potentially threatening indigenous populations, including environmental and social risks (Chukhrova & Johannssen, 2021).

Surprisingly, there is a lack of an adequate strategy for enhancing risk management in indigenous financial institution with another critical concern, which is enterprise-wide. It entails pinpointing and managing threats in the credit, liquidity, operations, and strategy fields. The systematic approach facilitates an understanding of the different risks within the indigenous financial institutions and how risks are connected and affect one another.

To implement a comprehensive risk management framework, indigenous financial institutions should consider the following steps:

- Establish a clear risk governance structure: Disseminate clear information on what is expected of personnel in managing risk with particular focus on the operational managers, the senior management, and the board of directors.
- Develop robust risk policies and procedures: Develop definite procedures for recognition, evaluation, and regulation of risks which should address the requirements of institutions and its risk characteristics. Fig. 14.2 illustrates the best use cases of AI in finance. Below is the detailed explanation of each use case.

4. Transforming Customer Relations With AI and ML Technologies

Integrating AI and ML technology into the indigenous banking systems is highly likely to have vast potential to change the face of relationships in client, as well as the engagement rate within indigenous communities. Through the use of this technology, indigenous financial institutions can develop a better appreciation of their clients' needs, wants, and behaviors to ensure that they provide them with services that are not only customized to their needs but remain sensitive to their pursuit of change.

One of the ways of differentiating the customer relations with help of AI and ML is to introduce intelligent chatbots and virtual assistants. These technologies use natural language processing and conversational AI pattern to help customers any time even get back information, make transactions, or get some advice from

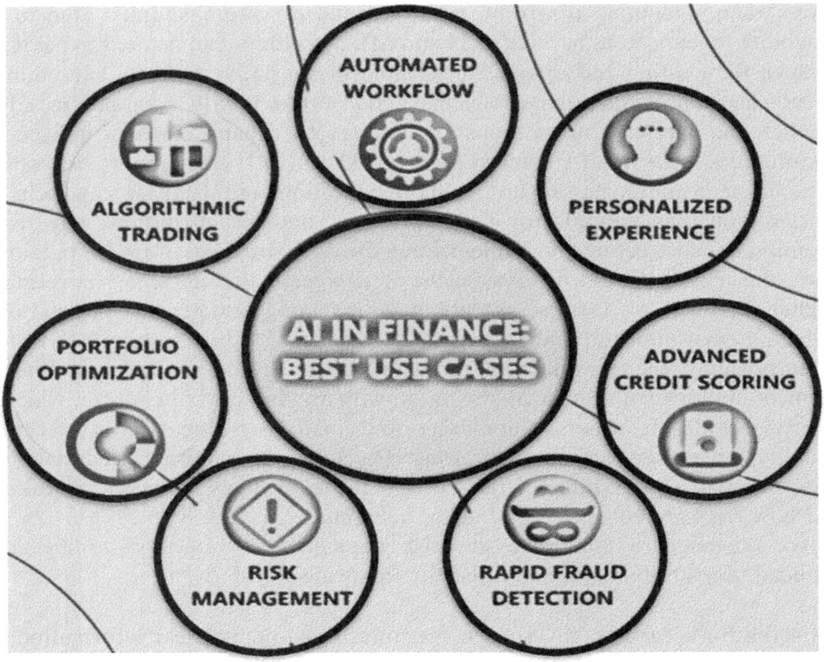

Fig. 14.2. Use Cases of Artificial Intelligence (AI) in Finance.

their created AI. Not only does it help enhance the expense per customer satisfaction in the long run but also it releases human jobs for more complicated tasks, thus enhancing the efficiency of employees and reducing costs.

Furthermore, with regard to analyzing massive amounts of data originating from various sources, such as transaction pasts, interactions on the web and feedbacks from consumers, it was identified that AI and ML might help local financial institutions in gaining more insightful knowledge of their clientele. These can then be utilized for designing more explicate appealing marketing strategies or deployment of more appropriate products or services or for being more sensitive to the needs and demands of the customers. For instance, an indigenous bank is in a position to incorporate adaptive learning and try to pin point out customers that would run into some forms of financial distress and then offer them propositions such as restructuring of the debt or even monetary advice that would prevent that from even happening.

Besides making an effective change in the definition of the way companies and other financial entities satisfy and interact with their customers, AI and ML can also be helpful in making the indigenous financial systems more secure and trustworthy. Through the application of refined fraud detection algorithms, these technologies are capable of detecting and eliminating frauds within an institution as well as its customers. This can be especially helpful concerning the indigenous population where they lost confidence in banking systems resulting to past incidences or social issues (Pnevmatikakis et al., 2021).

In addition, LI AND ML can help bring enhanced financial services from indigenous financial institutions closer to the people. For instance, through data mining, such technologies can give out recommendations in investments, and give out loans based on the customers' wants and needs, and can also act as beacons of education on what the customer ought to do financially based on their risk taking abilities. Such level of customization can't really enhance endorsing and satisfaction to the clients but can also enhance the financial experience of indigenous people. However, the realization of this dream through adoption of AI and ML/in indigenous financial systems is not without some of the disadvantages as highlighted in this paper. Some challenges that may affect Indigenous financial institutions include: availability of data, quality of data, and the need to engage professionals in the engineering of such technologies for its effective implementation. For the new advantages based on the new technologies to be fairly utilized, several questions that relate to the moral uses of AI besides prejudice and discriminating have to be answered. In trying to overcome these challenges therefore, indigenous financial institutions should adopt the strategies that are elaborated below. This might entail proper relationships with the technology vendors, universities, and other relevant stakeholders to implement specific AI and ML solutions to the existing needs of customers embracing certain cultural differences. Further, these institutions should facilitate the development of capacities and frameworks for governing the proper uses of these technologies in the ways that reflect indigenous peoples' values and autonomy.

Hence, the incorporation of the AI and ML technologies into the indigenous financial systems has the potentials of revolutionizing customer relations and increasing interaction with a customer within the indigenous communities. This way, micro financial institutions of indigenous peoples can adopt these sophisticated technologies to improve the efficiency of their services, obtain fuller and more comprehensive knowledge, strengthen safety, and provide a more effective and accessible banking services. However, it is imperative to ensure that the numerous technologies are applied in a well-coordinated but a planned manner that will address the special difficulties, not to mention the cultural barriers that are evident in the indigenous populations (Meng et al., 2022). Fig. 14.3 illustrates various use cases of AI and ML in Customer Relationship Management (CRM) systems.

5. Overcoming Challenges and Barriers to Adoption

There are some challenges and impediments that require to be solved for the proper adaptation and implementation of the AI and ML into native financial systems. All these issues cut across the technological, organizational, and cultural realms and cannot be solved individually but systematically. The use of data is important factor and indigenous financial institutions' key problem is access to

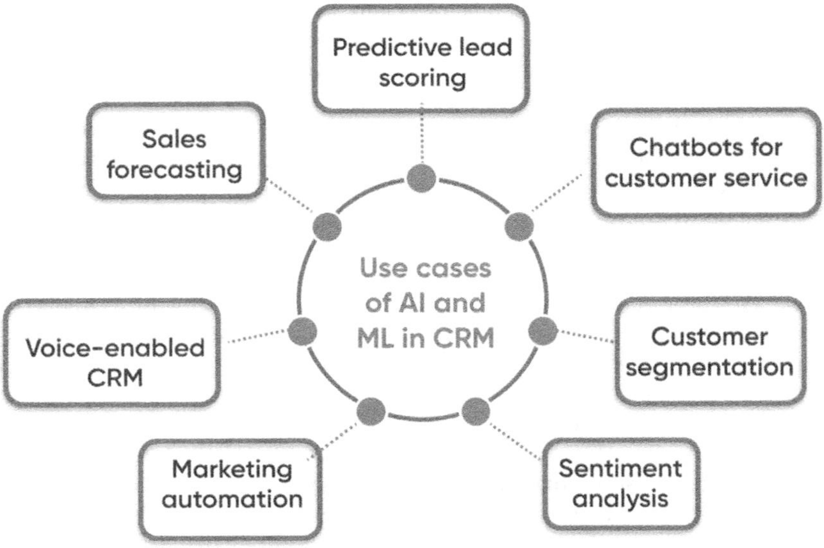

Fig. 14.3. AI and ML in CRM.

quality data. AI and ML are based on the availability and accuracy of big and clean datasets for the proper functioning of the model. However, regarding disclosing rich and adequate financial information, many indigenous communities may not have access to adequate financial data or lack the necessary means of converting and normalizing their records. This affects AI and ML as these technologies rely on high quality of data for the models to perform as expected (Bouhia et al., 2022).

To overcome this issue, it is suggested that indigenous financial institutions should carry out data searches with the help of external suppliers of data, IT companies, or research institutions. They could also have to incur on personnel as well as infrastructure for data management to ensure that their data resources are accurate, secure, and easily retrievable. This might involve implementing policies and procedures for managing data, creating awareness among the employees regarding data management and following new techniques for digital & collection of data fitting the needs & challenges of indigenous institutions.

The last core factor hindering the implementation of AI and ML in indigenous financial systems is the shortage of IT specialists within these institutions. AI and ML solutions need a combination of data scientists, software engineers, and subject matter experts capable of grappling with challenges surrounding and implementing sustainable AI and ML for financial organizations' functional systems.

To this end, the indigenous financial institutions should consider capacity enhancements activities including training, partnerships with learning institutions, offering employment opportunities for specialized personnel. They may also consider the possibilities of cooperation when they may partner with technology vendors or FinTech, or research institutions to co-design and co-implement the solutions based on the AI and ML technologies.

Also, deeper cultural and organizational set-up in the native financial institutions can work as a hindrance to deploy AI and ML technologies. Perhaps certain indigenous people still uphold some or many cultural beliefs and norms which might not be compatible with the seemingly efficiency of the AI systems to make decisions or with the perceived warfare of automation to conventional financial practices (Iyer et al., 2024a, b).

The last important issue of AI and ML technology implementation to the indigenous financial systems is the question of adequate governance structure and ethical frameworks. As the technologies become more prevalent, there are increasing concerns of topics like the algorithmic bias, the rights of data ownership, and other negative externalities. To achieve this, it may be required to establish framework for data management, educate people about data management strategies, and explore other innovative approaches to document and collect data that are relevant to Indigenous people and where they are. In this regard, it is vital that indigenous financial institutions engage policymakers, regulators, and Indigenous rights organizations to ensure the appropriate use of AI and ML by drafting a powerful governance structure that will include ethical aspects of AI and ML. This may require the creation of policies on data protection, customer and service provider's rights, as well as equitable use of these technologies in the development of indigenous financial structures. Therefore, it is clear that the implementation of the recommended AI and ML technologies must pass through several barriers in indigenous financial systems; the barriers include technological barriers, organizational barriers, and cultural barriers. The indigenous financial institutions should leverage the services of external partners, build capacity, stay people-centered, and ensure that the implementation of such technologies is done with the right governance systems in a bid to maximize on the impact that the technologies bring while at the same time ensuring that it meets the indigenous people's needs and standards.

There are a number of dangers and difficulties that are attached to the utilization of machine learning and AI for economic decisions. The first one is inherent bias – where biased data are fed into an AI model to train the model or where bias sneaks into the algorithm by designers. This may lead to prejudice of some groups, hinder provisions of fairly services in organizations, and expose the financial institutions to cases of noncompliance with set laws. Secondly, AI models might be complex making it hard to understand how AI makes decisions, which brings up issues of responsibility, openness, and, sometimes, even compliance issues (Tepe et al., 2022).

Lastly, operational risks are also an issue since the application of these technologies brings new risks such as the failure of systems and data leak risks, which poses a risk to the stability of the financial systems and in periods of crisis. In order to avoid these risks, the financial institutions and the regulators should collaborate on formulating good governance around the usage of AI and ML in effective decision-making for financial institutions.

6. Case Studies: Successful Implementation of AI and ML in Indigenous Financial Systems

Due to the increased adoption of the indigenous financial integrated systems involving AI and ML skills, it is possible to realize the financial inclusion to empower the neglected communities. Even though the implementation of the said sophisticated technologies is not without its peculiarities, indigenous financial systems have begun experiencing drastic positive changes attributable to AI and ML. Hence, it can greatly improve the degree of personalized financial consultancy services for the small businesses. It is in this front that such advanced technologies are capable of helping the small business in such ways as credit scoring and risk evaluation. With the help of AI and ML it is possible to use a large number of nonfinancial data such as payment history, social media activities, and industry analysis for better credit scoring. It thus results in higher loan delivery and improved credit quality through effective management of risks and allowing small business outfits, to get the funding they require for growth. Also, financial advisory services that involve the use of AI can also incorporate these technologies to provide an individualized approach to the management and investment hints given. These include specific investment recommendations and budgeting solutions for a small business, based on the business's financial data, goals, and risk (Iyer et al., 2024a, b).

It is possible to continue the use of robo-advisors that, with the help of AI, allow clients to get individualized financial advice at any time convenient for them, which will be useful for representatives of the small business with limited access to a financial advisor.

Furthermore, AI and ML can upgrade the advisory services for those areas and simultaneously provide more effective and efficient operations as a result of such. Thus, boring tasks like data entry, report generation, and checking for compliance help are relieved by these tools, freeing up the financial advisers' time for strategy development and customer interaction. Real time financial analysis providing by AI technology also contributes to the improved decision-making and implementation of decisions for the small businesses as well as provide timely responses to the occurrences in the market and the emerging opportunities (Li et al., 2021).

In addition, ML algorithms have a great potential of improving the risk management to the small enterprises. Through analyzing different large data inputs, these algorithms are able to detect potential dangers in investments and provide this information to the financial advisory so that potential threats could be nipped in the bud and small business manufactured portfolios protected. This can comprise of issues, such as tracing and reducing instances of embezzlement which remains a major problem affecting many small business organizations. And last, but not least, advanced technologies like AI and ML in the mentioned sphere can enhance client interaction and personalized approach within the standards of financial advising. Small business customers may receive individual experiences that are closer and more allied with their financial advisors if financial advisors employ AI-dependent tools to help them comprehend the patterns of behaviors and financial goals of their customer. Therefore, this may lead to improvement in small company clients' overall satisfaction, commitment, and confidence with the services offered. Applying the ideas of the AI and ML health continuous, financial advisory services can be personalized, productive, efficient, and beneficial to small business by offering them the precise tools that will help them urge their financial objectives and reach out sustainable advancement. A well-known interesting case is the Grameen Bank in Bangladesh, which applied AI/ML in its microfinancing operations since 2001. New data sources, which include telecommunications check and social relationships of borrowers, have assisted Grameen Bank develop AI credit scoring solutions that reflect the creditworthiness of borrowers who lacked any recognition in usual credit databases. This has made the bank in a position to advance money to many people who had been deemed uncreditworthy and the issue of recovery stands at 99% (Mittal & Agrawal, 2022).

Another successful example is Kenya where the M-Pesa, the mobile money transfer service, utilizes AI and ML in the fight against fraud and to improve clients' satisfaction. The AI system for detecting fraud in M-Pesa constantly examines the records of the transactions and alerts the management within minutes of any wrongdoing. From this kind of preventive measure that has been implemented in the operations of M-Pesa, incidences of fraud have been minimized, thus building credibility among the users, many of whom are from the less privileged backgrounds (Kristof & Virag, 2022).

Currently in Latin America, Banco Azteca, a loan company that targets the underbanked and low-income customers, has adopted the use of AI and ML in its operations to control its lending business. Using nontraditional credit checks aiming at loan applicants' bills and phone records, Banco Azteca's credit scoring utilizing AI efficiently evaluates creditworthiness of consumers with no credit history. Thus, the bank has been lending money to millions of people who previously had no access to the official financial market, which increases the rate of financial inclusion and economic freedom (Lux, 2015).

These case studies demonstrate the transformative potential of AI and ML in indigenous financial systems. By leveraging alternative data sources, automating processes, and enhancing customer experience, these technologies have been able to overcome traditional barriers to financial inclusion and empower underserved communities. However, it is essential to note that the successful implementation of AI and ML in indigenous financial systems requires a comprehensive approach that addresses technological, organizational, and cultural challenges.

7. Future Directions and Opportunities for AI and ML Integration

As AI and ML keep on advancing, inclusion of such sophisticated features into indigenized financial frameworks opens up a great potential for positive impacts with regards to indigenous people's financial rights and opportunities to unlock sustainable financial inventions.

Another topical area of AI/ML application is the improvement of the financial product and services' personalization. With the help of ML and real-time data analysis, these technologies can help indigenous financial institutions to gain more profound understanding of their customers' requirements, inclinations, and patterns. This can help to design suitable products and services of which would fit the situation and future plans of individual consumers, building up a better consumer relations and improving consumers' welfare (Akakpo et al., 2019).

For instance, robo-advisors which are in the category of the AI-assisted tools can provide indigenous community members with an investment plan and other important decisions regarding the financial future. Likewise, the credit-scoring models based on the ML approach can assess the user's creditworthiness using nontraditional data such as call detail records or social connections to offer credit services to customers who are underserved by traditional financial institutions. Another facet of future integration is the implementation of the AI and ML models to further improve the indigenous financial institution risk management and fraud detection. Through the constant analysis of large numbers of data and the understanding of complicated correlations, these technologies assist the institutions to prevent various risks such as credit losses, cyber threats, and compliance failures. This in turn can contribute to improved and more efficient operations, decrease in cost and loss, and positive perception and understanding from indigenous peoples' side (Khan et al., 2021).

However, the synergistic application of AI and ML in the platforms used by microfinance institutions can transform the delivery of these financial services especially in the remote and under-banked regions. AI-generated chatbots and virtual assistants can be effective in responding to customers' inquiries around the clock; as a result, indigenous people can also have fast access to banking, insurance, and other financial services regardless of the location. Furthermore, through embracing the blockchain and distributed ledger technologies together with AI, and ML, people will be able to conduct their financial transactions safely and efficiently, hence improving financial accessibility and control (Pattnaik et al., 2023).

In conclusion, we can state that the prospects of the further development of the integration of AI and ML with the indigenous financial systems remain extensive and relatively open. Therefore, by leveraging on the advanced technological frontiers, the indigenous FIs can promote the objective of including the indigenous people in financial systems, as well as improve on risk management and transform the delivery of financial services to support the indigenous people to gain sound economic base. However, one has to acknowledge that the integration of these technologies is not a simple technical matter which addresses only technological factors and proposes mere technical solutions to extant problems; it is a complex process that essentially encompasses technological, organizational, and cultural changes for which responsibilities and ethical issues cannot be left out (Baker et al., 2022).

8. Conclusion

This chapter has also revealed how the application of AI and ML in indigenous financial systems can help increase financial inclusion, support indigenous minority, and spearhead sustainable economic development. Through adoption of these advanced technologies, the indigenous financial institutions can make the decision-making process better, and the customer relations enhanced enabling them to deliver better services to their customers in the indigenous communities.

The opportunities outlined in this chapter regarding the organization of operations presented by successful biometric systems and focused on instances like Grameen Bank in Bangladesh, M-Pesa in Kenya, and Banco Azteca in Latin America demonstrated how emerging implements including AI and ML can help break generic barriers to inclusion. These institutions have proved that through examining additional data sources, setting autopilot, and improving customers' satisfaction, the technologies would help the previously unbanked populations and small and medium businesses to receive financing.

But, introduction of AI and ML, in indigenous financial systems, has its own advantages and advancement disadvantages. This chapter has brought out the fact that for indigenous financial institutions, adoption of such technologies is complicated because of the various technological, organizational, and cultural factors. Effective data governance, strategic partnerships, capacity enhancement, and coordination with relevant stakeholders, along with ongoing evaluation of the methods, are fundamental in addressing these challenges and promoting the effective and ethical implementation of the AI and ML.

From the outlook, the possibilities of applying AI and ML in the indigenous systems of financial management and control are enormous and encouraging. From customized solutions on investment and risk management to appropriate and efficient solution distribution and even improved financial position, these technologies can reshape the financial services and bring positive change in indigenous peoples' lives. This transformational potential can thus assist indigenous financial institutions in achieving the goals of the United Nation's Sustainable Development particularly those concerning poverty, women's rights, employment adequacy, and sufficient economy.

To fully exploit this potential, it is imperative that various players in the indigenous financial industry consisting of financial institutions, policymakers, regulators among others, devise a strategy of AI and ML with dedicated efforts from all parties involved. This involves promoting regional and international engagement, improving the capacity of organizations at various levels, and setting up sound governance structures geared toward facilitating the wishes and needs of indigenous populations. Thus, it is possible to achieve a change in the machineries, structures, and standards of financial practices and policies to better serve indigenous peoples and foster development based on equity and sustainability.

In conclusion, it is possible to agree with the statement that integration of the AI and ML into indigenous financial statements is a crucial step in attaining peoples' financial inclusion and self-sufficiency. Thus, by advancing the ideas about these transformative technologies and discussing the challenges and obstacles encountered during the process of their implementation indigenous financial institutions are able to open the new opportunities for development and innovations that can bring the positive impacts to society.

References

Akakpo, R. M., Xia, M., & Polansky, A. M. (2019). Frequentist inference in insurance ratemaking models adjusting for misrepresentation. *ASTIN Bulletin*, *49*(1), 117–146. https://doi.org/10.1017/asb.2018.41

Azzutti, A. (2022). AI trading and the limits of EU law enforcement in deterring market manipulation. *Computer Law & Security Review*, *45*, 105690. https://doi.org/10.1016/j.clsr.2022.105690

Baker, H. K., Kumar, S., Pattnaik, D., & Pandey, N. (2022). The journal of accounting and public policy at 40: A bibliometric analysis. *Journal of Accounting and Public Policy*, 107003. https://doi.org/10.1016/j.jaccpubpol.2022.107003

Bouhia, M., Rajaobelina, L., PromTep, S., Arcand, M., & Ricard, L. (2022). Drivers of privacy concerns when interacting with a chatbot in a customer service encounter. *International Journal of Bank Marketing*, *40*(7), 1159–1181. https://doi.org/10.1108/IJBM-09-2021-0442

Canhoto, A. I. (2021). Leveraging machine learning in the global fight against money laundering and terrorism financing: An affordances perspective. *Journal of Business Research*, *131*, 441–452. https://doi.org/10.1016/j.jbusres.2020.10.012

Cao, T., Cook, W. D., & Kristal, M. M. (2022). Has the technological investment been worth it? Assessing the aggregate efficiency of non-homogeneous bank holding companies in the digital age. *Technological Forecasting and Social Change*, *178*, 121576. https://doi.org/10.1016/j.techfore.2022.121576

Chukhrova, N., & Johannssen, A. (2021). Stochastic claims reserving methods with state space representations: A review. *Risks*, *9*(11), 198. https://doi.org/10.3390/risks9110198

Farbmacher, H., Low, L., & Spindler, M. (2022). An explainable attention network for fraud detection in claims management. *Journal of Econometrics*, *228*, 244–258. https://doi.org/10.1016/j.jeconom.2020.05.021

Goodell, J. W., Kumar, S., Lim, W. M., & Pattnaik, D. (2021). Artificial intelligence and machine learning in finance: Identifying foundations, themes, and research clusters from bibliometric analysis. *Journal of Behavioral and Experimental Finance*, *32*, 100577. https://doi.org/10.1016/j.jbef.2021.100577

Herrmann, H. (2023). What's next for responsible artificial intelligence: A way forward through responsible innovation. *Heliyon*, *9*, e14379. https://doi.org/10.1016/j.heliyon.2023.e14379

Iyer, R., Maralapalle, V., Mahesh, P., & Patil, D. (2024a). 13 Generative AI and LLM: Case study in E-commerce. In *Generative AI and LLMs: Natural language processing and generative adversarial networks* (pp. 253–272). De Gruyter. https://doi.org/10.1515/9783111425078-013

Iyer, R., Maralapalle, V., Patil, D., & Irfan, M. (2024b). The future of smart contracts: Pioneering a new era of automated transactions and trust in the digital economy. In *AI-driven decentralized finance and the future of finance* (pp. 229–256). IGI Global. https://doi.org/10.4018/979-8-3693-6321-8.ch010

Khan, A., Ahmad, A., & Shireen, S. (2021). Ownership and performance of microfinance institutions: Empirical evidences from India. *Cogent Economics & Finance*, *9*(1), 1930653. https://doi.org/10.1080/23322039.2021.1930653

Kristof, T., & Virag, M. (2022). EU-27 bank failure prediction with C5.0 decision trees and deep learning neural networks. *Research in International Business and Finance*, *61*, 101644. https://doi.org/10.1016/j.ribaf.2022.101644

Li, Q., Xu, Z., Shen, X., & Zhong, J. (2021). Predicting business risks of commercial banks based on BP-GA optimized model. *Computational Economics*, *59*(4), 1423–1441. https://doi.org/10.1007/s10614-020-10088-0

Lux, T. (2015). Emergence of a core-periphery structure in a simple dynamic model of the interbank market. *Journal of Economic Dynamics and Control*, *52*, A11–A23. https://doi.org/10.1016/j.jedc.2014.09.038

Meng, S., Gao, Y., & Huang, Y. (2022). Actuarial intelligence in auto insurance: Claim frequency modeling with driving behavior features and improved boosted trees. *Insurance: Mathematics and Economics*, *106*, 115–127. https://doi.org/10.1016/j.insmatheco.2022.06.001

Manahov, V. (2015). Can high-frequency trading strategies constantly beat the market? *International Journal of Finance & Economics*, *21*(2), 167–191.

Mittal, D., & Agrawal, S. R. (2022). Determining banking service attributes from online reviews: Text mining and sentiment analysis. *International Journal of Bank Marketing*, *40*(3), 558–577. https://doi.org/10.1108/IJBM-08-2021-0380

Moutinho, L., & Smith, A. (2000). Modelling bank customer satisfaction through mediation of attitudes towards human and automated banking. *International Journal of Bank Marketing*, *18*(3), 124–134. https://doi.org/10.1108/02652320010339699

Nagurney, A. (2015). A multiproduct network economic model of cybercrime in financial services. *Science*, *7*, 70–81.

Pattnaik, D., Hassan, M. K., Dsouza, A., & Ashraf, A. (2023). Investment in gold: A bibliometric review and agenda for future research. *Research in International Business and Finance*, *64*, 101854. https://doi.org/10.1016/j.ribaf.2022.101854

Pnevmatikakis, A., Kanavos, S., Matikas, G., Kostopoulou, K., Cesario, A., & Kyriazakos, S. (2021). Risk assessment for personalized health insurance based on real-world data. *Risks*, *9*(3), 46. https://doi.org/10.3390/risks9030046

Polychroniou, P. V., & Giannikos, I. (2009). A fuzzy multicriteria decision-making methodology for selection of human resources in a Greek private bank. *Career Development International*, *14*(4), 372–387. https://doi.org/10.1108/13620430910979853

Prix, J., Loistl, O., & Huetl, M. (2007). Algorithmic trading patterns in Xetra orders. *The European Journal of Finance*, *13*(8), 717–739. https://doi.org/10.1080/13518470701705538

Singh, K., & Best, P. (2019). Anti-money laundering: Using data visualization to identify suspicious activity. *International Journal of Accounting Information Systems*, *34*, 100418. https://doi.org/10.1016/j.accinf.2019.06.001

Tepe, G., Geyikci, U. B., & Sancak, F. M. (2022). Fintech companies: A bibliometric analysis. *International Journal of Financial Studies*, *10*(1), 2. https://doi.org/10.3390/ijfs10010002

Vorobyev, I., & Krivitskaya, A. (2022). Reducing false positives in bank anti-fraud systems based on rule induction in distributed tree-based models. *Computers & Security*, *120*, 102786. https://doi.org/10.1016/j.cose.2022.102786

Chapter 15

Empowering Indigenous Communities Through Education

Meeta Siddhu and Madhuri Singh

Symbiosis International (Deemed University), India

Abstract

The present study investigates the impact of digitalization in artisan communities from the perspectives of emerging economies and was conducted in the Chitrakoot craft cluster of Uttar Pradesh, India. It was an ethnographic cum exploratory study utilizing a case study approach to focus on the One District One Product (ODOP) Scheme launched by the government for the artisanal communities in Chitrakoot, Uttar Pradesh, India. Purposive sampling was used for recruiting the cases for primary data collection. Secondary data were collected from government documents, reports, policy papers, etc. Thematic analysis of the data generated valuable insights which were systematically presented in the result section. Findings revealed the multifaceted impact of digitalization on various aspects of the scheme, including market access, product development, skill development, and socioeconomic enhancement. Furthermore, the study highlighted the opportunities as well as challenges that have aroused for the artisan community in contemporary times. The study will thus have implications for development sector professionals, policymakers, craft researchers, and designers. It will also be useful in designing effective skill development programs for the artisans as per their requirements and lead to effective utilization of institutional resources.

Keywords: Artisans; Chitrakoot; craft cluster; digitalization; emerging economies; skill development

Indigenous Empowerment through Human-Machine Interactions, 253–272

Published under exclusive licence by Emerald Publishing Limited
doi:10.1108/978-1-83608-068-820251015

1. Introduction

Uttar Pradesh, India launched the "One District One Product" (ODOP) program in 2018. The initiative aims to boost artisans' welfare, increase local products' visibility, and generate employment Ministry of Commerce and Industry, Government of India (2021). The ODOP Program is a district-specific initiative aimed at revitalizing local crafts through market assistance, skill development, and establishment of a common facility center (Nisa, 2024). The main objectives of the program are to preserve and promote indigenous products, generate local employment, check migration, build a competitive ecosystem, improve product quality, boost exports, and foster a global platform for indigenous crafts (Department of MSME and Export Promotion, 2024).

Many emerging economies have rolled out ODOP. The program has been implemented in all the states and union territories of India too. However, there have been fewer studies that have tried to explore the effectiveness of this program (Tripathi & Agrawal, 2021). The present study is an endeavor to measure the effectiveness of ODOP, and it was conducted in the Chitrakoot district, Uttar Pradesh, known for its handcrafted wooden toys. These have been declared as the ODOP product of the district, and the government has been making consistent efforts for the development of this sector. Through ethnographic and exploratory approach, the study brings forward the effects of education in general and digital and skill education in particular on the artisan community of Chitrakoot, Uttar Pradesh, India.

The findings of the study will have theoretical implications for researchers and academicians. It will also have practical implications for the policymakers, government, and private organizations working for grassroots artisans and development sector.

2. Review of Literature

2.1 Genesis of ODOP Program in India

The "One District One Product" program (hereafter called ODOP in the chapter) is on the lines of the "One Village One Product" (OVOP) movement of Japan which was introduced in the 1970s there. The main aim of OVOP was revitalization of rural community crafts in Japan for which local products with unique identity were earmarked and provided all assistance for small-scale production (Mukai & Fujikura, 2015).

The OVOP began in 1970s when some women in Oita, Japan, started selling homemade biscuits in the local market. The demand for biscuits led the women to skill up-gradation as well as improved quality and packaging owing to a robust demand. This led to a foundation of sustainable livelihoods, entrepreneurship in the women artisans, and also initiation of the ODOP program in Japan. This also aided local employment creation, building artisan readiness, and preservation of traditional skills and cultural heritage in respective villages (UNDP, 2017).

The OVOP operates on three fundamental principles: (1) utilizing the region's potential and the local community's inventiveness to achieve self-reliance. (2) Creating a local product that may be used globally to promote export revenue while reflecting the culture and sense of pride. (3) Ensuring human resource development at local level (Mukai & Fujikura, 2015; Claymone, 2011).

2.2 ODOP Program of India and Its Components

The ODOP program was first launched in India by the Uttar Pradesh government in 2018–2019 and was subsequently launched in all the Indian states and union territories (Tripathi & Agrawal, 2021). It is aimed at providing an identity and boost sales of cultural and traditional products from each district (Sharma & Kumar, 2024). The program comprises four schemes.

2.2.1 Common Facility Center (CFC) Scheme

The CFCs under the ODOP are designed to provide solutions to all problems encountered by artisans and provide solution for raw material testing, a design development and training center for developing contemporary products, raw material bank for provision of quality raw material, packaging, labeling, and bar-coding. It also has information and broadcasting center for dissemination of information related to government schemes and programs.

2.2.2 Marketing Development Assistance (MDA) Scheme

The MDA scheme assists artisans in achieving a fair price for their products through utilization of CFC and better marketing of their products. Under this scheme of ODOP, the artisans are also provided with subsidies and financial assistance to participate in various national and international fairs/exhibitions.

2.2.3 Financial Assistance Scheme (Margin Money Scheme)

Under this scheme or the margin money scheme of ODOP, the government provides approximately 25% subsidy on project cost of the craft business to new entrepreneurs for setting up a business under ODOP (Bajpai, 2020).

2.2.4 Skill Development Scheme

Under this component of ODOP program, the artisans are provided with skill development to equip them with the knowledge of contemporary tools, product ideas, color schemes, and market demands among other things (Tripathi & Agrawal, 2021).

2.3 Present Status of Chitrakoot Artisans

Cultural changes and urbanization have led to a shrinking market for Chitrakoot's wooden toys (see Fig. 15.1) and a decline in traditional toymakers; however, the artisans continue to produce these (Mishra, 2018) as it is linked to their cultural heritage and is a source of decent livelihood.

The jungles of Chitrakoot supply Dudhi Sundari and Koraiya woods (see Fig. 15.2), with which the handicrafts are made. The tapping, sawing, and drilling sounds associated with woodwork that rang through the village have diminished much (see Fig. 15.3). Making and selling handicrafts that sustained more than 200 families earlier lacks patronage now due to urbanization and the better appeal of modern toys. The district, designated as backward by the Uttar Pradesh government, has reasonable employment opportunities for the educated since there are two government universities and a few schools. However, due to a fall in demand, artisans' livelihoods have been affected (Sengupta, 2023). Under this backdrop, India has launched several economic revitalization programs at both the central and state levels. In this series, the ODOP scheme launched by the government of Uttar Pradesh, India, in 2018, is regarded as one of the most important schemes for regional economic revitalization (Gaur, 2023).

2.4 Implementation of ODOP in Chitrakoot

Chitrakoot is a remote place (Sengupta, 2023) spread across Uttar Pradesh and Madhya Pradesh, India. It has mythological significance and is a vast tribal hinterland. It is inhabited by *Mawasi*, *Kol*, *Gond*, and other *Adivasi* communities who are untouched by modern technological developments and are simple, hardworking forest dwellers (Mishra, 2018). Since ancient times, Chitrakoot has enjoyed vibrant ethnic and biological diversity and the forests of Chitrakoot are rich in medicinal

Fig. 15.1. Wooden Toys and Products From the Chitrakoot Craft Cluster. *Source:* Authors' own photographs.

Fig. 15.2. Tools and Raw Material for Making Wooden Toys in Chitrakoot. *Source:* Authors' own photographs.

Fig. 15.3. An Artisan Making Wooden Toys in His Traditional Home Based Workshop. *Source:* Authors' own photographs.

plants (Dar, 2019) and wood-yielding trees which are used by the tribal for food, fodder, fiber, medicine (http://nopr.niscpr.res.in/handle/123456789/2391NIScPR), and wooden utilitarian products. Chitrakoot is also known for its wooden handicrafts such as idols, bangle stands, *lattoo*s or spinning tops, key rings, particularly toys. Wooden toymaking with locally sourced wood and simple tools is an age-old tradition in this small town (see Fig. 15.2). Owing to urbanization and better appeal of modern toys, the artisans of Chitrakoot are increasingly finding it difficult to sustain making and selling the traditional wooden toys (see Fig. 15.1) (Mishra, 2018; Singh et al., 2015).

2.5 Prospects

Known for its mythological significance from the time of Ramayana, Chitrakoot in present times is an important pilgrimage and tourism hub (cite). The government has also developed a district export plan for Chitrakoot under the "District Export Hub Scheme" initiated by the Ministry of Commerce and Industry, Govt of India and state Government of Uttar Pradesh. A District Level Export Promotion Committee has been formed by the office of District Industries Centre, Chitrakoot under the chairmanship of District Magistrate. The state government has also been emphasizing on boosting additional, nonfarm income other than agricultural incomes through training in entrepreneurship and the formation of income-generating Self-Help Groups (SHGs) (Nisa, 2024; Tripathi & Agrawal, 2021).

The wood carving handicraft is among one of the famous Indian crafts (Yadav & Mahara, 2018). With increasing sustainability concerns and government initiatives, the Chitrakoot craft cluster has several prospects and can serve as a sustainable livelihood option to the traditional as well as first generation artisans.

2.6 Research Gap

The review of literature revealed limited research on "One District One Product (ODOP)" scheme. Those that are there present narrow scope of the scheme and lack comprehensive deliberations on the skill development component of ODOP. Also there has been a gap in studies pertaining to usage of digital tools among artisan community, specifically among the artisans of Chitrakoot wooden toy cluster. Thus, a study was undertaken to explore the effectiveness of ODOP programme in bringing changes in skills of the artisan. The study also involved documentation of lives, work and socio-economic status of the artisans of Chitrakoot wooden toy making craft cluster which would also aid in providing policy nudges for effective policy formulation.

3. Methodology

Though the government has implemented several welfare schemes including the "One District One Product" scheme for the benefit of the indigenous toymaking artisans of Chitrakoot, there is a lack of research on the various aspects of the craft and its community. Field visits also revealed lack of disaggregated data on the subject. Thus, an exploratory cum descriptive research with a mixed-methods approach was used for the present study to explore the effects of skill and digital education imparted to Chitrakoot artisans under the ODOP program. As defined by Tashakkori and Creswell (2007) and cited by Dalal et al. (2023), the mixed-method research makes use of qualitative and quantitative methodologies to collect and analyze data to make conclusions and is highly desirable in social sciences as it helps in getting a comprehensive picture of the problem as compared to a standalone quantitative and qualitative study. Both primary and secondary data were collected from multiple data sources which also helped in achieving data triangulation.

The secondary data were collected from books, policy documents, research papers, blogs, and official documents. The primary data were collected with the help of a self-constructed, interview schedule and was administered face to face after taking informed consent from the respondents. In-depth interviews were also conducted with the government officials and resource persons from local nongovernmental organizations (NGOs). A total of 34 artisans were selected by snowball and purposive sampling for the primary data collection.

The information collected from various primary and secondary sources was subjected to detailed qualitative analysis. The same was then methodically arranged into themes and linked to the literature reviewed.

Participatory observation was utilized to delve deeply into the ground situations. The researcher verified the validity of the data at the time of information collection from the respondents by monitoring their facial expressions, speech patterns, emotions, and attitudes. Besides, the researcher could also acquire untold information by studying the condition of the surroundings, available amenities in-house/workplace, environment, and lifestyle. Sociologists and anthropologists make use of participant observation in ethnographic research and studies. These types of studies involve a researcher interacting with the subject of their study, living with them and participating in the day-to-day activities of the lives of the respondents while simultaneously collecting the required information (Oruc, 2017). Such an observation gives crucial information which the respondents might have hidden during the interviews. Field notes and video recording helped in ensuring that all the recorded information was duly available for analysis. Use of multiple tools for data collection also helped in achieving data triangulation.

4. Findings and Observations

The present study was conducted to explore the changes in the lives of traditional wooden toys artisans after providing them with skill education and financial assistance through the "One District One Product" program launched by the Government of Uttar Pradesh State, India.

A careful thematic analysis of the responses yielded following information:

4.1 Background of the Respondents

Traditionally practiced by males, the craft is also pursued by women now. However, women artisans mostly undertake the assembling and packaging of the products. It is the male artisans who perform the heavier work of cutting and smoothing the wood, shaping, and coloring it (see Table 15.1). This increase in participation of women was attributed to:

- Mounting economic pressure.
- Desire for a better lifestyle.
- Availability of skill development and financial assistance programs such as ODOP by state as well as central government.

Table 15.1. Demographic and Background Information of the Artisans.

S. No.	Name	Age (in years)	Gender	Years of Association With ODOP	Did You Receive Any Digital Training Under ODOP Program
1	A1	24	Female	More than 3 years	Yes
2	A2	40	Female	Less than 1 year	No
3	A3	39	Female	Less than 1 year	Yes
4	A4	37	Female	Less than 1 year	No
5	A5	29	Female	Less than 1 year	Yes
6	A6	45	Female	Less than 1 year	Yes
7	A7	39	Male	Less than 1 year	No
8	A8	33	Female	Less than 1 year	No
9	A9	36	Female	Less than 1 year	Yes
10	A10	48	Male	1–2 years	Yes
11	A11	49	Male	1–2 years	Yes
12	A12	28	Female	More than 3 years	Yes
13	A13	60	Male	More than 3 years	Yes
14	A14	27	Male	More than 3 years	Yes
15	A15	32	Male	More than 3 years	Yes
16	A16	48	Male	More than 3 years	Yes
17	A17	59	Male	More than 3 years	Yes
18	A18	52	Male	More than 3 years	Yes
19	A19	57	Female	More than 3 years	Yes
20	A20	33	Male	More than 3 years	Yes
21	A21	30	Female	More than 3 years	Yes
22	A22	38	Male	More than 3 years	Yes
23	A23	27	Female	More than 3 years	Yes
24	A24	38	Male	More than 3 years	No
25	A25	48	Female	More than 3 years	No
26	A26	40	Male	More than 3 years	No
27	A27	50	Female	Less than 1 year	No
28	A28	46	Male	More than 3 years	No
29	A29	47	Male	More than 3 years	Yes
30	A30	28	Male	1–2 years	No
31	A31	58	Male	More than 3 years	Yes
32	A32	33	Male	1–2 years	Yes

Table 15.1. *(Continued)*

S. No.	Name	Age (in years)	Gender	Years of Association With ODOP	Did You Receive Any Digital Training Under ODOP Program
33	A33	37	Female	1–2 years	No
34	A34	38	Male	2–3 years	No
35	A35	35	Male	More than 3 years	No
36	A36	35	Male	Less than 1 year	No
37	A37	41	Male	1–2 years	Yes
38	A38	49	Male	1–2 years	No

4.2 Socioeconomic Status of the Respondents

Most of the respondents were from rural backgrounds with only 2% artisans having received a postgraduate degree and 15% artisans reporting receiving a graduate degree. None of the respondents possessed doctorate or higher professional degrees. Approximately 5% of artisans could only attend high school. Another 10% had received education till higher secondary that is 12th. Four percent artisans revealed dropping the studies during school, whereas approximately 64% artisans were illiterate. The educational attainment of males was better as compared to females. Observation and in-depth interviews further revealed the presence of sociocultural barriers which affected the education of women artisans, early marriages, and early motherhood being the most common factors. These barriers also restricted their mobility due to which they were found working from home.

The earnings of the artisans have increased as a result of various intervention programs by private and government organizations (see Figs. 15.4 and 15.5), but they are still not sufficient for a decent living. The monthly incomes of the respondents ranged between 8,000 Rs and 20,000 approximately. The earnings of females were considerably lesser than the males. It was noticeable that most of the artisans practiced the craft as a part-time or secondary activity. Primary activities of the artisans involved selling ritualistic items for religious ceremonies as Chitrakoot is a popular Hindu pilgrimage. A considerable number of artisans were also found pursuing agriculture as their primary livelihood.

4.3 Entry Into the Wooden Toymaking Handicraft

The craft has been passed down as oral tradition with no formal documentation of the process, raw material and tools. The exact details about the craft are not very clear, and most of the artisans also had no or little awareness about the exact origins of the craft. However, it is reported in the literature that an abundance of forests and availability of wood could have been the probable reasons for the

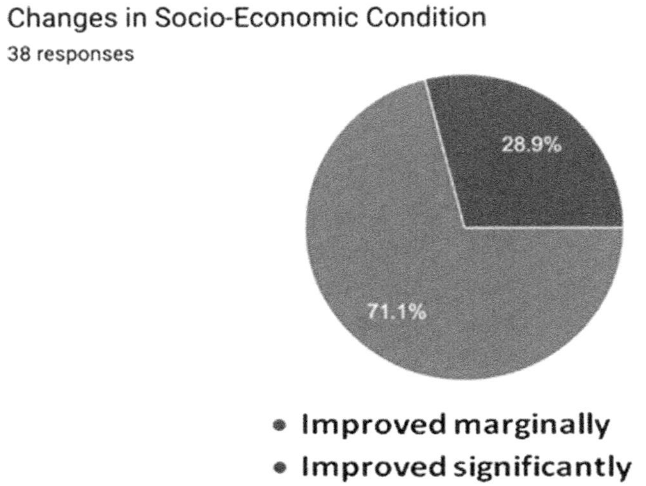

Fig. 15.4. Changes in Socioeconomic Status of the Respondents After Joining the ODOP Program.

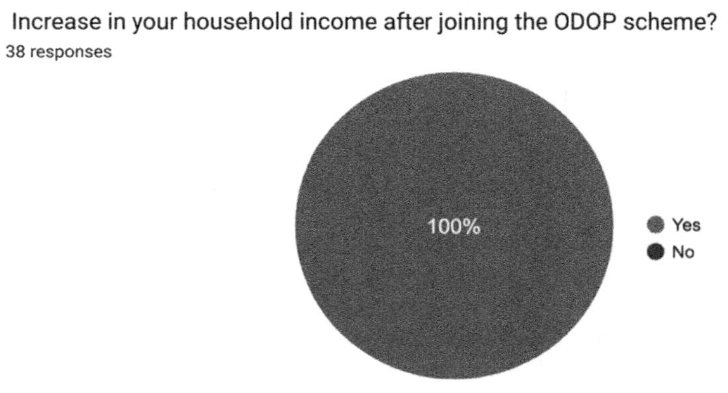

Fig. 15.5. Increase in Household Income After Joining ODOP Program.

genesis of wooden toymaking in Chitrakoot. An increasing number of first-generation artisans, mostly women, were also found to be engaged in the craft in the study which can again be attributed to the increasing number of skill development programs and it being seen as a source of secondary or additional income. In the post-COVID times, youth have also entered the craft as they had to shift back to their native places.

4.4 Advantages of the Wooden Toymaking Sector

The artisans, especially the women artisans, acknowledged the importance of the craft of toymaking in upliftment of their socioeconomic status and saw it as an example of a cottage industry. It was also seen as having a huge potential to generate local employment. It has minimal capital, electricity, and infrastructure requirements and mostly utilizes locally available raw material. Being an activity that can be pursued as a household activity, it can also help in women empowerment. The toys are handmade which also makes it a sustainable product.

4.5 Issues and Challenges Faced by the Artisans

In the past decade, the Chitrakoot craft cluster has received Government attention, but there remain some unresolved issues that need urgent attention the foremost being artisan readiness despite training and interventions. Despite obvious advantages, the craft and the community are facing challenges such as lack of market linkages, poor usage of technology, and threat from cheaper machine-made alternatives. Most of the artisans have been working in the toymaking handicraft since generations in Chitrakoot as there are abundant forests yielding quality wood. However, as revealed in the unstructured interviews, the younger generation of the toymakers is not very keen in continuing the craft. They pursue it only if they don't succeed in other jobs. The un-ergonomic working conditions, expensive raw material, and tedious work have pushed them toward services such as construction, driving, hotel industry, shop keeping, and other such jobs.

Indarpal (59 years) reported that all his three sons have migrated to cities as the working conditions and pay are better.

Despite challenges, the local craft community has managed to keep the craft alive due to factors such as cultural significance and ritualistic usage, traditional family craft, and local employment. If the challenges could be mitigated, the craft can be an excellent source of income for youth, women, and marginalized persons.

4.6 Government Interventions

The state government of Uttar Pradesh has launched numerous schemes in Chitrakoot such as the Vishwakarma scheme, free tool kit distribution, artisan card and ODOP among several others. The central government is also supporting the cluster in artisan upliftment through schemes such as design development workshop, product development, free tool kit distribution, Artisan Insurance scheme, margin money, and many other such schemes. ODOP is however the chief initiative of the government for the upliftment of the artisans of Chitrakoot. The scheme was launched in the Chitrakoot craft cluster in 2019.

4.7 Association with ODOP Government Scheme

The scheme was implemented in 2019 in Chitrakoot craft cluster, and almost 54 % of the respondents revealed being associated with the scheme ever since its implementation. 18 % artisans reported having enrolled in the program from the last 1–2 years. A considerable 28 % of artisans had enrolled fairly recently, less than a year before the interviews were conducted.

Results of the in-depth interviews with master craftsmen, state awardee artisan, and national awardee artisan revealed that the majority of artisans are still ignorant about the government programs even after almost five years of the inception of one district one product program. According to the state awardee artisan for the Chitrakoot craft cluster, the welfare programs only reached a small number of deserving artisans for a variety of reasons that included favoritism, discrimination, and ignorance. The master craftsmen also revealed that several factors restricted the craftsmen' access to these schemes, with presence of middlemen and contractors cited as the major reason. Most of the respondents opined the need for a better promotion campaign and strategies for the government welfare schemes so that it could reach the actual beneficiaries. Some artisans even suggested a door-to-door promotion campaign of the schemes.

4.8 Benefits from ODOP Scheme

A significant number of artisans responded having availed the benefit from the ODOP scheme and cited skill development and free tool kit as of immense help in earning their livelihood. 71.1% of artisans revealed that they had seen considerable improvement in their skill as well as earnings which led to positive changes in their socioeconomic status (see Fig. 15.4), and remaining 28.9% artisans cited marginal improvement in their socioeconomic status. None of the respondents chose 'no change or negative change' in their incomes and socioeconomic status (see Figs. 15.4 and 15.5). Almost all the artisans (100% artisans) indicated at achieving some increase in their monthly incomes after attending the ODOP program.

4.9 Digital and Skill Training during Skill Development Scheme Implementation of ODOP Program

The artisans enrolled in the ODOP program undergo a skill development module; however, there is no prescribed syllabus for the same. The respondents did reveal being trained for 10 days which included sessions with designers on theoretical and practical skills such as history and significance of the craft, design development, understanding market trends and product development. Resource persons from craft industry, banking, and retail were also invited to give a comprehensive picture of the contemporary trends and working style. The artisans also revealed being made familiar with digitalization and various digital tools for business (see Fig. 15.6).

However, artisans with better education and digital literacy showed more improvement as compared to illiterate artisans. Approximately 39.5% artisans revealed undergoing training for digital payments (see Fig. 15.7), use of social media for product promotion, and getting orders. Before the launch of ODOP in Chitrakoot, majority of the artisans were doing cash transactions but 21.1% respondents in the study informed of shifting to digital transactions (see Figs. 15.8 and 15.9) after attending the ODOP.

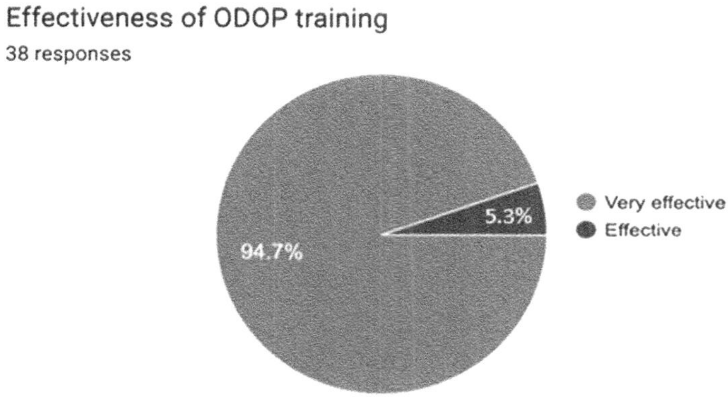

Fig. 15.6. Effectiveness of ODOP Program in Raising the Artisan Readiness.

Fig. 15.7. Scan Code of Various Payment Gateways Displayed on the Wall at a Shop Selling Wooden Toys in the Local Market Indicating the Increasing Use of Digital Payment. *Source:* Authors' own photographs.

4.10 Online Market Assistance Through ODOP Program

The artisans were given training to use digital payment gateways such as BHIM, UPI, PhonePe, Paytm (see Fig. 15.10). The artisans were also provided hand-holding with selling their products on e-commerce websites such as Amazon, gem (government-enabled marketplace) portal, eBay, and ODOP mart.

Fig. 15.8. Mode of Receiving Orders Used by the Artisans.

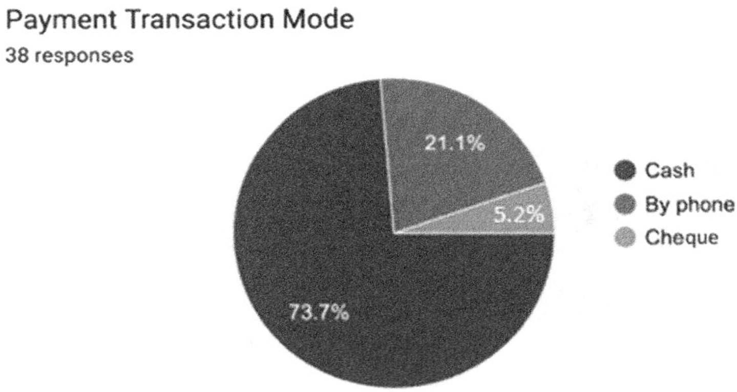

Fig. 15.9. Mode of Transaction.

4.10.1 Offline Market Assistance Through ODOP Program

Offline marketing of the products through national and international exhibitions and fairs was also offered to the artisans; however, only 22% of artisans revealed taking advantage of national fairs and exhibitions. None of the artisans could attend the international fair. Under this component, artisans reported being provided with exhibition stall, travel, and lodging expenses.

All the respondents, that is, 100% artisans reported an increase in their household incomes (see Fig. 15.1); however, the benefits were limited by education and awareness.

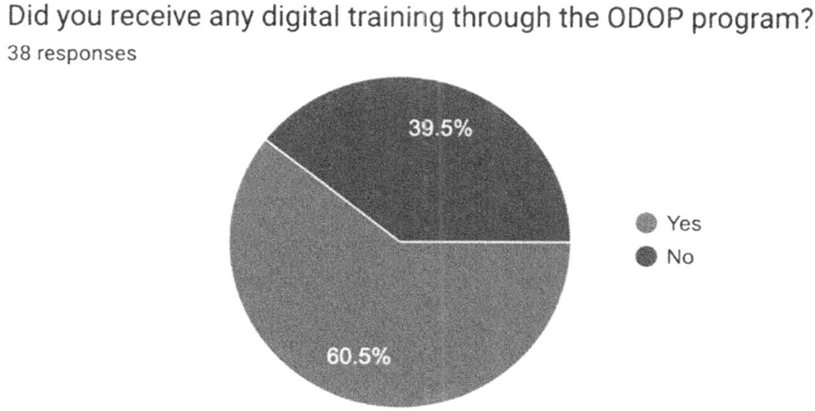

Fig. 15.10. Digital Training Through ODOP Program.

4.11 Unresolved Issues of Chitrakoot Toy Artisans

Despite numerous efforts by central and state governments, private organizations, academic institutions, the results have shown limited success. Moreover, certain issues were reported to be unresolved in the program (see Fig. 15.11). Most of the artisans reported that benefit was limited to only artisans who were educated or were familiar with the government working and officials. A few artisans also reported that the skill development workshops should be of longer duration as 10 days were not sufficient for developing expertise in crucial topics such as digital payment gateways, marketing trends, and customer demand.

Un-ergonomic working condition was another response cited by most of the artisans. Since the craft is pursued from home, the artisans worked from dingy, dim lit workshops with a lack of comfortable seating. Power fluctuation and frequent power cuts were also reported by artisans which resulted in lesser production and break in their workflow.

4.12 Challenges

Though the scheme has been effective in uplifting the socioeconomic status (see Fig. 15.4) and also an increased monthly income of the artisans (see Fig. 15.5), there have been challenges that still need to be resolved (see Fig. 15.11). These challenges were attributed to subjectivity in the craft trainers. No structured syllabus could be traced, and thus, the effectiveness of the schemes under ODOP programs was found to be highly dependent on the instructor or trainer.

The field interaction and participant observation (see Fig. 15.12) revealed loss of productivity and workflow due to frequent power cuts. High cost of wood was another deterrent in taking additional or surplus work.

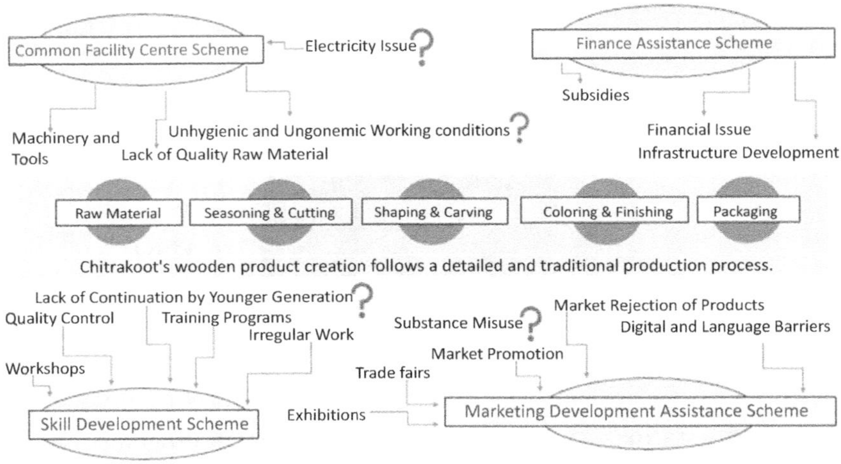

Fig. 15.11. Craft Loop of Various Issues and Challenges and Role of ODOP in Mitigating These. The Question Marks Represent Unresolved Issues. *Source:* Created by authors based on the results of the study.

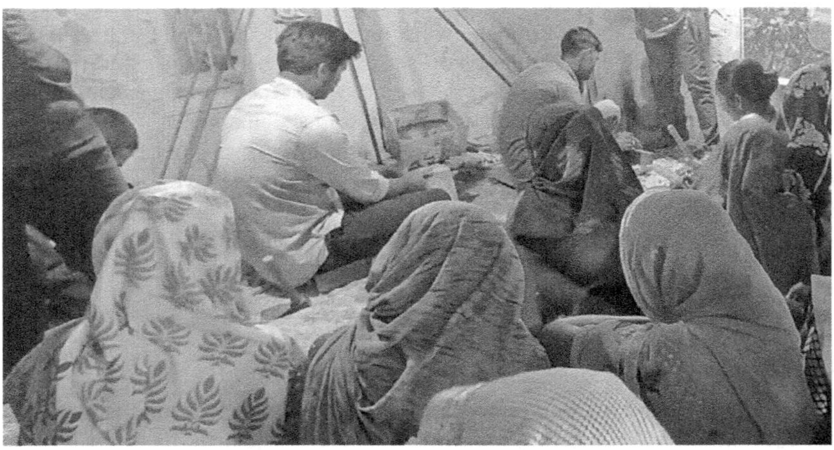

Fig. 15.12. ODOP Training Program in Progress in Chitrakoot Craft Cluster, Uttar Pradesh, India.

Lack of education and English language also paved hurdles in optimum use of digital platforms and e-commerce marketing in a considerable number of artisans.

5. Discussions and Conclusion

Traditional crafts and products serve as a backbone for economic development in emerging economies (Azharunnisa et al. 2024). The Indian handicraft sector is the second largest employment generating sector of India (Shah & Patel, 2017). In Uttar Pradesh also, the handicraft and cultural industries serve as the foundation of its industrial growth and economic development.

However, many handicraft industries are suffering a slow down due to the inability of the traditional artisan community in understanding modern markets and contemporary product demand. Most of the artisans in the study revealed that until recently, the artisans of the Chitrakoot craft cluster conducted their craft businesses traditionally, primarily through middlemen and cash transactions. It is only since past 4–5 years that the younger artisans started using digital transactions that too after *training. Majority of older artisans are still skeptical of digital payment and* transactions.

After the introduction of the ODOP program in 2019, the artisans have shown improvements in their skills (Tripathi & Agrawal, 2021). The findings revealed that the efforts taken up by the government for strengthening of the craft cluster have yielded success though it is limited than desired. It is important to measure that the skill development scheme under ODOP has yielded good results, and many artisans are now using digital tools for enhancing their craft businesses and earnings (refer to tables). This further indicated that all efforts to strengthen the local craft cluster through state government initiatives such as seed money for craft business, provision of free and subsidized stalls in fairs and exhibitions, and skill education have proven to be immensely helpful for the upliftment of the artisans. Thus, the present study has practical implications for other Indian states as well as emerging economies of the world for successful implementation of skill education which could be transformative in harnessing the true potential of the creative and cultural industries for sustainable economic development. The findings also have theoretical implications for scholars and academicians who can take the scheme as a benchmark for policy formulation and implementation.

The study also has implications for designers and artisan for improved product development and solution of productivity barriers.

The researchers thus conclude that:

- The government is making concerted efforts for preservation and revival of the indigenous wooden toymaking craft.
- The skill development undertaken in the ODOP scheme has helped the artisans in upgrading their digital skills in terms of e-commerce marketing, digital transactions, and getting orders.
- The partner institutions, such as Amazon, eBay, have proven to be of immense help to the local artisans in leveraging the digital platform in scaling the businesses as well as accessing the local and global markets.

6. Recommendations

The ODOP program has been effective in bringing a positive change in the lives and work of the indigenous toymaking community of Chitrakoot by assisting them in marketing and sales, providing financial assistance for new venture, and bridging the skill gap. However, to make the program more sustainable and effective, authors recommend:

- A fixed prescribed syllabus so that dependence of the program outcomes on the trainer's capabilities is reduced and all the outcomes of the program are achieved. This could be framed as per the various levels in the National Skill Qualification Framework.
- Industry collaboration and internship as part of ODOP for artisans to build an awareness of the latest market demands and developments such as digital tools, Artificial Intelligence, UNDP goals.
- The artisan representation should be ensured during the policymaking stage so as their specific requirements and sociocultural barriers could be targeted.
- Scientific studies on making the work of artisans more ergonomic such as designing of ergonomic workstations, digital marketing apps with easy navigation, and regional apps should be mobilized.
- One of the biggest challenges in proper utilization of these schemes is lack of education and language barriers in artisans which is a hindrance in utilizing government schemes; hence, efforts should be made to educate them through community colleges or evening classes with incentives.

Acknowledgment

The researcher would like to acknowledge the cooperation extended by the artisans during the field visits.

Conflict of Interest

The researchers declare no potential conflict of interest.

References

Azharunnisa, A., Gupta, S., & Panda, S. (2024). Craft culture revival through a sustainable approach of integrating tourism with craft promotion: Case study of Puri, Odisha. *Journal of Cultural Heritage Management and Sustainable Development*, *14*(3), 397–418. https://doi.org/10.1108/JCHMSD-07-2021-0120

Bajpai, L. (2020). Odop scheme for local craft governance: 'Chikankari' entrepreneur's perspective. *Corporate Governance Insight*, *2*(1), 129–139. https://doi.org/10.58426/cgi.v2.i1.2020.129-139

Claymone, Y. (2011). A study on one village one project (OVOP) in Japan and Thailand, as an alternative of community development in Indonesia: A perspective on Japan and Thailand. *The International Journal of East Asian Studies*.

https://www.researchgate.net/publication/255801823_A_Study_on_One_village_One_Product_Project_OVOP_in_Japan_and_Thailand_as_an_Alternative_of_Community_Development_in_Indonesia_A_Perspective_on_Japan_and_Thailand

Dalal, A., Chattopadhyay, S., & Bhattacharya, S. (2023). Towards sustainable livelihood promotion for artisans – A holistic narketing framework for improving the Indian handicraft sector. In *Perspectives in sustainable management practices* (pp. 111–119). https://doi.org/10.4324/9781032640488-15

Dar, M. H. (2019). Folklore knowledge among the tribal communities of Chitrakaoot valley in Madhya Pradesh, India. *Our Heritage*, *67*(7), 808–819. https://www.researchgate.net/publication/341203408_Folklore_knowledge_among_the_tribal_communities_of_Chitrakaoot_valley_in

Department of MSME and Export Promotion, U. P. (2024). One district one product, Uttar Pradesh. https://odopup.in. https://odopup.in/en/page/introduction

Gaur, P. (2023). *Promoting districts as export hubs in the export policy India's experiment with the one district one product programme. India habitat centre; research and information system for developing countries.* Research and Information System. https://www.ris.org.in/sites/default/files/Publication/DP-285-Pankhuri-Gaur.pdf

Ministry of Commerce and Industry, Government of India. (2021). One district one product scheme. https://pib.gov.in/PressReleasePage.aspx?PRID=1694805. Accessed on February 14, 2021.

Mishra, J. (2018). Toymakers of Chitrakoot struggle with declining demand. *Village Square*. https://www.villagesquare.in/toymakers-of-chitrakoot-struggle-with-declining-demand/

Mukai, K., & Fujikura, R. (2015). One village one product: Evaluations and lessons learnt from OVOP aid projects. *Development in Practice*, *25*(3), 389–400. https://doi.org/10.1080/09614524.2015.1020763

Nisa, Z. (2024). The role of entrepreneurship in promoting sustainable development through the ODOP programme in Uttar Pradesh. In P. K. Rai, A. Singh, B. Beingachhi, & S. K. Singh (Eds.), *Natural resources management and sustainable livelihoods in the mountainous region. Advances in Geographical and Environmental Sciences*. Springer. https://doi.org/10.1007/978-981-97-2100-9_17

Oruc, I. (2017). Participant observation as the data collection tool and its usage in the CSR researches. In *Handbook of research methods in corporate social responsibility*. https://doi.org/10.4337/9781784710927.00025

Sengupta, A. (2023). "This is our ancestral land, respect us" – Advisasis of Chitrakoot. *The Citizen - Independent Journalism | Indian News | The Citizen*. https://www.thecitizen.in/in-depth/this-is-our-ancestral-land-respect-us-advisasis-of-chitrakoot-895936

Shah, A., & Patel, R. (2017). Problems and challenges faced by handicraft artisans. *Voice of Research*, *6*(1), 57–61.

Sharma, A. K., & Kumar, P. (2024, June). A study on the role of the ODOP (one district, one product) scheme on strengthening the metal craft industry in Moradabad region, Uttar Pradesh. *Asia-Pacific Journal of Management Research and Innovation*. https://doi.org/10.1177/2319510X241239796

Singh, R., Agarwal, S., & Modi, P. (2015). Market separations for BOP producers: The case of market development for the chanderi cluster weavers in India. *International Journal of Rural Management*, *11*(2), 175–193.

Tashakkori, A., & Creswell, J. W. (2007). Editorial: The new era of mixed methods. *Journal of Mixed Methods Research*, *1*(1), 3–7.

Tripathi, A. P., & Agrawal, N. (2021). Impact of an economic revitalization programme on export and employment generation: A case study of one district one product programme in Uttar Pradesh, India. *Public Administration and Policy: An Asia-Pacific Journal*, *24*(3), 306–319. https://doi.org/10.1108/PAP-03-2021-0018

United Nations Development Programme (UNDP). (2017). One village one product: How a Japanese idea is changing lives and helping rural communities. https://reliefweb.int/report/afghanistan/one-village-one-product-how-japanese-idea-changing-lives-and-helping-rural

Yadav, R., & Mahara, T. (2018). An exploratory study to investigate the value chain of Saharanpur wooden carving handicraft cluster. *International Journal of System Assurance Engineering and Management*, *9*, 147–154. https://doi.org/10.1007/s13198-016-0492-5

http://nopr.niscpr.res.in/handle/123456789/2391

Chapter 16

Integrating Artificial Intelligence in Islamic Financial Management: Opportunities and Challenges in Maintaining Shariah Compliance

Early Ridho Kismawadi[a], *Mohammad Irfan*[b] *and Isnaini Harahap*[c]

[a]IAIN Langsa, Indonesia
[b]Christ University, India
[c]Universitas Islam Negeri Sumatera Utara, Indonesia

Abstract

The objective of this study is to examine the potential incorporation of Artificial Intelligence (AI) technology into financial management practices that are based on Islamic principles, with a particular emphasis on ensuring Shariah compliance. The literature analysis methodological approach is employed to identify the opportunities and challenges associated with the adoption of AI in the Islamic Finance (IF) environment. The results indicate that the implementation of AI can enhance the efficacy, transparency, and precision of IF operations. However, there are numerous challenges associated with Shariah compliance and ethics. The findings of this study emphasize the importance of establishing a regulatory framework that is consistent with Shariah principles in order to ensure the successful implementation of AI in Islamic financial institutions (IFI). The need for collaboration between finance experts and academics to ensure that the technology is implemented in accordance with Shariah principles, as well as the expansion of training for IF (Islamic Finance) practitioners regarding the implications of AI, are among the recommendations. Future research should examine the influence of more specific AI implementation strategies on Islamic conformance and operational efficacy in the context of IF.

Indigenous Empowerment through Human-Machine Interactions, 273–288

doi:10.1108/978-1-83608-068-820251016

Keywords: Artificial Intelligence; IF; financial management; technology integration; big data; natural language processing

1. Introduction

The contemporary financial era presents significant opportunities for the incorporation of Artificial Intelligence (AI) technology into the financial system, particularly within the framework of Islamic Finance (IF) (Abbas & Hafeez, 2021; Kismawadi et al. 2023; Ridho Kismawadi et al., 2023) Amid rapid technological advancements, research on how AI technology can be applied in financial practices that adhere to Islamic principles is becoming increasingly important. This integration is expected to improve efficiency, transparency, and adherence to Shariah principles, which are at the core of a financial system rooted in cultural and religious values. The incorporation of AI technology into authentic financial systems, namely those grounded on Islamic principles, is intriguing as it amalgamates tradition and modernity. The ethical basis and adherence to Shariah principles in IF create a distinctive context for the application of AI technology. With the help of AI, native financial institutions can improve operational efficiency, improve decision-making, and better manage risk, while still adhering to Shariah principles (Irfan et al., 2023; Sheela et al., 2023).

The market for AI is showing significant growth in the coming decade, according to multiple sources. According to Statista data, the size of the AI market (Fig. 16.1) is expected to increase from \$241.8 billion in 2023 to nearly \$740 billion by 2030, representing a compound annual growth rate of 17.3%. Meanwhile, according to Next Move Strategy Consultants, the value of AI around 208 billion US dollars in 2023 is expected to increase ninefold to around

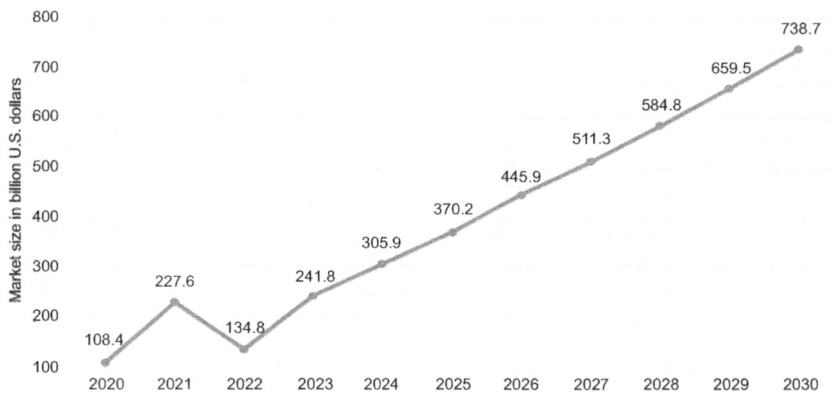

Fig. 16.1. Market Size and Revenue Comparison for AI Worldwide From 2020 to 2030 (in Billion US Dollars). *Source:* Statista, May 2024.

1.85 trillion US dollars by 2030. The AI market spans a wide range of industries, including health care, education, finance, media, and marketing. Trends such as chatbots, image-generating AI, and mobile applications are anticipated to drive growth in the coming years, as the use and implementation of AI technology are increasing worldwide. Data administration is one of the primary obstacles to AI-related infrastructure. AI companies face difficulties in acquiring specific data and organizing existing data, which is compounded by data storage restrictions imposed by international bodies such as the EU, US, and China. These restrictions present significant barriers for data-dependent AI companies.

Despite these challenges, AI has the potential to boost productivity growth, especially in the United States. Although the adoption of AI can lead to workforce eviction, effective management can facilitate the transition of workers to more productive and value-added industries. This shift is expected to contribute to a more productive economy in the long run. However, the impact of AI on productivity growth will depend on factors such as the capabilities of next-generation AI systems, the complexity of the tasks they can perform, and the number of workers displaced (Shaikh et al., 2024).

This research is of great importance in the context of IF and the development of AI technology. The IF industry's expansion necessitates the integration of AI to improve its efficacy and competitiveness. This research contributes to the sustainability and expansion of the IF sector by providing a more comprehensive comprehension of the integration of AI into Shariah-compliant financial practices. Secondly, the study identified and addressed the distinctive ethical and Shariah challenges that the use of AI in IF presents by emphasizing the necessity of a regulatory framework that is in accordance with Shariah and ethical principles. Third, the incorporation of AI can enhance transparency and accountability in Islamic financial institutions (IFI), which is crucial for the establishment of public trust and the adherence to Shariah principles. The implications of this study affirm the urgent need to develop a regulatory framework aligned with Shariah principles, thereby ensuring ethical and Shariah-compliant applications of AI in IF.

2. Opportunities and Challenges of AI in IF

The application of AI in IF offers significant opportunities that can improve efficiency, transparency, and accuracy in operations (Himeur et al., 2023; Soori et al., 2023; Yigitcanlar et al., 2021). AI can automate manual processes that are time-consuming and prone to human error. For example, in the financing approval process, AI can be used to verify documents and analyze prospective customers' financial data quickly and accurately. This not only speeds up the approval process but also reduces operational costs (Elkholy et al., 2024). AI can help ensure compliance with Shariah principles through deep and automated data analysis. AI-based systems can be used to monitor transactions and ensure that no elements of usury, gharar, or maysir are involved. For example, Islamic banks in the Middle East have used AI technology to monitor Shariah compliance in their transactions, thereby increasing transparency and customer trust.

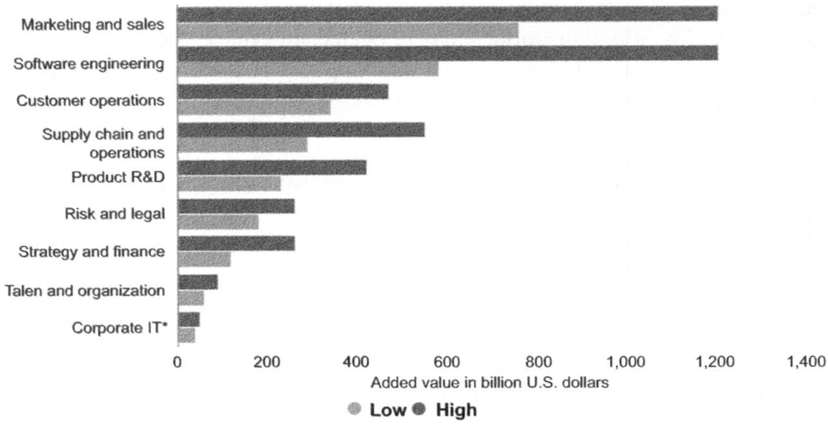

Fig. 16.2. Potential Impact of Generative AI on Productivity Worldwide in 2023, by Business Functions (in Billion US Dollars). *Sources:* McKinsey & Company, S&P Global, Oxford Economics, IHS Markit, Graphics by Statista.

Generative AI has the potential to significantly enhance the productivity of a variety of business functions, as evidenced by an analysis of 63 use cases by 2023. The marketing and sales functions have the potential to generate the most significant benefits, with an additional value of 760 to 1,200 billion US dollars. Likewise, the discipline of software engineering has a high potential for value addition. This demonstrates that the implementation of generative AI has the potential to significantly enhance productivity and efficiency in a variety of business sectors (Fig. 16.2), particularly in software engineering, sales, and marketing. In the field of supply chain management, generative AI assists in the more efficient planning and administration of the supply chain by predicting demand, optimizing inventory, and identifying potential disruptions. In general, the potential productivity impact of generative AI is substantial and encompasses a broad spectrum of business operations. Companies can enhance operational efficiency, reduce costs, and generate substantial added value across a variety of business functions by incorporating these technologies. Despite the initial investment and modifications to business processes required for the implementation of generative AI, the long-term benefits are expected to result in significant increases in productivity and competitiveness.

AI-based chatbots and virtual assistants can provide 24/7 customer service, answer questions related to Islamic financial products and services, and provide recommendations based on customer data analysis. For example, Bank Islam Malaysia has launched a chatbot capable of providing information about Islamic products and assisting customers in the financing application process. AI algorithms can also analyze big data to identify patterns and trends that may not be visible to humans. This is very useful in risk analysis and financing decision-making (Kismawadi et al., 2023). AI can assist IFI in evaluating financing risks and determining customer eligibility based on more accurate and up-to-date data. AI enables the development of financial products that are more sophisticated and in accordance with market needs. For example, blockchain-based smart contracts combined with AI can be used to create Islamic financial contracts that automatically execute agreed terms, reducing potential disputes, and increasing efficiency.

The application of AI in IF offers a great opportunity to improve efficiency, transparency, and accuracy (Canhoto & Clear, 2020; Han et al., 2023). However, challenges related to Shariah compliance, algorithm transparency, the role of Shariah experts, regulation, and data integrity must be carefully addressed. With the right collaboration between technologists and Shariah experts, as well as support from regulators, AI can be a powerful tool to drive innovation and growth within the IF sector. Successful implementation in various countries shows that these challenges can be overcome, paving the way for a more inclusive and efficient future of IF.

3. Ethics and Justice in the Development of AI

The application of AI in IF must consider the ethical aspects inherent in the Shariah-based financial system. Ethics in IF includes principles such as fairness, transparency, and adherence to Islamic law that avoids *riba* (interest), *gharar* (uncertainty), and *maysir* (gambling). To ensure that AI does not violate these ethical values, there needs to be a cautious and responsible approach.

First, AI must be developed with transparent and auditable algorithms. The transparency of this algorithm is important to ensure that the decisions taken by the AI system can be accounted for and do not harm any party. Second, AI must be programmed to adhere to Shariah principles. This means all decisions taken by AI must ensure that there are no elements of usury, gharar, or maysir in financial transactions. Third, data privacy and security are crucial ethical aspects. AI relies heavily on data, in this case the personal data of Islamic finance customers. Data misuse or data leakage can result in huge losses for customers and erode public trust. Therefore, AI systems must be equipped with sophisticated security mechanisms to protect customer data and ensure that personal information is not misused.

AI has great potential to increase fairness and transparency in Islamic financial transactions. One way AI can improve fairness is through objective, unbiased data analysis. In the traditional financial system, financing decisions are often influenced by subjective judgments that can lead to discrimination. However, with AI, decisions can be made based on objective data analysis, reducing the potential for bias and discrimination. Transparency is another important aspect that can be improved with AI (Fukuda-Parr & Gibbons, 2021; Lo Piano, 2020; Varma et al., 2023). Transparency is essential in IF to guarantee that all parties involved in a transaction comprehend all aspects of the transaction. AI has the potential to enhance transparency by generating comprehensive and lucid reports on each transaction. For instance, AI systems can produce reports that delineate the process of financing decisions, the data that are employed, and the analysis that is conducted. This not only enhances customer confidence in financial institutions but also guarantees that all transactions are conducted in accordance with Shariah principles.

4. The Future and Innovation in AI in IF

The use of AI in IF is expected to experience significant developments in the future (Di Vaio et al., 2020; Dwivedi et al., 2021). Collaboration between Islamic FinTech and AI is one of the primary trends that will emerge. AI-based FinTech platforms will provide more advanced products and services that are consistent with Shariah principles. For example, equity crowdfunding platforms and peer-to-peer lending platforms will employ AI to more precisely evaluate the viability of projects and customers. AI will facilitate the development of financial services that are more personalized and customized to the unique requirements of each customer, including investment recommendations that are customized to the customer's risk profile and preferences. AI will also be instrumental in risk management, enabling IFI to more effectively identify and manage risks by conducting more in-depth data analyses. It will become increasingly crucial to ensure compliance with Shariah regulations in real time through the use of AI-based Regulatory Technology (RegTech). A significant trend that will emerge is the integration of AI with blockchain technology and smart contracts, which will result in more secure and transparent financial solutions (Kismawadi, 2024a).

To ensure that applied AI technologies remain relevant and beneficial to IF, continuous innovation is key. One way to achieve this is through investment in ongoing research and development (R&D). Collaboration between academia, industry, and financial institutions will foster continuous innovation, and dedicated research centers for FinTech and AI in IF can be set up to explore new solutions. Training and skill development for human resources is essential to ensure that professionals in the industry have the necessary skills to effectively implement and manage AI technology. International collaboration will also accelerate AI adoption and innovation, with the exchange of knowledge and best practices between IFI in different countries. Regulations that support innovation are also crucial, with regulators working with industry to develop flexible and adaptive regulatory frameworks (Sangwan et al., 2020; Vu et al., 2023; Yadav et al., 2022).

With the right approach, AI can be a powerful tool to drive broader financial growth and inclusion in the Islamic financial system (Behera et al., 2023; Dampitakse et al., 2021; Rjoub et al., 2023). Future trends in the use of AI in IF show great potential for transformation and improved efficiency, fairness, and transparency. To prepare for this trend, IFI must invest in research and development, skills training, and international collaboration. Continuous innovation supported by adaptive regulation and a vibrant technology ecosystem will ensure that applied AI technologies remain relevant and beneficial to IF. With the right approach, AI can be a powerful tool to drive broader financial growth and inclusion in the Islamic financial system.

5. Evaluation of the Impact of AI on Public Trust and the Use of Technology in IF

The application of AI in IF could bring significant changes in increasing public trust in IFI. One way AI can increase trust is through increased transparency in operations and decision-making (Hangl et al., 2023; Mohd Yusof et al., 2023). With AI, processes such as credit scoring, financing approval, and risk management can be carried out more objectively and transparently. AI enables more in-depth and fair data analysis, reducing the likelihood of bias and human error that can harm customers. When customers realize that decisions made by financial institutions are the result of objective data analysis and without discrimination, their trust in these institutions will increase.

AI can assist in real-time monitoring and auditing of financial transactions, ensuring that all activities comply with Shariah principles. This monitoring can be done automatically and continuously, so that any discrepancies with Islamic law can be immediately detected and dealt with. This provides additional assurance to customers that their funds are managed in a manner that complies with Shariah principles, which in turn increases their trust in IFI.

AI can also play an important role in financial inclusion (How et al., 2020; Mhlanga, 2020). In numerous countries, particularly those with substantial Muslim populations, there are still numerous individuals who lack access to formal financial services. AI can assist in the identification and provision of Shariah-compliant microfinance solutions to these groups. For instance, AI algorithms could be implemented to evaluate customer eligibility with respect to alternative data, including social media activity or mobile phone utilization patterns. This would enable individuals without formal credit histories to obtain financing.

6. Risks and Mitigations in AI Application

While AI has the potential to provide numerous advantages in IF, it also introduces a variety of risks that necessitate meticulous oversight. The application of AI in IF is associated with certain specific hazards. In order to make well-informed decisions, AI heavily depends on data. The primary risk is the possibility of data leakage, data theft, or cyberattacks that could compromise the confidentiality and integrity of consumer data (Fig. 16.3). For financial institutions, inadequate data security can lead to a loss of consumer trust and financial and reputational losses. Operational risks, including algorithm errors, system failures, and AI's incapacity to manage unusual situations, may arise as a result of AI implementations. Algorithm errors can result in inaccurate judgments, including errors in financial transactions or inaccurate credit scoring.

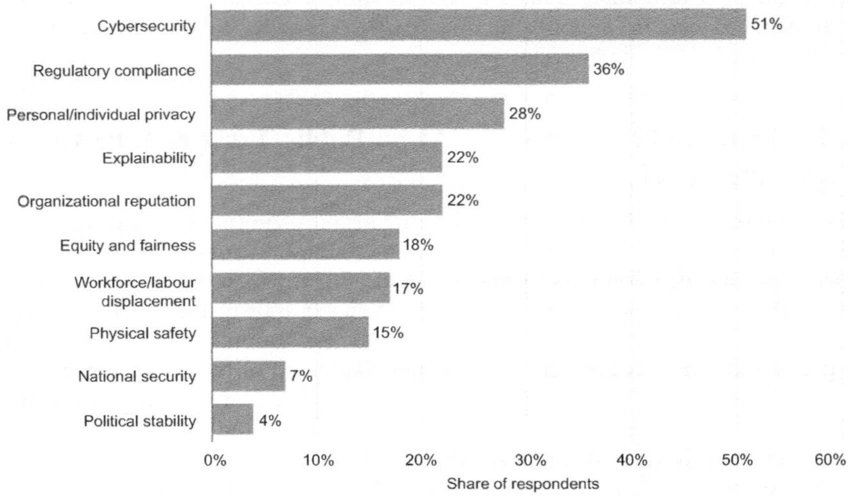

Fig. 16.3. Global Risks Being Mitigated by Organizations When Adopting AI in 2022. *Sources:* Stanford University; McKinsey & Company, Graphics by Statista.

In 2022, cybersecurity remains a key mitigation target for organizations adopting AI in their business. More than half of businesses surveyed stated that they are taking steps to reduce cybersecurity risks. This is likely due to the recent nature of AI along with a desire to keep corporate interests secret, which has led businesses to pay close attention to the risks of adopting new programs. Political stability and national security are the least of concern for businesses, with less than 10% of businesses seeking to address those issues when adopting AI.

By 2023, most organizations consider inaccuracies in the use of generative AI to be a major risk when using this new form of AI. Nevertheless, most organizations are taking steps to mitigate risks from cybersecurity (Fig. 16.4), which is only considered the second riskiest generative use of AI. Most companies do not seek to address political, environmental, or national security stability risks by 2023, likely considering these areas outside the scope of individual businesses.

To mitigate the risks associated with the application of AI in IF, a number of mitigation strategies can be implemented. First, the development of strong security protocols is crucial, with the adoption of sophisticated data encryption techniques to protect customer information and the use of multi-factor authentication (MFA) to strengthen system access. The use of AI in monitoring and conducting real-time audits of system activities is important to detect and mitigate cyber threats quickly. Regular security training for employees is also needed to raise awareness of cybersecurity practices.

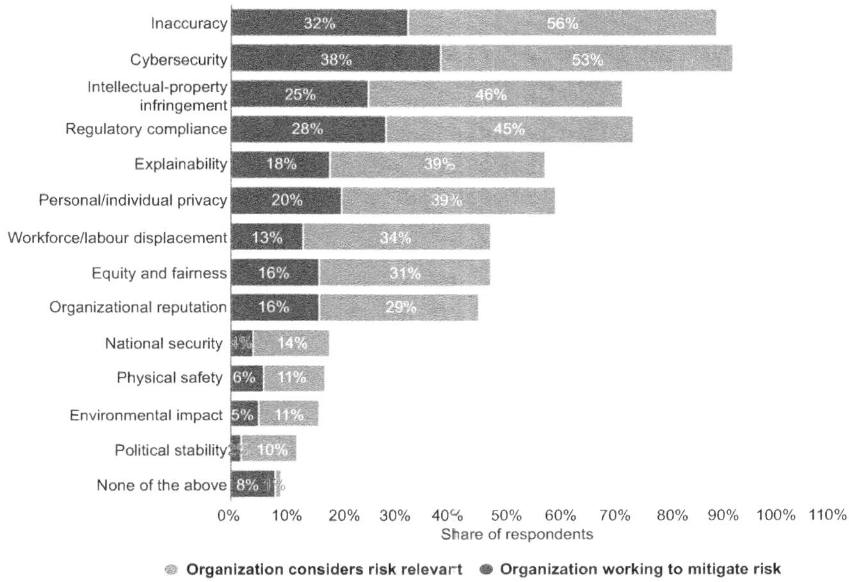

Fig. 16.4. Generative AI–Related Risks that Organizations Worldwide Consider Relevant and Are Working to Mitigate in 2023. *Source:* McKinsey & Company, Graphics by Statista.

In order to mitigate the risks associated with the application of AI in IF, it is crucial to implement effective supervision and regulation. Initially, it is imperative to establish unambiguous regulations and guidelines regarding the application of AI in IF. These regulations should encompass Shariah security, operational, and compliance standards that financial institutions are required to adhere to. Financial institutions can guarantee that AI implementation is conducted in compliance with Islamic principles and with regard to pertinent factors by establishing a defined framework. Secondly, it is imperative to establish a Supervisory Agency that is responsible for the regulation and oversight of the use of AI in IF. This supervisory body is responsible for minimizing the risks associated with the implementation of AI and assuring compliance with established regulations and standards. Financial institutions can be continuously monitored and provided with the requisite guidance to rectify inappropriate practices when a robust supervisory body is established. It is crucial to recognize the significance of transparency and accountability in the application of AI.

7. AI Ecosystem in IF

The AI ecosystem in IF is characterized by the presence of numerous key players who are instrumental in the development and application of this technology. It is crucial for these players to collaborate and work in tandem to ensure the successful implementation of the technology. Initially, regulators are responsible for the establishment of frameworks and regulations that govern the use of AI in IF. They are accountable for the establishment of Shariah security, operational, and compliance standards that must be followed by financial institutions and technology providers. Regulators can contribute to the reduction of risks and the enhancement of confidence in the application of AI by establishing explicit guidelines.

The AI ecosystem in Vietnam encountered a variety of challenges in 2018, with the absence of experts serving as the most significant impediment (Fig. 16.5). In Vietnam, the primary obstacle to the industry's growth was identified by as many as 59% of AI companies as a paucity of talent and expertise in the field of AI, according to a survey conducted that year. This challenge is not exclusive to Vietnam. One-third of organizations in Southeast Asia attribute their employment decisions to advancements in AI technology. The absence of experts is a significant impediment for a variety of reasons: The existing curriculum frequently fails to align with industry requirements, and the number of AI graduates remains restricted. Consequently, the increasing market demand has been unable to be met by existing education and training programs. The global demand for AI talent is high, and as a result, a significant number of experts are opting to work abroad in order to secure more favorable opportunities and compensation.

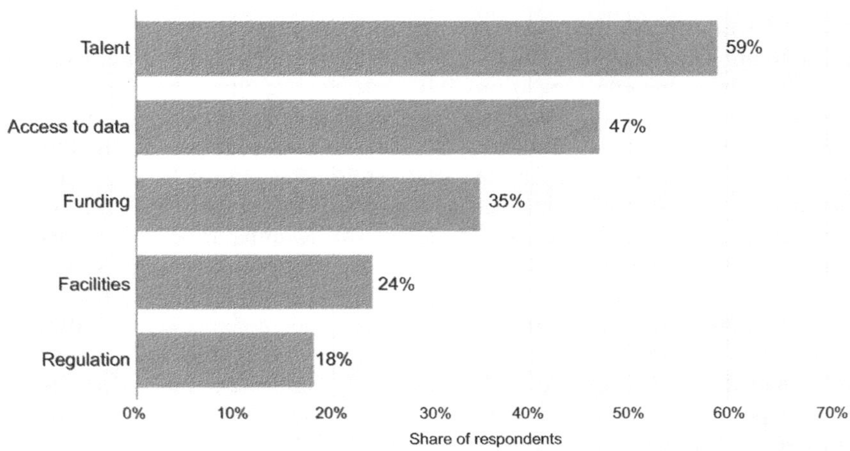

Fig. 16.5. Main Constraints for the AI Ecosystem in Vietnam 2018. *Sources:* Rubik AI, G&H Ventures, Graphics by Statista.

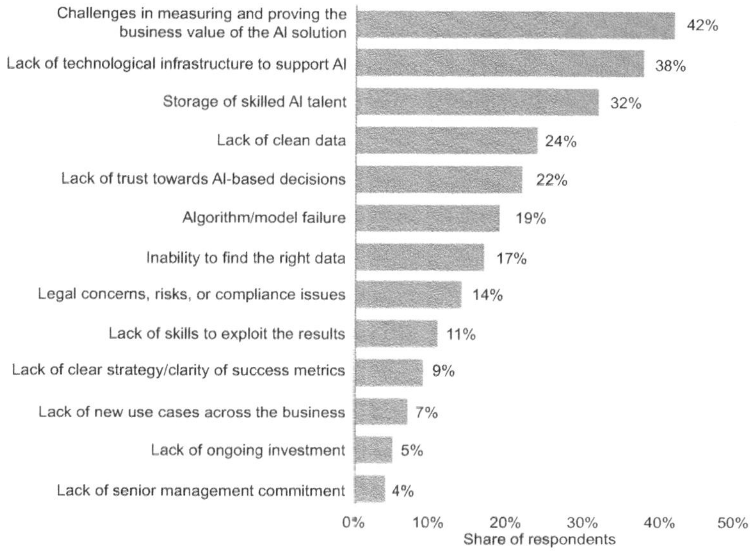

Fig. 16.6. Leading Challenges in Implementing AI Initiatives in Businesses in 2023. *Sources:* Rackspace; Dell; VMware; Coleman Parkes Research, Graphics by Statista.

The primary obstacle to the successful implementation of AI initiatives in the business sector in 2023 is the proof of their utility and benefits (Fig. 16.6). This is the most significant challenge that companies face when attempting to implement AI programs, as these technologies are relatively new, rapidly evolving, and frequently lack established metrics or require additional time to develop. There is a lack of clear metrics to assess the success of AI initiatives in many companies, and existing metrics are often not standardized and require time to develop in order to be able to measure results objectively. Many companies find it challenging to comprehend how AI can concretely contribute to their business objectives, as AI is a complex technology that frequently necessitates a profound comprehension in order to be successfully implemented. The adoption of AI technology in IF can also be accelerated by the synergies between various stakeholders. Key actors in the ecosystem can develop joint strategies to promote the widespread and sustainable use of these technologies by collaborating to identify challenges and opportunities in the application of AI.

8. Application of Big Data in Islamic Financial Risk Analysis

The application of big data in Islamic financial risk analysis offers a range of significant advantages (Jung et al., 2022; Kernchen, 2020; Pandey & Shukla, 2022). Big data is becoming into a useful tool in the digital age that helps

Islamic financial organizations perform more precise and efficient risk analysis. Islamic banks can create more comprehensive customer profiles by gathering information from a variety of sources, including financial transactions, purchasing patterns, payment histories, and social media activities (Kismawadi, 2024b, 2024c; Kismawadi et al., 2024). This information can then be gathered, stored, and analyzed. With the use of machine learning algorithms and this thorough profile, it is possible to assess customer performance in greater detail and create risk scoring models that are more accurate. Naturally, these models must adhere to Shariah's values of justice and openness.

Big data also enables the identification of market trends through sentiment analysis from social media, news, and other online platforms (Kumar et al., 2020; Menzies et al., 2024; Oo, 2022). IFI can use natural language processing (NLP) to glean insights about consumer preferences and market trends from unstructured material. Analysis of both historical and current data, such as changes in the price of halal assets, improves forecasts of market trends. Big data is crucial to the management of financing risk since it helps detect fraud by continuously observing transaction patterns and identifying irregularities. By comprehending the relationship between different assets and economic sectors, big data analysis additionally aids in portfolio diversification. It also enhances credit risk management by providing more profound insights into the repayment capacities of customers.

9. Market Sentiment Analysis

AI has opened up new opportunities in market sentiment analysis related to Islamic financial products, particularly through the use of NLP technology (Hassan et al., 2022; Saini & Sharma, 2022; Srivastava et al., 2023; Ullah et al., 2024). NLP enables automated analysis of text from a variety of unstructured data sources such as public opinion, news, and social media. In the context of IF, this sentiment analysis can provide valuable insights into people's perceptions and attitudes toward financial products that conform to Shariah principles.

AI can, first and foremost, gather and analyze data from a variety of internet platforms. This includes social media posts and comments, product reviews on websites, and pertinent news stories. AI is able to recognize terms, expressions, and situations that convey a favorable, negative, or neutral opinion of Islamic financial products by using NLP algorithms. AI, for instance, is capable of analyzing thousands of tweets to assess the likelihood of a good or negative general opinion regarding the introduction of a new Islamic investment product. Moreover, marketing plans can be informed by the sentiment analysis data. IFI can increase their marketing campaigns for a product if public opinion is favorable and they emphasize the features that the public finds most appealing.

10. Conclusion

The research findings indicate that incorporating AI into Islamic financial management presents substantial prospects for enhancing efficiency, transparency, and precision in financial operations that adhere to Shariah rules. AI has the capability to expedite the funding approval process, guarantee adherence to Shariah principles by conducting thorough data analysis, and enhance accountability by continuously monitoring transactions in real time. Nevertheless, there are obstacles associated with adhering to Shariah principles and ethical standards, including the need for algorithmic transparency and ensuring the integrity of data. This study highlights the need to establish a regulatory framework that aligns with Shariah principles in order to ensure the ethical and Shariah-compliant use of AI in IFI.

References

Abbas, K., & Hafeez, M. (2021). Will artificial intelligence rejuvenate Islamic finance? A version of world academia. *Hitit Theology Journal*, *20*(3), 311–324. https://doi.org/10.14395/hid.931401

Behera, R. K., Bala, P. K., & Rana, N. P. (2023). Creation of sustainable growth with explainable artificial intelligence: An empirical insight from consumer packaged goods retailers. *Journal of Cleaner Production*, *399*. https://doi.org/10.1016/j.jclepro.2023.136605

Canhoto, A. I., & Clear, F. (2020). Artificial intelligence and machine learning as business tools: A framework for diagnosing value destruction potential. *Business Horizons*, *63*(2), 183–193. https://doi.org/10.1016/j.bushor.2019.11.003

Dampitakse, K., Kungvantip, V., Jermsittiparsert, K., & Chienwattanasook, K. (2021). The impact of economic growth, financial development, financial performance and capital growth on the adoption of artificial intelligence in the Asean countries. *Journal of Management Information and Decision Sciences*, *24*(4), 1–14. https://www.scopus.com/inward/record.uri?eid=2-s2.0-85112838680&partnerID=40&md5=1c8999e82dc88622dc68b7a57dd5e2de

Di Vaio, A., Palladino, R., Hassan, R., & Escobar, O. (2020). Artificial intelligence and business models in the sustainable development goals perspective: A systematic literature review. *Journal of Business Research*, *121*, 283–314. https://doi.org/10.1016/j.jbusres.2020.08.019

Dwivedi, Y. K., Hughes, L., Ismagilova, E., Aarts, G., Coombs, C., Crick, T., Duan, Y., Dwivedi, R., Edwards, J., Eirug, A., Galanos, V., Ilavarasan, P. V., Janssen, M., Jones, P., Kar, A. K., Kizgin, H., Kronemann, B., Lal, B., Lucini, B., ..., & Williams, M. D. (2021). Artificial intelligence (AI): Multidisciplinary perspectives on emerging challenges, opportunities, and agenda for research, practice and policy. *International Journal of Information Management*, *57*. https://doi.org/10.1016/j.ijinfomgt.2019.08.002

Elkholy, M. H., Senjyu, T., Elymany, M., Gamil, M. M., Talaat, M., Masrur, H., Ueda, S., & Lotfy, M. E. (2024). Optimal resilient operation and sustainable power management within an autonomous residential microgrid using African vultures optimization algorithm. *Renewable Energy*, *224*. https://doi.org/10.1016/j.renene.2024.120247

Fukuda-Parr, S., & Gibbons, E. (2021). Emerging consensus on 'Ethical AI': Human rights critique of stakeholder guidelines. *Global Policy*, *12*(S6), 32–44. https://doi.org/10.1111/1758-5899.12965

Han, H., Shiwakoti, R. K., Jarvis, R., Mordi, C., & Botchie, D. (2023). Accounting and auditing with blockchain technology and artificial intelligence: A literature review. *International Journal of Accounting Information Systems*, *48*. https://doi.org/10.1016/j.accinf.2022.100598

Hangl, J., Krause, S., & Behrens, V. J. (2023). Drivers, barriers and social considerations for AI adoption in SCM. *Technology in Society*, *74*. https://doi.org/10.1016/j.techsoc.2023.102299

Hassan, M. K., Hudaefi, F. A., & Caraka, R. E. (2022). Mining netizen's opinion on cryptocurrency: Sentiment analysis of Twitter data. *Studies in Economics and Finance*, *39*(3), 365–385. https://doi.org/10.1108/SEF-06-2021-0237

Himeur, Y., Elnour, M., Fadli, F., Meskin, N., Petri, I., Rezgui, Y., Bensaali, F., & Amira, A. (2023). AI-big data analytics for building automation and management systems: A survey, actual challenges and future perspectives. *Artificial Intelligence Review*, *56*(6), 4929–5021. https://doi.org/10.1007/s10462-022-10286-2

How, M.-L., Cheah, S.-M., Khor, A. C., & Chan, Y. J. (2020). Artificial intelligence-enhanced predictive insights for advancing financial inclusion: A human-centric ai-thinking approach. *Big Data and Cognitive Computing*, *4*(2), 1–21. https://doi.org/10.3390/bdcc4020008

Irfan, M., Kadry, S., Sharif, M., & Khan, H. U. (2023). Fintech applications in Islamic finance: AI, machine learning, and blockchain techniques. In *Fintech applications in Islamic finance: AI, machine learning, and blockchain techniques*. https://doi.org/10.4018/979-8-3693-1038-0

Jung, K., Kim, D., & Yu, S. (2022). Next generation models for portfolio risk management: An approach using financial big data. *Journal of Risk and Insurance*, *89*(3), 765–787. https://doi.org/10.1111/jori.12374

Kernchen, R. (2020). Risk forecasting in the light of big data. *Journal of Risk Analysis and Crisis Response*, *10*(4), 160–167. https://doi.org/10.2991/jracr.k.201230.001

Kismawadi, E. R. (2024a). Blockchain technology and Islamic finance: Empowering small businesses for financial sustainability. In *Technopreneurship in small businesses for sustainability* (pp. 50–77). https://doi.org/10.4018/979-8-3693-3530-7.ch004

Kismawadi, E. R. (2024b). Integrating Islamic finance for sustainable development in Industry 5.0. In *Sustainable development in Industry and Society 5.0: Governance, management, and financial implications* (pp. 48–66). https://doi.org/10.4018/979-8-3693-7322-4.ch003

Kismawadi, E. R. (2024c). Sustainable Islamic financial inclusion: The ethical challenges of generative AI in product and service development. In *Exploring the ethical implications of generative AI* (pp. 237–258). https://doi.org/10.4018/979-8-3693-1565-1.ch013

Kismawadi, E. R., Irfan, M., Al Muddatstsir, U. D., & Abdulkarim, F. M. (2023). Fintech innovations: Risk mitigation strategies in Islamic finance. In *Fintech applications in Islamic finance: AI, machine learning, and blockchain techniques* (pp. 35–58). https://doi.org/10.4018/979-8-3693-1038-0.ch003

Kismawadi, E. R., Irfan, M., Haron, R., & Masitah, M. (2024). Bridging finance and sustainability: Islamic financial institutions' contributions to the transition to sustainable energy. In M. Irfan, K. Hussainey, S. A. C. Bukhari, & Y. Nam (Eds.), *Issues of sustainability in AI and new-age thematic investing* (pp. 138–161). IGI Global. https://doi.org/10.4018/979-8-3693-3282-5.ch009

Kumar, A., Shankar, R., & Aljohani, N. R. (2020). A big data driven framework for demand-driven forecasting with effects of marketing-mix variables. *Industrial Marketing Management*, *90*, 493–507. https://doi.org/10.1016/j.indmarman.2019.05.003

Lo Piano, S. (2020). Ethical principles in machine learning and artificial intelligence: Cases from the field and possible ways forward. *Humanities and Social Sciences Communications*, *7*(1). https://doi.org/10.1057/s41599-020-0501-9

Menzies, J., Sabert, B., Hassan, R., & Mensah, P. K. (2024). Artificial intelligence for international business: Its use, challenges and suggestions for future research and practice. *Thunderbird International Business Review*, *66*(2), 185–200. https://doi.org/10.1002/tie.22370

Mhlanga, D. (2020). Industry 4.0 in finance: The impact of artificial intelligence (AI) on digital financial inclusion. *International Journal of Financial Studies*, *8*(3), 1–14. https://doi.org/10.3390/ijfs8030045

Mohd Yusof, R., Zainuddin, Z., Musaddad, H. A. B., Harun, S. L., & Abdul Rahim, M. A. A. (2023). Democratizing Islamic home financing and reimagining fractional homeownership model: A conceptual framework via big data and blockchain technology. *Journal of Islamic Accounting and Business Research*. https://doi.org/10.1108/JIABR-02-2022-0033

Oo, M. C. M. (2022). Real-time predictive big data analytics system: Forecasting stock trend using technical indicators. *International Journal of Business Intelligence and Data Mining*, *21*(1), 1–22. https://doi.org/10.1504/IJBIDM.2022.123852

Pandey, K. K., & Shukla, D. (2022). Stratified linear systematic sampling based clustering approach for detection of financial risk group by mining of big data. *International Journal of System Assurance Engineering and Management*, *13*(3), 1239–1253. https://doi.org/10.1007/s13198-021-01424-0

Ridho Kismawadi, E., Irfan, M., & Shah, S. M. A. R. (2023). Revolutionizing islamic finance: Artificial intelligence's role in the future of industry. In *The impact of AI innovation on financial sectors in the era of Industry 5.0* (pp. 184–207). https://doi.org/10.4018/979-8-3693-0082-4.ch011

Rjoub, H., Adebayo, T. S., & Kirikkaleli, D. (2023). Blockchain technology-based FinTech banking sector involvement using adaptive neuro-fuzzy-based K-nearest neighbors algorithm. *Financial Innovation*, *9*(1). https://doi.org/10.1186/s40854-023-00469-3

Saini, A., & Sharma, A. (2022). Predicting the unpredictable: An application of machine learning algorithms in Indian stock market. *Annals of Data Science*, *9*(4), 791–799. https://doi.org/10.1007/s40745-019-00230-7

Sangwan, V., Prakash, P., & Singh, S. (2020). Financial technology: A review of extant literature. *Studies in Economics and Finance*, *37*(1), 71–88. https://doi.org/10.1108/SEF-07-2019-0270

Shaikh, Z., Irfan, M., Sarea, A., & Panigrahi, R. R. (2024). The emergence of Islamic Fintech and Bahrain: Prospect for global financial sectors. *Studies in Systems, Decision and Control*, *503*, 669–683. https //doi.org/10.1007/978-3-031-43490-7_52

Sheela, P., Kusuma, K., Panigrahi, R. R., & Shaikh, Z. H. (2023). Fintech and Islamic banking: A systematic view and future research agenda. In *Fintech applications in Islamic finance: AI, machine learning, and blockchain techniques* (pp. 1–23). https://doi.org/10.4018/979-8-3693-1038-0.ch001

Soori, M., Arezoo, B., & Dastres, R. (2023). Artificial intelligence, machine learning and deep learning in advanced robotics, a review. *Cognitive Robotics*, *3*, 54–70. https://doi.org/10.1016/j.cogr.2023.04.001

Srivastava, S., Pant, M., & Gupta, V. (2023). Analysis and prediction of Indian stock market: A machine-learning approach. *International Journal of System Assurance Engineering and Management*, *14*(4), 1567–1585. https://doi.org/10.1007/s13198-023-01934-z

Ullah, R., Ismail, H. B., Islam Khan, M. T., & Zeb, A. (2024). Nexus between Chat GPT usage dimensions and investment decisions making in Pakistan: Moderating role of financial literacy. *Technology in Society*, *76*. https://doi.org/10.1016/j.techsoc.2024.102454

Varma, A., Dawkins, C., & Chaudhuri, K. (2023). Artificial intelligence and people management: A critical assessment through the ethical lens. *Human Resource Management Review*, *33*(1). https://doi.org/10.1016/j.hrmr.2022.100923

Vu, N., Ghadge, A., & Bourlakis, M. (2023). Blockchain adoption in food supply chains: A review and implementation framework. *Production Planning and Control*, *34*(6), 506–523. https://doi.org/10.1080/09537287.2021.1939902

Yadav, S. P., Agrawal, K. K., Bhati, B. S., Al-Turjman, F., & Mostarda, L. (2022). Blockchain-based cryptocurrency regulation: An overview. *Computational Economics*, *59*(4), 1659–1675. https://doi.org/10.1007/s10614-020-10050-0

Yigitcanlar, T., Mehmood, R., & Corchado, J. M. (2021). Green artificial intelligence: Towards an efficient, sustainable and equitable technology for smart cities and futures. *Sustainability (Switzerland)*, *13*(16). https://doi.org/10.3390/su13168952

Chapter 17

Bridging Cultures: A Framework for Inclusive Indigenous Technologies

Abhijit Chandratreya

SCES's Indira Institute of Management, India

Abstract

The global environmental concern has led Modern Science to seek refuge in Indigenous technologies. Researchers across the world are implementing Indigenous science into Modern Science to achieve sustainable solutions. The proponents of Modern science have realized the importance of incorporating indigenous perspectives to create more inclusive and culturally sensitive technologies.

Frameworks developed by researchers often highlight the potential benefits of incorporating indigenous knowledge into technological development but fall short of providing actionable frameworks that guide this integration. There is a need for much more robust technology.

This chapter aims to address this gap by including holistic approaches to Indigenous technology that facilitate the integration of Indigenous knowledge and practices into technology development processes. This framework will not only serve as a guide for designers, developers, and policymakers but also contribute to the preservation and promotion of indigenous cultures and traditions in the digital age.

There is an urgent need for research that focuses on developing frameworks for inclusive indigenous technologies and preserving the indigenous culture. The chapter aims to fill this gap by providing a comprehensive framework that promotes the integration of Indigenous knowledge, thereby fostering cultural preservation, inclusivity, and innovation in technological development.

Keywords: Indigenous knowledge; technology integration; cultural inclusivity; sustainable innovation; frameworks; technological sovereignty

Indigenous Empowerment through Human-Machine Interactions, 289–312

Published under exclusive licence by Emerald Publishing Limited
doi:10.1108/978-1-83608-068-820251017

1. Introduction

In this chapter, a brief acquaintance with Indigenous technologies and their evolution with time has been discussed. Their importance in present-day technology has also been discussed as well (Andersen & O'Brien, 2020).

The term indigenous knowledge can be defined as the local knowledge held by indigenous peoples or local knowledge unique to a particular culture or society. Indigenous Science, Indigenous technology, and Indigenous Knowledge Systems (IKS) are synonyms (Banerjee & Linstead, 2021).

The Indigenous cultures developed technologies with the minimum available resources, which helped them to survive daily life in a better and more convenient way. They used their indigenous knowledge to produce food, clothing, and shelter (Battiste, 2005). They treasured them and passed them to their future generations through communication. The indigeneity mostly depended on the geographical distribution of the area and its biodiversity. The indigenous technologies practiced by them worked in parallel with the Ecosystem without any considerable damage to the Natural Resources (Brown & Strega, 2015). These indigenous technologies came to light with the colonization of foreign cultures which started a few centuries back. The world came to know about different technologies developed and practiced across the world (Cameron, 2012).

In the current scenario, much indigenous knowledge is endangered, and its authenticity is much debated (Cheryan et al., 2009). The IKS of many colonial countries were subjected to neglect and malign due to their dominancy. Modern Science considers Indigenous science very primitive, simple, and static. Indigenous science can't nurture the needs of the vast global population. Indigenous Science lacks the literature in written form (Coombes, 2018). The whole of the Indigenous Science is transferred to the future generations in the form of oral communication alone. The Indigenous communities share their experiences in the form of stories, rituals, and ceremonies. They use signs and symbols as inscriptions to pass their message to the forerunners (Davis, 2017). They followed unique methods of farming, protected the forests, built their dwellings naturally, and protected themselves from natural calamities. Indigenous communities are experts in the conservation of resources.

The increase in global threat to the environment has made scientists study Indigenous knowledge elaborately, and they have provided a detailed overview and a general explanation of Indigenous knowledge, with a particular emphasis on Environmental conservation (Dudgeon et al., 2014). The revival of indigenous science is much needed by the global community to bring back the lost glory of Mother Earth (Garrison & Monroe, 2016).

The rapid globalization in the last century helped the global mankind to win the battle to meet the basic needs of man but failed to create sustainability, putting the future generations' survival in danger (Walter & Andersen, 2013).

Native technologies can provide effective solutions to the environmental, social, and economic issues facing humanity today (Tsing, 2011). The integration of Indigenous science into modern science can provide sustainable development (SD) in the fields of agriculture, forestry, and biodiversity preservation. Sustainable Development Goals (SDGs) are set in the areas where this

Indigenous science is utilized (TallBear, 2013). In recent times, modern non-indigenous communities have started using Indigenous technologies and have been adopting and incorporating useful technologies developed outside of their nations (Smith, 2012). The increasing utilization of technologies for meaningful community-driven goals has paved the way for innovations in technology.

The introduction of Indigenous science in modern scientific education is indeed a great leap that will help mankind to attain sustenance. The concept is to make the curriculum greener. United Nations Environment Program (UNEP) is a developmental project that works for a greener economy. They aim for economic policies that reduce environmental pollution. By greening the curriculum, students also get the opportunity to share their knowledge, skills, and ideas for the green economy.

Basic engineering can be effectively used to "green" engineering education. The information and knowledge in science, technology, engineering, and math (STEM) fields should be perceived in the cultural and philosophical aspects as well rather than restricting them as purely technical and practical knowledge. Modern science is the reflection of Western philosophy and does not accept Indigenous ideologies. Indigenous knowledge is the outcome of the experiences of community members, which includes responsibilities generated by complex systems of relationships, age and gender, and other social contexts. Indigenous technology creates effective learning environments like engaging, valuing, and management.

Several Indigenous communities are adapting digital technologies and other forms of technology to form a more meaningful community (Rifkin, 2000). They are utilizing modern technology to reclaim their culture and protect the remains of culture and to improve their public relations, social conditions, innovations in energy production, and cultural and linguistic revival.

There is a growing realization around the world about really having to also integrate IKS into our technological futures. The researcher aims to fill a gap in the literature by proposing an approach for designing inclusive indigenous technologies with due emphasis on respecting and integrating underlying cultural values that underpin the practices of local communities.

1.1 Research Objectives

This research aims to find a universal framework for embedding the use of different native traditional knowledge systems and cultural approaches in the design (Hudson & Henwood, 2009), development, and implementation of technologies (McGregor, 2018). The framework aims to:

- What are the basic principles and practices that govern the development of cultures in technology?
- Provide practical guidelines for technologists, designers, and policymakers on how to incorporate indigenous perspectives in technology design.

- Assessing Indigenous technology initiatives to determine what works and what doesn't.
- Increase the inclusion of indigenous communities in the technological development process so that technologies are designed and implemented to be applicable and useful for them.

1.2 Research Questions

The study focuses on asking the following research questions to meet its objective:

- Why is the history of Indigenous communities being ignored in tech design?
- What changes could be made to technology design processes to account for indigenous landmarks?
- What challenges and opportunities can arise while co-designing technologies with indigenous communities?
- How can the proposed framework operationalize across different technology contexts to make sure that it remains sensitive to cultural diversity and inclusiveness?
- What is the evidence that inclusive technology development leads to more successful Indigenous millions using these technologies?

This research aims to inform the creation of technologies that empower and encourage indigenous communities, and in doing so contribute to a more culturally aware framework, which helps direct an inclusive digital future.

2. Literature Review

At least 370 million people globally are identified as Indigenous populations. Across cultures, Indigenous peoples often share deep social, cultural, and spiritual connectivity to their "lands" with hunting, fishing, herding, and farming as their main means of livelihood. They follow unique and effective soil and water conservation methods which were passed on to the forerunners through oral communication. Their ideology is that human beings are interconnected with nature (plants, animals, water, soil, mountains) and personify it. This stewardship has blessed them to be highly resilient, and their learning tendency is based on their experiences. They hold high ethical values giving importance to spirituality, mutualism, harmony, self-restriction, and determination. Many communities lack the script and communicate through oral communication, pictures, and symbols. Indigenous living is a holistic way of living. They are keen observers of natural changes and make accurate climate predictions.

The impact of climate change has hit the Indigenous communities in a harder way than the rest of the world. They must face hardships in the form of famines, floods, forest fires, epidemics, pest infestations, and critical climatic conditions. As a result, many people of the communities have migrated to the nearby towns and cities for their livelihood much at the loss of Indigenous knowledge (Johnston, 2020).

The importance of recognizing indigenous and local knowledge (ILK) systems emerged in 1987 from the World Commission on Environment and Development (WCED) report on SD. According to WCED, the disappearance of the ILK system would likely result in the loss to the larger society of the traditional skills, practices, and understandings necessary for the sustainable management of complex ecosystems. ILK has also contributed to demography, ecology, and evolutionary studies. Traditional knowledge is gained by experience, transferred among individuals (especially from elders to youth) by stories and direct teachings, and may evolve through experience and observations.

3. Methodology

The research consisted of several phases all set to incorporating indigenous viewpoints with modern technology design methods. The following steps were followed in the study:

The research was initiated with an in-depth background review of the literature on IKS, technology design, and inclusive innovation. This review examines the academic literature, books, and reports as well as case studies on technology in indigenous settings. The literature review aimed to:

The key themes and gaps in the extant research were:

- Know the challenges and opportunities to consider in incorporating indigenous worldviews into the design of technology.
- Develop insights into successful Indigenous technology initiatives.

3.1 Qualitative Case Studies

The research employed a qualitative case study of indigenous technology projects to better understand how these communities are interacting with technologies (Kovach, 2009). The cases were chosen according to their relevance, geographical diversity, and degree of indigenous participation in the design process. The case studies involved:

- Reading the project documentation, reports, and results.
- Interviews of project leaders, participants, and indigenous community members engaged in these initiatives.
- Finding out the best practices, challenges, and learning from these projects.
- Leading Indigenous voices & techMARK chats.

There are four challenges to sustainable education. These are: (1) creating better, more integrated science and quantifying the biophysical wealth; (2) engaging the people; (3) transforming societal value systems through "empathic education"; and (4) improving knowledge transfer around sustainability. The inclusion of Indigenous technology has gained importance in recent times since modern science failed in environmental sustainability. Human health, social justice, equity, economic development, ethics, and governance are the secondary issues associated with the challenges. Advocates of the inclusion of Indigenous science argue that all cultures have developed their unique form of science which is important to the overall diversity of human knowledge related to the biosphere (Kuokkanen, 2007).

The basic framework of "sustainable" development is achieved by the introduction of principles of Indigenous science into community education, planning, development, and policy. Global environmental threats have made modern education understand science not only as a body of knowledge but also as a way of thinking. Indigenous science is the knowledge that reflects the indigenous way of thinking about the physical world. Modern science education has urged learners to view science not as a body of knowledge but also as a way of thinking. The integration of modern science with Indigenous science is a constructive process that allows the learner to add something useful to the existing knowledge by their understanding. Fig. 17.1 gives the framework overview.

3.2 Traditional Ecological Knowledge (TEK)

TEK framework emphasizes the importance of local, place-based knowledge. It recognizes the value of long-term observations and practices and integrates the environmental, cultural, and spiritual aspects. It often focuses primarily on environmental knowledge, but potentially overlooks other aspects of Indigenous wisdom. It is "static" rather than evolving.

3.3 IKS Integration Model

This framework considers the holistic nature of Indigenous knowledge and supports the integration of IKS into various fields, including education and research. However, this system will not fully address the technological needs of Indigenous communities.

3.4 Two-Way Framework

Two-way framework considers the Indigenous and the modern science equally. It is flexible and collaborates with mutual respect. However, it lacks the guidelines for specific development. It works randomly.

3.5 Indigenous Research Paradigm

This considers Indigenous worldviews and methodologies in research. It promotes Indigenous self-determination and addresses issues of cultural appropriation and misrepresentation. It works mostly on research methodology but does not have any guidelines for the integration of modern technology. It is difficult to form cross-cultural collaborations.

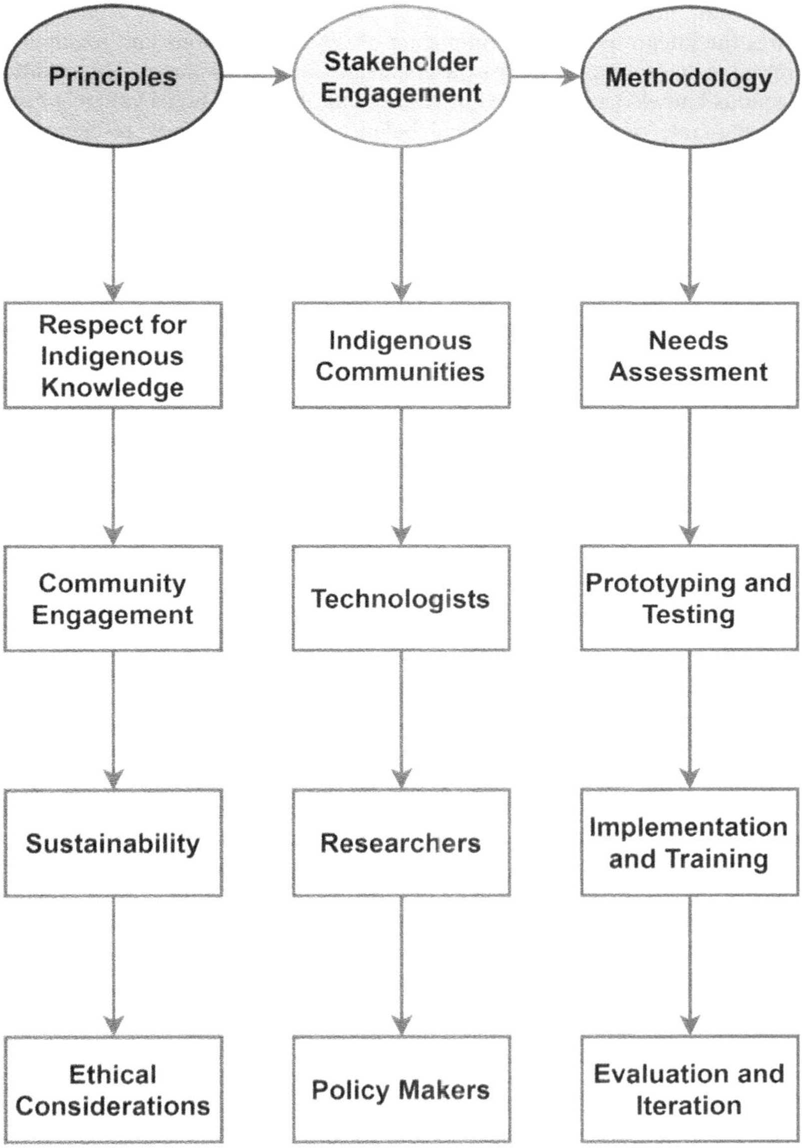

Fig. 17.1. Framework Overview.

3.6 Principles of Inclusivity

Human growth and development lie in SD that aims to fulfill the needs of the present without compromising the ability of future generations to meet their own needs. The objective is to have a society where living conditions and resources meet human needs without weakening the Earth's Integrity. Achieving SD requires the integration and harmonizing of diverse traditions and manipulating advanced technologies to address the challenges facing our planet. The traditional Indigenous knowledge embedded in various cultures often holds valuable insights into sustainable practices, ecological balance, and community resilience. Integrating this wisdom with modern technologies presents an opportunity to create innovative and contextually relevant approaches to SD. However, this integration also brings forth challenges related to cultural sensitivity, ethical considerations, and the need for collaborative frameworks.

3.7 Principle of Epistemological Similarity

This theory states that "different philosophies have elements that have the same goal of producing knowledge." This principle facilitates communication among different cultures, which offers the option of crossing philosophical boundaries, starting from possible similarities that allow exchanges between different philosophies. This theory is used to identify similar practices, values, devices, and techniques, in Modern science and traditional knowledge systems.

3.8 Integration of Indigenous Perspectives

They point out four major concerns of research and development associated with integrating Indigenous knowledge into science education.

These are: (a) Students from modern science should be impartial toward the outcome of the collaborative studies. (b) Indigenous knowledge should contribute to the knowledge base of modern science, (c) collaboration should aim for Environmental concern over sustainability, and (d) inclusion of the Philosophical ideology into the science.

It developed a model based on methodologies to incorporate modern and Indigenous knowledge and philosophy into ecological studies and pedagogy. Métissage offers a diversity of views and experiences about nature which is required for the development of environmental education research for future generations. Environmental education researchers from all cultural backgrounds are encouraged to acknowledge and engage with indigenous knowledge, philosophies, and methodologies.

The integration of Indigenous knowledge in education should recognize Indigenous frameworks and methodologies to give more attention to their history, politics, cultural beliefs, and philosophical views as well as to balance the Western modern perspective.

The youth of the Indigenous population have less exposure to new technologies, and educators should develop innovative teaching methodologies using technology and encourage Indigenous students to integrate their learning of linguistic skills with their increasing use of social media and technology. Some workers have developed the Epistemological bridge Framework to include traditional epistemologies in teaching. This framework allows the knowledge exchange among different philosophies and learning by the students. Through the EB, students can actively participate in the domain of every epistemology ultimately resulting in explanations of the same phenomenon using each epistemology. Indigenous communities have distinct communication systems among different communities. The system is dynamic and is robust represented by artists who influence the system with the help of their voices and acting skills to settle the communities' issues. Communication in the form of folksongs and plays reaches the community in a better way than anything. The integration of this communication system into the present digital communication system is proven to be highly effective.

3.9 Collaboration With Indigenous Communities

Though primitive, indigenous knowledge is refined and flexible. It is based on the observations and experiences of the communities gathered through generations practically. This knowledge worked following environmental and social change. However, researchers studying Indigenous science should also have a deep understanding of the indigenous culture and its depth. This helps them to increase the project productivity. Tolerance, humility, and understanding will go a long way in creating fruitful relationships between the researchers and the indigenous community members. Researchers should try their best to prevent conflict creation and manipulate their values and sentiments in all possible ways.

The incorporation of Indigenous knowledge paved the way for the increasing involvement of the indigenous communities in many research programs across the world thereby guaranteeing the safeguard of the Indigenous culture across time. The incorporation also has led to the exploration of Indigenous knowledge from the scientific point of view. The application of the IKS can be applied or theoretical.

An instance demonstrated that closures of Dungeness crab (*Metacarcinus magister*) fisheries, ordered by Haíłzaqv (Heiltsuk), Kitasoo/Xai'Xais, Nuxalk, and Wuikinuxv (First Nations in the province of British Columbia of Canada), increased the abundance and body size of crabs within protected areas. Indigenous Law imposed the fisheries closures as a remedial measure when the Canadian federal government failed to act on observed declines in catch rates. IK has also contributed to understanding related to evolutionary studies in the ecosystems. The observation of differences in various phenotypes is the result of many interspecific and intraspecific variations that form the basis of Evolution. More recently, IK has been used to identify life-history characteristics and population genetic structure. It described how the Sahtú Dene tribe and Métis peoples of northern Canada distinguished among genetically different populations of boreal, mountain, and barren-ground caribou based on unique behaviors, habitat preferences, and morphology. However subsequent genetic analyses provided evidence of distinct caribou subpopulations closely related to the Dene tribe.

Collaborative research involving the indigenous participants requires the same respect and recognition as the Modern scientists. The collaborative research may impose some harm to the IK and may result in the disregard of IK and inappropriate use of information of IK. Initially, the research scholars involved in the research programs with Indigenous participants should develop research agreements in a locally appropriate form independent of any institutional ethics approval. The indigenous participants should be educated on the project thoroughly and transparency should be maintained. A strong and enduring research is achieved by the two-way collaboration of the indigenous participants and the research scholars. Sometimes even research with good intentions will result in a failure when the IK is reproduced inappropriately.

3.10 Case Studies in Bridging Cultures

With the industrialization of the world that started a century ago, humanity had to face the hard reality of ecological imbalance. Humanity is experiencing hard realities of climate change ranging from global warming, melting of glaciers, rising sea levels, extreme rainfall, floods, drought, food scarcity, epidemics, pests, etc.

Given below are a few of the most visible case studies the world has noted (Larsen & Johnson, 2017).

3.11 Examples of Successful Inclusive Indigenous Technologies

3.11.1 Case Study 1

Farmers in many African nations were actively involved in planting many trees that helped in creating forests. Modern tree planting technologies in the form of "Drone Seed automation tree planting" are used. The adaptation of land and water resources to climate change was effective. According to experts, agroforestry is the most sensible solution for creating a balance between farming and forestry.

In Southern Africa, game farming instead of livestock farming is practiced· Agriculturists reaped much more economy by game farming than by Livestock Farming. In addition to this, agroforestry is practiced to revive the dry conditions of soil. They practice soil conditioning (Feeding organic matter content and mulch to improve the soil). The over-exploitation of land is reduced by these measures. This Practice helps in reviving soil fertility and its conservation.

3.11.2 Case Study 2

The prediction of weather events and seasons with indigenous knowledge is an interesting aspect especially when such predictions come true. However, there is a high possibility of uncertainty associated with these predictions especially in today's world as many anthropogenic activities have altered environmental processes. As per Indigenous Knowledge, the flying of sparrows indicated the onset of rain. But modern science contradicts it as a sign of pollution. However, the Indigenous knowledge cannot be neglected. Modern science goes hand in hand with Indigenous knowledge by incorporating smarter monitoring tools based on artificial can be integrated in the flying of sparrows to get more accurate results in weather. Weather events can be better predicted, and appropriate climate change mitigation and adaptation strategies can be designed for implementation by developing such integrated technologies.

3.12 Challenges and Lessons Learned

The integration of Indigenous science with modern science is not a walk in the cake but a herculean task. Both Indigenous Science and Modern Science have their advantages and disadvantages. Indigenous science is conservative and qualitative whereas Modern Science is skeptical. The integration of Indigeneity and modernity to obtain susceptibility must overcome many challenges.

Given below are a few notable challenges:

- Integrating traditions with technologies requires a deep understanding of Indigenous culture and traditions. Conflicts can arise between modern and traditional sciences while information is exchanged. Showing respect and tolerance toward cultural perspectives while implementing technological interventions can be a sensitive task.
- Caution should be maintained to preserve their values, and they should not be distorted by modern science.
- Confidential information of the communities should be safeguarded along with the intellectual property rights of modern science. Most of the communities have no access to modern science.
- Even when Indigenous communities are educated about modern Science, they are reluctant to take it up due to the fear of losing their customs.
- Many communities lack the resources for integrated studies with Modern Science.
- Using indigenous science without proper understanding leads to ecosystem damage.
- Integration of Indigenous Science with Modern Science should be nondominant and should benefit the Indigenous Communities also.

4. Results

4.1 Ethical Considerations and Responsibility

Indigenous Science and Modern Science are two sides of a coin. Both are unique and similar in their aspects. Ethical values are of the highest order for Indigenous Science, but for Modern Science, it is much more compromised. Indigenous Science operates on a holistic approach giving utmost Importance to spirituality, culture, and respect to the supernatural forces (Nakata, 2007). It is bound by the metaphysical elements. Indigenous Science always worked for conservation and sustainability. They practiced mutualism and reciprocity socially. Modern Science is based on capitalism aiming at progress and profitability. Modern science emphasizes individual growth.

4.2 Ethics in Indigenous Technological Development

Indigenous Science across the world considers that Mother Earth owns life. So, they live in harmony with Mother Nature and its elements. They also follow community living with mutual benefit. Indigenous living is a purposeful living and a contented one. It adds much more value to human life. It has been the basis for the sustenance of life on Earth ever since human life started.

4.3 Respect for Cultural Intellectual Property Rights

Digitalization of the world has made it easily accessible to indigenous knowledge across the world. The integration of the sciences for the betterment of mankind requires the protection of their cultural rights. Modern science has a well-established intellectual property protection policy while Indigenous Science has got any. Their protection is in their community law and owned by the community as a whole. Their infringement only affects the cultural integrity of the indigenous community.

The commercialization of the world has led to the inappropriate use of indigenous cultural heritage at risk of cultural loss. The randomized exposure to Indigenous culture results in the loss of ancestral knowledge and their sanctity, authenticity, livelihood of the indigenous people, and confidentiality. The misuse of their sacred symbols and artifacts impedes the emotional sentiments of the people. Figs. 17.2–17.5 have been developed by the author to allow the readers to visualize the review.

An international collaboration by the name "The Intellectual Property Issues in Cultural Heritage (IPinCH) project" is founded for the protection of Indigenous Cultural rights and their sovereignty. This project is participated by archaeologists, lawyers, anthropologists, museum specialists, ethicists, and 25 partnering organizations actively working on the Project. They work on the intellectual property of cultural heritage, analyze the situations of infringement, and make policies for the Aboriginal communities' development. They follow the community-based participatory research methodology. So far, this organization has protected many artifacts and databases of the indigenous communities and collected them in museums for public use.

4.4 Hypothetical Variables (Fig. 17. 6)

Belief in Mother Earth (BME): Represents the extent to which a community believes life is owned by Mother Earth.

Reverence for Nature (RFN): Represents the reverence for unseen forces of nature.

Community Living (CL): Represents the extent of community living with mutual benefit.

Environmental Sustainability (ES): Represents the level of environmental sustainability achieved by the community (Fig. 17. 7).

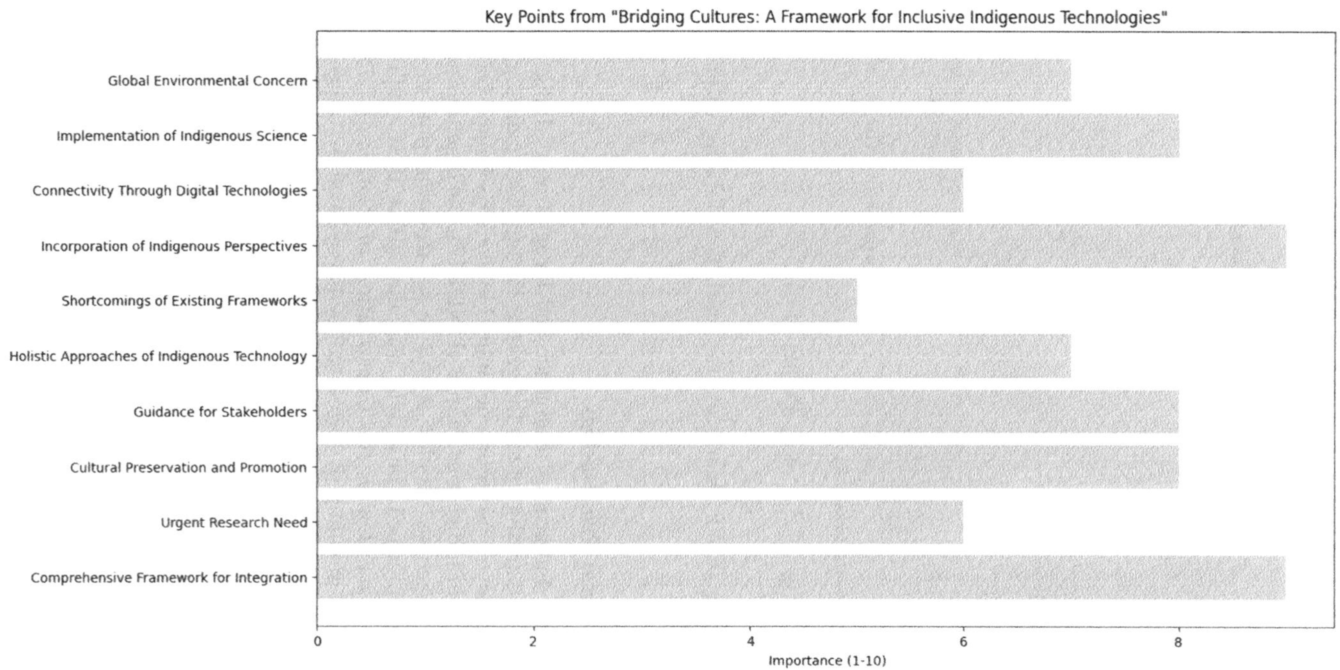

Fig. 17.2. To Visualize the Key Points Mentioned in the Provided Text Can Help Highlight the Emphasis on Various Aspects of Integrating Indigenous Knowledge With Modern Science.

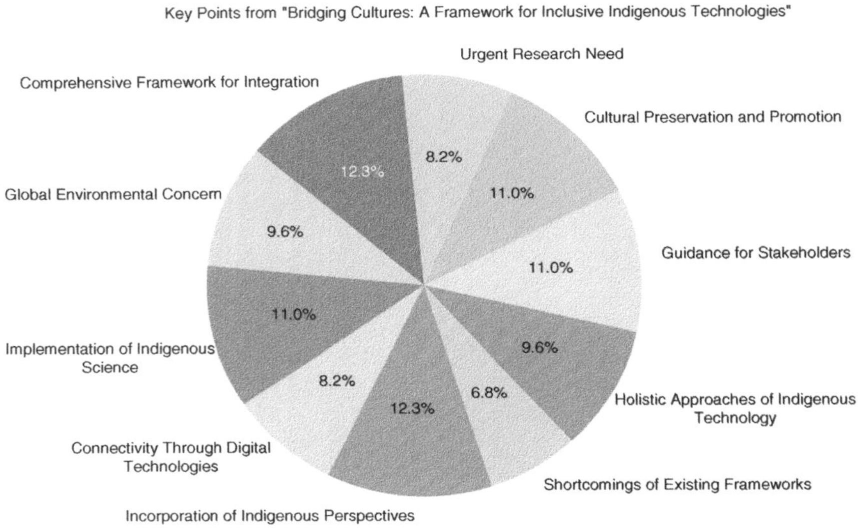

Fig. 17.3. Key Points of %.

Community Well-being (CW): Represents the level of well-being within the community (Fig. 17. 8).

4.5 Establishing Long-Term Relationships

This is the period for the Indigenous resurgence and recognition of Indigenous rights. Areas under the stewardship of the Indigenous Peoples have gained global attention due to the urgency of protecting declining biodiversity. United Nations Declaration of the Rights of Indigenous Peoples (UNDRIP), 144 countries across the world recognized Indigenous Peoples' Rights to self-determination, cultural identity, and free prior informed consent to uses that affect their traditional Territory. The International Union for Conservation of Nature (IUCN) is another organization that monitors the protected areas governed by Indigenous Peoples. The Convention of Biological Diversity (CBD) is also an organization that works toward the creation of protected areas, assisting the indigenous communities in conservation.

The Indigenous Protection and Conservation Area (IPCA) initiative was started by many countries. They safeguard the Indigenous cultural, spiritual, and livelihood values associated with IPCAs. With cooperation, Indigenous Peoples can develop traditional territories. Many policies were developed for the proper functioning of IPCAs some were designed by Indigenous groups themselves, while others were jointly initiated with state agencies and nongovernment organizations (NGOs).

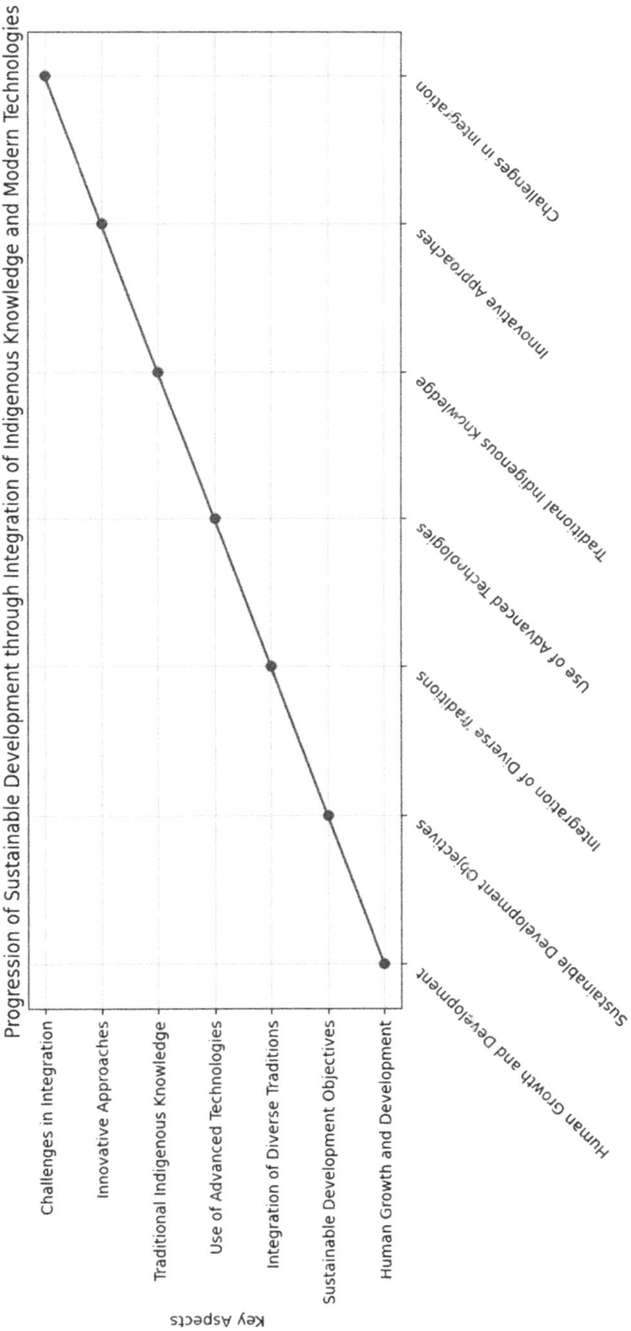

Fig. 17.4. Components of Inclusivity.

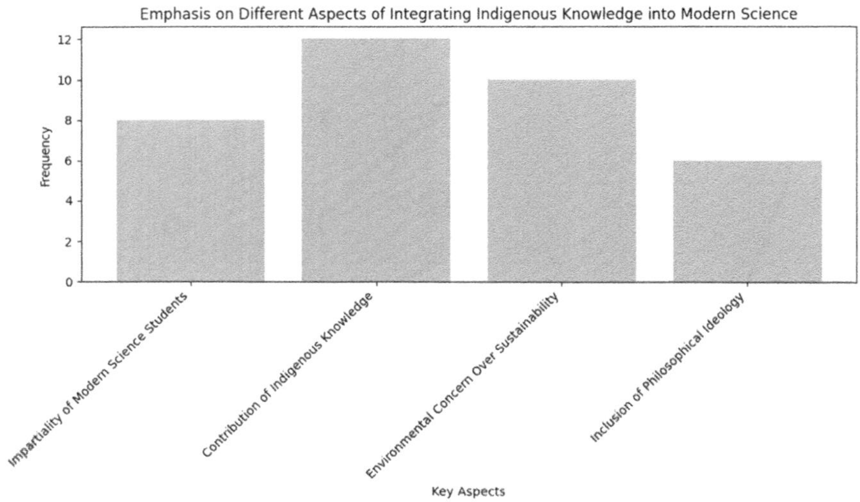

Fig. 17.5. The Different Aspects and Their Emphasis as Mentioned in the Provided Text, We Can Represent the Key Aspects as Different Categories and Assign a Hypothetical Frequency or Weight to Each Aspect Based on Their Perceived Importance.

4.6 Practical Steps for Incorporating Indigenous Perspectives

The incorporation of Indigenous perspectives can be achieved through the introduction of Indigenous Science in the Modern Education system. The students can use their ideas and beliefs in alternate ways of knowing nature, which contributes to a better understanding of social, cultural, economic, political, and natural aspects of local environments (Table 17.1). By integrating multiple ways of knowing into science classrooms, students can learn the value of traditional ways of knowing (Table 17.2).

Scientists have identified many practices of indigenous peoples in indigenous soil taxonomies; soil fertility; and agronomic practices (terracing), such as contour banding, fallowing, organic fertilizer application, crop-rotation, and multi-cropping; conservation of soil and water beneficial to the Modern scientific point of view also. UNESCO created the Local and Indigenous Knowledge System (LINKS) (UNESCO, 2002) program. Under this program, the exploration of ILK systems contributions to understanding, mitigating, and adapting to climate change, environmental degradation, and biodiversity loss are deeply assessed. In addition, as part of its Education for a Sustainable Future project, UNESCO launched the Teaching and Learning for Sustainable Future: A Multimedia Teacher Education Program (UNESCO, 2002). It provides education to the teachers and students about local cultures, their wisdom, and ethics, and

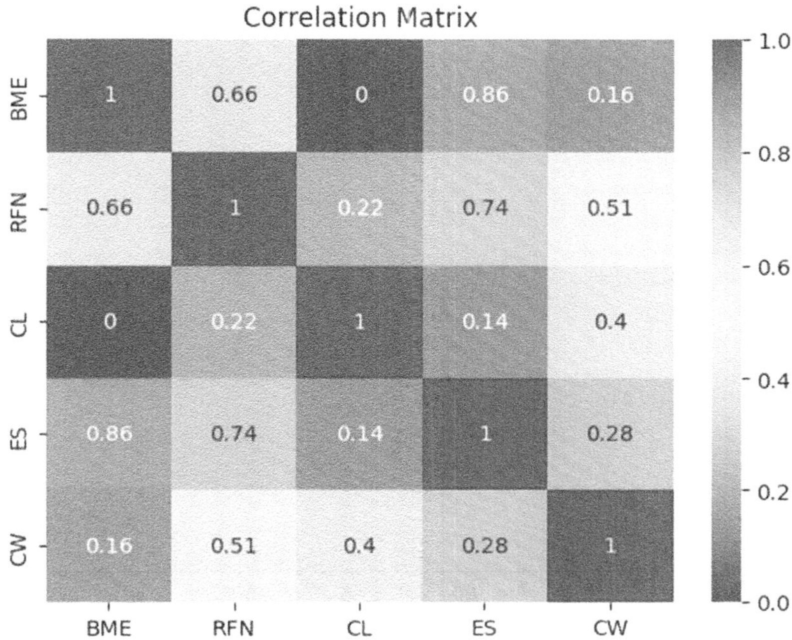

Correlation Matrix:

	BME	RFN	CL	ES	CW
BME	1.000000	0.664364	0.000000	0.859776	0.156174
RFN	0.664364	1.000000	0.217443	0.740924	0.512936
CL	0.000000	0.217443	1.000000	0.136845	0.403239
ES	0.859776	0.740924	0.136845	1.000000	0.281425
CW	0.156174	0.512936	0.403239	0.281425	1.000000

Fig. 17.6. Correlation Matric Analysis.

suggests ways of teaching and learning locally relevant knowledge and skills thereby supporting the professional development of student teachers, teachers, curriculum developers, education policymakers, and authors of educational materials. The aim is to create respect for the indigenous communities among the learners (Table 17.3).

5. Discussion

A major part of the world's biodiversity was maintained by Indigenous Communities. They are the rightful owners of the ecosystems across the world. Indigenous survival is based on the intimate knowledge of local ecosystems. For centuries, Indigenous communities passed their ecological knowledge through

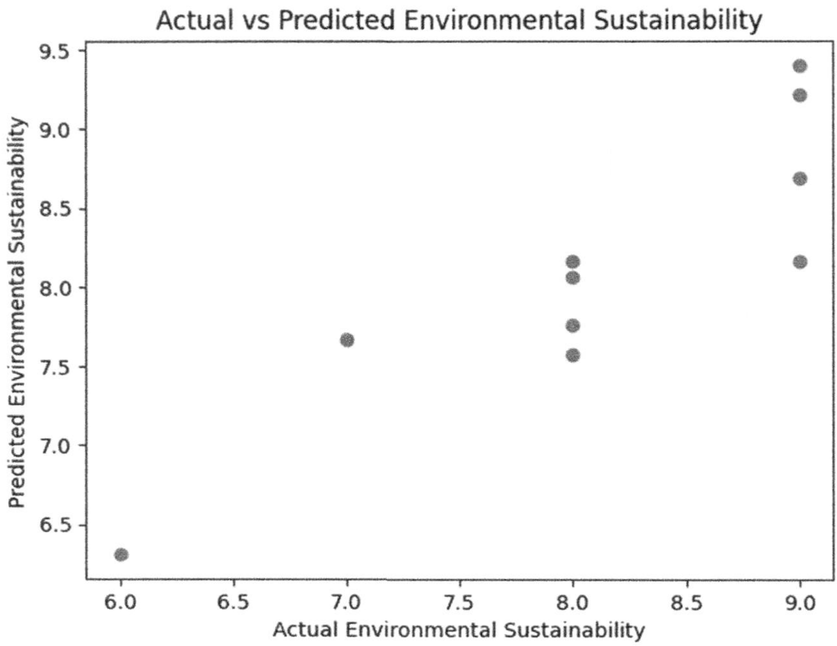

Fig. 17.7. Regression Analysis for Environmental Sustainability.

oral communication to their forerunners (Nardi, 2010). TEK is the knowledge held by the Indigenous communities related to these ecosystems. With colonization, the TEK and the Indigenous freedom were in jeopardy.

The United Nations has declared 2021–2030 as the UN Decade on Ecosystem Restoration. Under this program, the UN has urged the indigenous culture and people to restore their respective ecological landscapes to repair the degraded ecosystems. As per Article 31 of the UN Declaration on the Rights of Indigenous Peoples, it is the right of Indigenous peoples to maintain, protect, and control their culture and TEK. Attentive listening to Indigenous people and exhibiting humility and respect are all needed to engage with such communities (Pacey, 1983). There is an urgent need for the non-Indigenous Communities to restore the rights of Indigenous peoples. By doing this, global conservation is speeded. Giving Indigenous Leadership for restoration is of high priority now.

5.1 Overcoming Barriers to Implementation

Many countries must come out of the colonial system of education, which is not possible at a faster pace. Even the UN policies for the tapping of Indigenous knowledge and their conservation policies are yet to be more effective. The Indigenous populations are at the risk of their culture being infringed. The digital

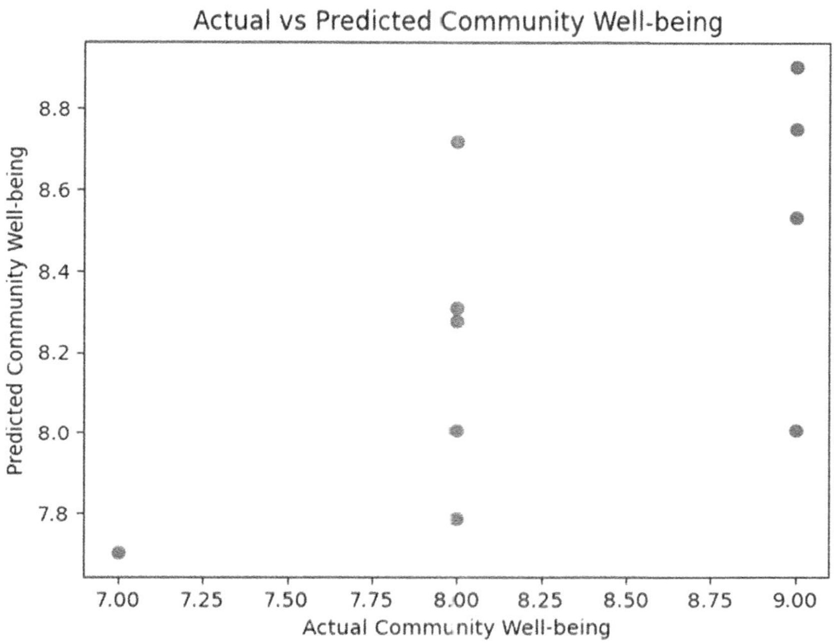

Fig. 17.8. Regression Analysis for Community Well-being. *Source:* Developed by the author.

Table 17.1. Summarizing the Key Points of Aspects and Description.

Aspect	Details
Tree planting	Farmers in many African nations are involved in planting trees to create forests.
Modern technologies	Use of "Drone seed automation tree planting" for efficient tree planting.
Climate change adaptation	Effective adaptation of land and water resources to climate change.
Agroforestry	Sensible solution for balancing farming and forestry, as stated.
Game farming versus Livestock farming	In Southern Africa, game farming is practiced instead of livestock farming, yielding better economy.

(Continued)

Table 17.1. *(Continued)*

Aspect	Details
Soil reviving practices	Agroforestry and soil conditioning (adding organic matter and mulch) to improve soil in dry conditions.
Soil conservation	Measures reduce land over-exploitation, helping to revive soil fertility and conservation.

Source: Developed by author.

Table 17.2. Summarizing the Key Points Case Study 1.

Aspect	Details
Prediction of indigenous knowledge	Indigenous methods predict weather events and seasons accurately at times.
Uncertainty in predictions	High possibility of uncertainty in predictions due to altered environmental processes by anthropogenic activities.
Traditional indicators	Flying of sparrows indicates the onset of rains according to indigenous knowledge.
Modern science view	Modern science sees the flying of sparrows as a sign of pollution.
Integration of knowledge	Combining modern science with indigenous knowledge using smarter monitoring tools and artificial intelligence.
Improved prediction	Integrated technologies can enhance weather event predictions.
Climate strategies	Better weather predictions can lead to effective climate change mitigation and adaptation strategies.

Source: Developed by author.

technology is yet to reach many communities across the world. Indigenous issues like gender equality and empowerment, lifelong learning opportunities, innovation promotion, and global partnerships have a high probability of further research. United Nations policies alone are not sufficient to tackle the global environmental threat. A framework that suits all the countries irrespective of their sociocultural and cultural disparities is also needed.

Table 17.3. Summarizing the Key Points Case Study 2.

Integration	Holistic	Understanding	Transformative	Traditional Knowledge
8	9	7	8	9
7	8	8	7	8
9	9	9	8	9
6	7	6	6	7
8	8	8	9	8
5	6	5	5	6
7	8	7	8	7
9	9	8	9	9
6	6	6	6	6
8	7	8	7	8

Basic Statistics

	Integration	Holistic	Understanding	Transformative
Count	10.000000	10.000000	10.000000	10.000000
Mean	7.300000	7.700000	7.200000	7.300000
Std	1.337494	1.159502	1.229273	1.337494
Min	5.000000	6.000000	5.000000	5.000000
25%	6.250000	7.000000	6.250000	6.250000
50%	7.500000	8.000000	7.500000	7.500000
75%	8.000000	8.750000	8.000000	8.000000
Max	9.000000	9.000000	9.000000	9.000000

Traditional Knowledge

Count	10.000000
Mean	7.700000
Std	1.159502
Min	6.000000
25%	7.000000
50%	8.000000
75%	8.750000
Max	9.000000

Correlation Matrix

	Integration	Holistic	Understanding	Transformative
Integration	1.000000	0.852592	0.905571	0.875776

(Continued)

Table 17.3. *(Continued)*

Correlation Matrix				
	Integration	**Holistic**	**Understanding**	**Transformative**
Holistic	0.852592	1.000000	0.748357	0.852592
Understanding	0.905571	0.748357	1.000000	0.770411
Transformative	0.875776	0.852592	0.770411	1.000000
Traditional knowledge	0.924239	0.917355	0.826311	0.780946
Traditional Knowledge				
Integration	0.924239			
Holistic	0.917355			
Understanding	0.826311			
Transformative	0.780946			
Traditional knowledge	1.000000			

Source: Created by author.

6. Implications and Conclusion

6.1 Implications

Collaborating with indigenous communities has the potential to question scientific approaches and substantially enrich them while leading to sustainable transformation. But this transformation should not create the neglect of indigenous communities and sustain Modern science supremacy. Recent research indicated the lack of understanding of transformation from Indigenous knowledge. This can be rectified by collaboration of the Indigenous Scholars who already worked hand in hand with the indigenous knowledge holders in sustainability transformation. It designed an approach that included five tasks, i.e., mobilize, translate, negotiate, synthesize, and apply. This helps in the better knowledge transformation and understanding.

6.2 Empowering Indigenous Voices in Technology

Though the Indigenous communities' population is predominant across the world, they were deprived of basic scientific education due to their economic conditions like poverty, nonaccessibility, etc. Recent advances in information and communication technology (ICT) have greatly contributed to the digitalization of Indigenous Communities. Indigenous youth are incorporating ICT into their daily lives. A study reported that it is the belief of the Indigenous elders, that the use of digital technology opened many opportunities for young people, who in turn play a crucial role in maintaining cultural traditions bringing contemporary innovations into the community and helping amplify Indigenous voices across time and distance. Digital technology has led to linguistic evolution of Indigenous communities, which has increased the literacy rate and their communication skills. The receptivity of the Indigenous youth to digitalization shows that they are highly resilient.

ICT have enabled self-authoring and significantly transformed these public representations. In addition, several studies have suggested that Indigenous peoples are utilizing ICT to focus on the development of online communities and interfaces to tell stories and construct representations of self. Further, communities have been developing innovative uses of wireless access for intercultural dialogue, political mobilization, and action related to civil and human rights issues. In short, tribal communities are increasingly asserting their right to self-determination and are innovating with technologies to assert this right.

7. Conclusion

The global threat to the world led to the exploration of the researchers to find more sustainable solutions. A deep look into the Indigenous communities provides a ray of hope to protect the environment right now. Many researchers across the globe extensively studied the Indigenous communities, understood their practices, learned about their resilience and their conservation strategies, and identified Indigenous Science as the most effective tool available to mankind now. In this chapter, the importance of the Indigenous communities, their unique culture, and their scientific knowledge were discussed. The importance of including them in Modern science is discussed. Many frameworks to integrate the IS in Modern Science and the challenges faced during integration are discussed. This chapter also discussed the case studies, cultural rights, and future studies of the Indigenous Communities.

Forces of globalization, colonialism, political oppression (Noble, 2018), and economic interests on the territories of Indigenous Peoples and local communities have put ILK systems at risk of extinction. These pressures can generate incidents of intergenerational trauma and reduced cultural engagement, leading to declines in people's physical and mental well-being, including feelings of shame and insignificance. Supporting the Indigenous communities in conservation, helping them in building developmental policies, protecting their sovereignty, and respecting their culture and values are the few notable solutions to preserve them for sustainable future generations.

References

Andersen, C., & O'Brien, J. (2020). *Indigenous data sovereignty: Toward an agenda*. Australian National University Press.

Banerjee, A., & Linstead, S. (2021). Technology and indigeneity: Deconstructing the modernity discourse. *Journal of Information Technology*, *36*(3), 193–210.

Battiste, M. (2005). *Indigenous knowledge: Foundations for first nations*. UBC Press.

Brown, L., & Strega, S. (2015). *Research as resistance: Critical, indigenous, and anti-oppressive approaches*. Canadian Scholars' Press.

Cameron, E. (2012). Securing indigenous knowledge: From rights to responsibilities. *Social & Cultural Geography*, *13*(6), 585–603.

Cheryan, S., Plaut, V. C., Davies, P. G., & Steele, C. M. (2009). Ambient belonging: How stereotypical cues impact gender participation in computer science. *Journal of Personality and Social Psychology*, *97*(6), 1045–1060.

Coombes, B. (2018). *Indigenous knowledge and cultural sustainability in the digital age*. Springer.

Davis, M. (2017). The digital divide: Its impact on indigenous communities and policy initiatives. *Journal of Information Policy*, *7*, 92–116.

Dudgeon, P., Milroy, H., & Walker, R. (2014). *Working together: Aboriginal and Torres Strait Islander mental health and wellbeing principles and practice*. Commonwealth of Australia.

Garrison, V., & Monroe, C. (2016). Indigenous technological practices and sustainable development. *International Journal of Technology and Design Education*, *26*(4), 555–574.

Hudson, M., & Henwood, W. (2009). Embedding indigenous principles in computer science education. *ACM Transactions on Computing Education*, *9*(3), 1–16.

Johnston, P. (2020). *Indigenizing the digital: The challenge of weaving technology with tradition*. Routledge.

Kovach, M. (2009). *Indigenous methodologies: Characteristics, conversations, and contexts*. University of Toronto Press.

Kuokkanen, R. (2007). *Reshaping the university: Responsibility, indigenous epistemes, and the logic of the gift*. UBC Press.

Larsen, S. C., & Johnson, J. T. (2017). *Being together in place: Indigenous coexistence in a more than human world*. University of Minnesota Press.

McGregor, D. (2018). Digital inclusion for indigenous communities: Barriers and opportunities. *Journal of Information, Communication and Ethics in Society*, *16*(2), 115–130.

Nakata, M. (2007). *Disciplining the savages, savaging the disciplines*. Aboriginal Studies Press.

Nardi, B. (2010). *My life as a night elf priest: An anthropological account of World of Warcraft*. University of Michigan Press.

Noble, S. U. (2018). *Algorithms of oppression: How search engines reinforce racism*. NYU Press.

Pacey, A. (1983). *The culture of technology*. MIT Press.

Rifkin, J. (2000). *The age of access: The new culture of hypercapitalism, where all of life is a paid-for experience*. Tarcher/Putnam.

Smith, L. T. (2012). *Decolonizing methodologies: Research and indigenous peoples*. Zed Books.

TallBear, K. (2013). *Native American DNA: Tribal belonging and the false promise of genetic science*. University of Minnesota Press.

Tsing, A. L. (2011). *Friction: An ethnography of global connection*. Princeton University Press.

UNESCO. (2002). *Bridging cultures: A framework for inclusive indigenous technologies*. UNESCO Publishing. https://www.unesco.org/en/articles/cutting-edge-indigenous-languages-gateways-worlds-cultural-diversity

Walter, M., & Andersen, C. (2013). *Indigenous statistics: A quantitative research methodology*. Routledge.

Chapter 18

Cultural Perspectives on Wealth and Prosperity in Indigenous Communities

Indira Bhati[a], *Amit Joshi*[b] *and Rishabh Jain*[c]

[a]Maharishi Markandeshwar (Deemed to Be University), India
[b]Guru Nanak Khalsa Institutes of Technology and Management, India
[c]Symbiosis International (Deemed University), India

Abstract

The Human–machine interaction (HMI) refers to a technology that can fill the gap between the traditional indigenous traditions and the latent needs of contemporary society. It does this through preserving and transferring cultural knowledge, generating long-term economic activity, and improving the community's well-being. HMI can be utilized to overcome the challenges faced by the Indigenous groups as they have experienced devastating effects from colonization, land dispossession, and forced assimilation. It is important to address these challenges in order to engage with the technology and business development opportunities. Using HMI solutions, it is quite possible to support indigenous communities in a number of ways like archiving oral histories and cultural knowledge along with revitalizing endangered languages and selling traditional crafts directly to a global audience. The HMI solutions can positively contribute to the Indigenous communities by preserving the traditional know how, building sustainable economies, and improving the general well-being. Community based governance structures are needed to be established to ensure the ownership and control of Indigenous communities over HMIs. At the same time, stronger data security arrangements are also required along with security protocols, that are updated regularly and are able to assess potential vulnerabilities.

Keywords: Indigenous societies; community; HMI; wealth; resource; generation

Indigenous Empowerment through Human-Machine Interactions, 313–328

doi:10.1108/978-1-83608-068-820251018

1. Introduction

Earning wealth and social success are the most common benchmarks of an ideal person in Indigenous societies (IS) globally. Western economic theories are mostly centered on monetary benefits and materialism. These global trends also have a profound impact on indigenous culture. Communities are also built around cultural heritages, respect toward nation, and the manifestation of wealth.

Nathan Grandjambe, a specialist in Indigenous economic development, postulated that the economic prosperity in IS is not measured only in financial terms. It is based on the presence of a system which allows the IS citizens to engage willfully in exploring economic opportunities as per their native traditions instead of the pressures by western or any other communities.

Local cultural considerations weigh heavily in practices defining and deriving wealth. "Wealth" is not clearly defined in most of the indigenous languages. These cultural manifestations also mark a deviation from the well-established western economic practices, i.e., individualism or selfishness and resource intensity. The IS citizens usually define the economic processes around the principles of reciprocity, togetherness, shared cultural norms, and social and environmental benefits to be employed and enjoyed by everyone (Diaz-Reviriego et al., 2016).

Turner et al. (2008) have gauged this relationship between the land and its natural resources and documented this as traditional ecological knowledge (hereafter Traditional Ecological Knowledge [TEK]). The interdependence between human and nature is also illustrated by Richards (2019), which discussed cultural transference in forms of stories and folklore depicting this relationship.

IS histories globally are filled with tales of discrimination and deprivation by colonists. Brosius et al. (2018) and Tauli-Corpuz (2019) have documented the various battles waged by IS communities against colonial practices of exploitation and destruction of environment, demonstrating the resilience of IS citizens. Human–machine interaction (HMI) can be an effective intervention to overcome various challenges faced by IS citizens. HMI can effectively bridge the gaps between the economies of development, and the cultural perspectives about wealth and success in IS citizens, while trying to balance both the approaches. It is hereby posited that HMI can achieve the objectives of economic development, while preserving IS culture and general well-being of people.

2. The Objectives

The following objectives are postulated for the study:

- Exploring the role of HMI in economic progress, preserving culture and social development in IS.
- Identifying the major roadblocks such as digital divide and cultural sensitivity in HMI implementation in IS communities.
- Recommending strategies for the designing and implementation of HMI in IS communities.

3. Literature Review

The current study is a review study focusing on role of HMI technologies to gauge the concepts of wealth and prosperity in IS. The current study intends to fill the gap between the concepts of traditional values in IS and business-oriented approaches while instilling the wealth creation objective using traditional mechanisms.

3.1 The Western and Indigenous Views of Wealth

Western scholars have defined wealth as a result of business efforts (Smith, 2008). The definition in IS community is somewhat broad inclusive of cultural wealth (Daigle, 2010). The wealth of a culture is rooted in caring and community feeling instead of material possession (Cajete, 2000). This social care and togetherness are in true sense the prosperity that people strive (Wilcox, 2008). As per Cruikshank (2005), in such a society, the more valuable assets are the TEK and artistic skills. This TEK gives rise to cultural continuity and environmental governance (Nakata, 2000). In many IS cultures, prosperity is derived from this synchronization and spiritual connect with nature (Kawagley, 1995). The spirit of resource conservation and sustainability in future generations are a result of such connect (Pascoe, 2018).

3.2 Prosperity and Cultural Values

The connection between IS cultural values and prosperity has been discussed in multiple works. Coulthard (2012) conducted a study in South American IS communities who practice the concept of "Buen Vivir" (Living Well) which postulates the superiority of social unity and ecological awareness over material possessions. In contrast, the studies by Acosta and Martínez (2010) established the supremacy of individual wealth in Western societies. Simpson (2014) explored the IS communities' approach toward wealth which is holistic and inclusive which contrasts with major economic development models.

3.3 Opportunities and Challenges

Sanders (2003) demonstrated the ill effects of colonization which tried to impose the western wealth theories to IS communities, thereby leading to land deprivation and forced annexation if IS properties. In addition to disrupting traditional ways of life, these historical injustices severed Indigenous connection to the land and undermined Indigenous knowledge systems (Alfred & Corntassel, 2005). In order to understand how Indigenous communities might engage with technology and business development opportunities, it is crucial to understand these challenges (Da Silva, 2018).

It is possible to support Indigenous communities in various ways through HMI technologies. Using digital storytelling tools, Todd (2016) explores ways to archive oral histories and cultural knowledge, ensuring their preservation for future generations. In a similar vein, McCarty (2002) examines the use of language learning apps to revitalize endangered languages. E-commerce platforms also provide Indigenous artisans with the opportunity to sell traditional crafts directly to a global audience, fostering economic development opportunities that are aligned with cultural values of self-reliance and artistic expression (Natcher & Davis, 2013), as Anderson (2014) discusses.

The benefits of HMI must be weighed against ethical considerations. Smith (2019) highlights the importance of Indigenous communities owning their data and cultural heritage within the digital sphere. A second issue that Cameron (2012) emphasized was the need to respect and integrate Indigenous knowledge systems into the development of HMIs in order to avoid cultural appropriation.

4. The Traditional View of Wealth and Prosperity: Beyond Material Possessions

Wealth and success are much more than tangible possessions for Indigenous cultures around the world. As part of its holistic approach, it considers social well-being, spiritual connection, and environmental balance. Here are five key components of Indigenous wealth, illustrated by unique cultural examples.

4.1 A Network of Strong Relationships and Reciprocity

There is much more to prosperity within Indigenous societies than collecting tangible items. To be economically and socially successful, they need strong social networks and reciprocity. We will discuss the relevance of this fundamental idea and present cultural examples in this section.

Indigenous communities frequently function as interconnected entities that prioritize collaboration and social cohesiveness. There is an inextricable link between individual prosperity and community health. According to "Nativegov.org," wealth is measured by the strength and interconnection of a person's social network rather than personal possessions.

A key component of Indigenous cultures' social connections is reciprocity. As individuals, we are expected to use our talents, resources, and labor for the greater good and to receive assistance when needed (Smith & Lightfoot, 2014). It develops social relationships, encourages a sense of shared responsibility, and meets everyone's fundamental needs.

To demonstrate the importance of social networks and reciprocity in Indigenous wealth creation, the following examples can be studied:

Among Indigenous groups of the Pacific Northwest coast, such as the Tlingit and Haida, the potlatch ceremony exemplifies this concept. In these extravagant events, blankets, food, and prestige items were distributed. Rather than acquiring stuff, the focus was on developing social relationships, exhibiting kindness, and completing communal tasks.

In Tanzania, the Hadzabe people live off a sharing economy based on social cooperation. In the tribe, food is freely distributed, so everyone has access to sustenance regardless of hunting success. Bliege Bird and Coppinger (1995) have demonstrated that this strategy strengthens social bonds and boosts group survival.

In East Africa, the Maasai have a complex system of communal land management. In this, the community grazing areas are divided and rotated so their survival can be ensured. This technique reemphasizes putting long-term social well-being over individual interests (Homewood, 2004).

To achieve sustainable growth and resilience, the values as trust and commitment, social gregariousness, and togetherness provide the foundation. To develop HMI solutions for IS communities, these values must be embedded in the social fabric. To empower IS communities, technology must embrace and amplify these values by promoting native behavior such as knowledge dissemination, resource sharing, and building strong social communities.

4.2 Indigenous Wealth and Its Living Legacy

The collective expertise passed over from one generation to another which includes not only the ancient wisdom but also artisanal skills is the most important wealth in IS communities. This collected wisdom is known as TEK. TEK also includes the understanding of flora and fauna for food and medicinal purposes. Such a knowledge also serves as the bridge between man and nature and promotes a symbiotic relationship culminating into sustainable practices which ensures resource availability to future generations. It instils a sense of responsibility for maintaining the environment (Berkes, 2008).

To maintain the relevance of TEK, it is vital that it is transferred from generation to generation. Stories, songs, and rites passed down through generations are among the strategies traditional civilizations use to do this. Hands-on experience is often used to teach skills like hunting, fishing, and plant identification. The younger generation learns from seniors who have this expertise by observing and collaborating with them. Often, rituals have an ecological component, as they support traditional resource management practices while connecting participants to the natural world.

4.3 The Dissemination of TEK by Indigenous Cultures Must Be Studied in Some Specific Instances

A complex system of environmental knowledge transfer exists among the Yolngu people of Arnhem Land, Australia. Seasonal changes, animal behavior, and medicinal herbs are discussed in stories, chants, and ceremonies. As a result, it is guaranteed to be preserved and passed down from generation to generation.

The Maori Whakatauki are a set of proverbs which convey short and interesting phrases about the natural systems. These proverbs have been very effective to propagate a feeling of care toward nature and resources across generations.

A major challenge posed in today's time is the hindrance in the transference of TEK on account of formal education mechanisms which replace TEK with modern science knowledge.

We postulate that HMI can be employed to preserve and revive TEK. We can digitally record and share archives containing traditional knowledge in form of oral accounts, music, and stories, thus preserving them for generations to come. These technologies can be employed to create interactive learning interventions for early to advance age students and for societies as well. For this, HMI needs to integrate and develop language learning algorithms. In this way, HMI can support IS communities greatly.

4.4 The Connection Between Spiritual Well-Being and Land

Indigenous cultures view the land as more than just a physical location; it is inextricably linked to their spiritual well-being. This section discusses the significance of this link and its contribution to Indigenous wealth. The land, in some cultures, is believed to have a spirit or life force. Land itself, sacred sites, and natural characteristics are frequently associated with the spirit world. Indigenous cultures often link spiritual well-being to living in harmony with nature. A balanced partnership promotes spiritual fulfillment and belonging.

But rampant resource exploitation has disturbed this synchronization and damaged the native communities in process.

4.5 Community Harmony and Balance

To bring harmony, the IS communities have been practicing various festivities and rituals which brings the community together. The process is self-fulfilling and brings a lot of cohesiveness, trust, and a sense of shared cultural perspectives. IS communities practice their own code of ethics and governance, providing a mechanism based on shared responsibilities fostering acceptability. Such an acceptance leads to the members sharing the common gains to the community, thereby reinforcing this behavior of subordinating individual good to shared benefits.

This brings the members closer to the community enhancing togetherness in IS communities.

4.6 Cultural Values and Prosperity Beyond Material Wealth

In the studies of IS communities, it is essential to study the value systems and definitions of success, as these form the parameters of prosperity in IS communities. Focus on these attributes is more prominent than material gains. The prosperity in IS communities is defined in terms of reciprocity among the members, communal ownership, and spiritual connection to the land (Fig. 18.1).

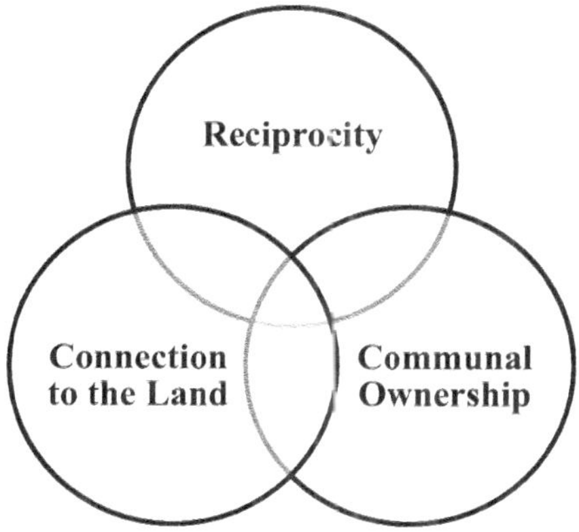

Fig. 18.1. The Basic Components of Wealth in IS Communities. *Source:* Author's compilation.

5. Wealth and Values of Indigenous People

The conception of wealth in IS communities is significantly different from western ideas. The culture defines the idea of wealth in terms of:

- Social network of a person.
- Reciprocity.
- Giving and receiving in a balanced manner. An example of such practice is Northwest coast's "potlach" ritual, in which the wealth is distributed to enhance social bonding.

HMI has the potential to add to this wealth by providing mechanisms to:

- The documentation and conservation of TEK.
- Develop instructional tools that promote cultural identity and environmental sustainability.
- Facilitate IS community's communication and collaboration.

But such HMI solutions must be designed while giving due importance to IS values, thereby contributing to the prosperity concept.

6. The Challenges to Traditional Wealth and Prosperity

The crux of colonization was based on enhancing material possessions as most of the colonist's viewed wealth in this perspective only.

Their methods were coercive and manipulative at times which contradicted and somewhat replaced the prosperity conception in IS communities.

The various methods employed by colonists to displace natives and capture their lands were:

- Signing of treaties using coercive methods and misleading narratives about the consequences, which actually led to the IS communities losing major chunks of land (Brosius et al., 2018).
- At times, the land was captured by forcing natives from their parental properties, destroying important parts of their traditional life style, and severing the spiritual connection from land (Tauli-Corpuz, V., 2019).
- Exploration and possession of native land due to presence of lumbar and mineral resources without proper procedure and compensation to the actual holders of the land (Brosius et al., 2018).

This dispossession also resulted in loss of livelihood to IS communities as they lose lands rich in marine, fauna, and agricultural resources. This led to situations where these displaced communities became dependent on colonists for their livelihood. Continuous exposure to these practices weakens the cultural identity due to the loss of TEK. Such a loss is then carried forward by future generations, and they find themselves incapable of managing their resources effectively.

The colonial government's forced assimilation efforts were also a key problem for Indigenous communities due to dispossession of land and resources. During the colonial period, indigenous civilizations were often viewed in a negative light and undermined in a variety of ways. In these programs, Indigenous languages, knowledge systems, social structures, and spiritual beliefs were eradicated and replaced with dominant colonial ideals (Tauli-Corpuz, 2019).

Excellent examples of these policies are residential schools, which were established in several nations with large Indigenous populations. Children were forced to leave their families and communities, sent to harsh regimes that repressed indigenous languages and customs while indoctrinating them into mainstream culture by these institutions. Indigenous cultures disintegrate, eroding their fundamental foundations. Cultural practices, languages, and knowledge systems were disrupted or destroyed, reducing social cohesion and limiting communities' ability to pass on their distinctive wealth.

The sustainability of the environment has always been intrinsic to the wealth and success of Indigenous communities throughout history. During colonial and modern development initiatives, extractive methods have severely degraded the environment and exploited its resources, threatening Indigenous wealth. The lands possessed by IS communities attract mining and exploration corporations, when some deposits such as minerals, fuel, and lumbar resources are discovered on such lands.

These corporations exploit the core of land excessively without looking at environmental impacts of their activities. The members of local communities are usually alienated, and they are not paid the reasonable compensations. The members appointed on such projects due to agreements are subjected to subpar treatment.

Sustainable resource management strategies have been passed down through generations by indigenous groups. In TEK, the emphasis is on living in harmony with the land and ensuring that resources will be available for future generations (Turner et al., 2008). As opposed to the exploitative techniques used by extractive businesses, this strategy offers a sharp contrast.

The environmental repercussions of such an excessive exploitation are:

- Native food gathering communities such as hunters, fishermen, and scavengers find it hard to earn livelihood through their traditional methods.
- Skyrocketing pollution is another outcome due to negligent exploration practices. Depletion and pollution of groundwater, soil degradation, and fine particles leading to air pollution other than the deafening noise of machines used in these processes add to the vows of society.
- Other than methods of livelihood, the avenues of biological diversity also diminish due to such practices, making it even harder to live traditional life style.

Ecological, social, and cultural degradation are derivatives of environmental damage. A disruption in traditional resource management practices, as well as the loss of sacred sites, can undermine social cohesion, cultural identity, and Indigenous communities' spiritual connection to the land. Resource exploitation jeopardizes the ability of future generations to fulfill their fundamental needs and maintain their cultural traditions when it prioritizes immediate economic gain over long-term sustainability. Indigenous wealth emphasizes intergenerational responsibility and living in harmony with nature as its foundational ideas.

Alternative development options that respect Indigenous rights and environmental sustainability are critical. Here are a few promising avenues:

- Indigenous communities can make sustainable decisions and manage their resources through community-based natural resource management (CBNRM).
- The development of renewable energy sources such as solar and wind power can create economic opportunities while minimizing environmental damage.
- HMI can be used to track environmental consequences, map traditional areas, and assist Indigenous tribes in advocating for their rights.

Those businesses that operate in Indigenous areas can cause severe environmental damage, altering traditional ways of life and access to resources. In these activities, short-term economic gains are prioritized over long-term ecosystem sustainability, putting future generations in danger (Brosius et al., 2018). Inextricably linked to the health and sustainability of the natural environment is the basic base of Indigenous wealth (Cajete, 2000).

7. HMI for Indigenous Empowerment: Integrating Traditional and Modern Approaches

A HMI system can bridge the gap between traditional indigenous traditions and the needs of contemporary society. By strategically using technology, indigenous communities can maintain traditional knowledge, generate long-term economic activity, and improve community well-being. Here's an overview of how HMI can help with these goals:

7.1 The Preservation and Transfer of Cultural Knowledge

A variety of formats can be used to document and archive traditional knowledge using HMI tools, including audio and video recordings of elders sharing stories, songs, and ceremonies (National Geographic Society, 2018), digital systems for recording languages, traditional activities, and ecological knowledge (microsoft.com). Microsoft Translator delivers Artificial Intelligence-powered translation to endangered indigenous languages.

Through interactive games and simulations, schools and communities can learn traditional ecological knowledge and cultural practices in a fun and immersive manner using the latest technology. Another technological solution suggested to engage communities with traditional knowledge is a virtual reality experience that allows users to visit sacred locations or experience traditional lifestyles (Council of Tribal Colleges and Universities, n.d.).

7.2 The Development of Sustainable Economic Activities Includes the Following

Ecotourism activities – There are online booking platforms for ecotours with Indigenous guides who can offer their expertise and viewpoints. There are apps which provide culturally sensitive information about the places that are sacred and local ecosystems (Langford, 2019). The resource management can be monitored and managed with the help of HMI tools in a sustainable way. The environmental factors like water quality and animal numbers may be monitored using sensors (Jensen et al., 2018).

8. Community Welfare and Social Capital

Digital Alliance provides the online platforms/forums for the exchange of information, conservations, and project collaborations. The HMI solutions/technologies can be used to increase communication and collaborative skill sets within the Indigenous groups. Social media platforms are used to promote the cultural news, events, and educational content.

Revitalization of the language can be done by using the HMI solutions. This can help in revival of the language that can be accomplished by using HMI. Some mobile apps can also be used to translate between indigenous and mainstream languages.

9. Case Studies Demonstrating the Application of HMI to Preserve Cultural Knowledge

The following case studies show how HMI solutions/technologies can preserve cultural knowledge:

9.1 Case 1: Saving Yup'ik Eskimo Culture

It is known that the Yup'ik people of Alaska have a rich cultural tradition. It is based on a close connection and dependence on the land and a complicated verbal tradition. However, due to some societal changes, environmental problems, and less use of their language, there was a threat of extinction of this knowledge.

The Yup'ik Elders' Council and Alaska Native Heritage Center collaborated with each other to establish the Yup'ik Elders Digital Archive to handle this issue. The Yup'ik elders recounted all the oral histories, cultural melodies, folk music, and TEK in their original language on this digital platform. It helps in making it an important resource for Yup'ik culture (Jacobsen, 1970). The stories of Elders' from this community are captured in their original form using audio and video recordings. This HMI solution preserves the substance along with the storytelling style and pronunciation also as they used the audio and video recordings. Search option was available by topic and language. This technology ensures that the future generations have access to the saved knowledge for a longer period of time.

Through this TEK, the permanent record of the Yup'ik culture was reserved. This encouraged the cultural preservation of this community and opened its reach for the future generations. This technology and the available Yup'ik' recordings serve as an excellent example of reviving the language loss of this community. The Yup'ik Elders Digital Archive illustrates how technology may influence Indigenous cultures.

Yup'ik people are responsible for owning and controlling the material, due to which the archive has been seen so successful. The Yup'ik Elder Council worked with the Alaska Native Heritage Center to ensure traditional standards were followed throughout the process (MacArthur, 2017). Besides promoting long-term sustainability, the initiative teaches community members about proper archival procedures and uses digital storage options.

9.2 Case 2: New Zealand's Mori Language Faces a Serious Problem in Everyday Use due to Its Decline. Encouraging Mobile Applications, like Drops, to Become Effective Language Revitalization Tools (Cassels & Farr, 2019; Grenoble & Whaley, 2021)

Aside from offering interactive tools that facilitate language learning, Drops is ideal for people of all ages. Drops simplifies the process of learning languages by emphasizing essential terms and phrases.

There is a considerable influence of such applications. The organization supports language revitalization by fostering informal and participatory engagement with Te Reo Māori. In addition to improving vocabulary and understanding, this also contributes to the development of cultural identity among Māori communities.

It is important to realize, however, that mobile applications are only one component of the solution. Integrated language immersion programs and community activities that support Te Reo Mori use in everyday life are necessary for the full revitalization of Te Reo Mori.

9.3 Case 3: Arnhem Land's Economic Empowerment Through E-commerce (Altman, 2003)

A traditional barrier to selling Indigenous artisans' wonderful works in Arnhem Land was the lack of online presence. Internet markets such as Artland are reshaping the landscape. By using this platform, they can reach a worldwide audience, eliminating geographical restrictions and expanding their market.

As a result of Artland's direct relationship with the buyers, these artists benefited majorly in two ways. Firstly, the platform empowers the artists to exhibit their skills and earn a stable/sustainable income from their creative artworks/activities. This ultimately leads to economic growth.

Secondly, Artland contributes in maintaining Arnhem Land's distinctive artistic heritage through bringing indigenous culture to a broader market/ audience. This helps in preserving its survival for future generations.

The HMI can contribute positively and provide good potential to Indigenous communities. At the same time, it should be used in a deliberate way. These HMI can help in preserving the traditional know-how, building the sustainable economies, and improving the general well-being of indigenous societies by stressing more on community control, cultural sensitivity, and bridging the digital gaps. The latest technology must be used for this empowerment and self-determination. This is possible through collaboration among the Indigenous societies, technologists, and researchers.

10. Conclusion and Recommendations

The following recommendations may be considered for culturally relevant HMI solutions in Indigenous communities:

10.1 Ownership and Control of the Communities

For sustainable development process, Indigenous communities should be engaged in the design and development process. The HMIs should hold seminars, presentations, and consultations to understand the specific needs, goals, and concerns of the communities. In addition, community-based governance structures should be established. These should ensure the ownership and control of Indigenous communities. The ownership includes ownership of data and the decision-making power, and the control means the control over the cultural content. The Indigenous Knowledge Systems (IKS) should be incorporated into the HMI development to recognize their value. It should also be ensured that the used technology is consistent enough with cultural conventions. The traditional methodology should be recorded by consulting the elders and other persons who have the knowledge.

10.2 The Culturally Sensitive Design

The HMI solutions should be used in the community's primary languages, if possible. Multimedia features can help to accommodate different learning styles and accessibility requirements, like audio recordings and visual assistance. In this process, the inclusive design principles should be adhered to, to ensure the solution is accessible to the individuals easily. The user interface of this design should be culturally appropriate. The design should be visually appealing/pleasing and reflect the community's aesthetics. To ensure this thing, the community members should be involved.

10.3 Sovereignty and Security of the Data

The ownership of the data should be clearly established including the approval to collect, store, and use these data from the community. The HMI solution should ensure the protection of sensitive cultural information and personal information obtained through this collected data. For this, strong data security measures must be taken and security protocols be updated regularly and assess vulnerabilities. The data governance framework should be established for data collection, access restrictions, and data anonymization or deletion procedures, if required.

10.4 Capacity-Building and Sustainability

Community-wide training programs to be developed to teach the individuals the skill set they need to use and maintain. In case the individuals face technical issues, the technical support procedures should be framed for timely resolution and ensure the HMI system's long-term operations. Moreover, the financial and other resources should be assigned to support continuous development, maintenance, and capacity-building activities that are related with the HMI solution. This will ensure the overall sustainability of the solution.

10.5 Ethical Consideration

The cultural protocols and sensitivities should be respected throughout all the phases of HMI solutions like, development, design, and deployment phases. The cultural appropriation should be avoided here. In endeavor to form relationships and collaborations that benefits the indigenous communities, the exploitation of cultural heritage for the purpose of making money should be avoided. It is important throughout the entire HMI implementation process to maintain transparency in the implementation. The HMI solution should take the responsibility and accountability on its own to the community for the development and success of the solution.

If the HMI technology follows these best practices, it may prove to be an effective tool to empower indigenous communities. Here, prioritization of community ownership, culturally sensitive design, data sovereignty, and capacity-building are important for the HMI solution. The focus is to use the technology that preserves the culture, economic growth, and ensures the better future for the communities in the digital age.

References

Acosta, A., & Martínez, E. (Eds.). (2010). *Soberanías*. Abya Yala.

Alfred, T., & Corntassel, J. (2005). Being Indigenous: Resurgences against contemporary colonialism. *Government and Opposition*, *40*(4), 597–614.

Altman, J. (2003). People on country, healthy landscapes and sustainable Indigenous economic futures: The Arnhem Land case. *The Drawing Board: An Australian Review of Public Affairs*, *4*(2), 65–82.

Anderson, J. (2014). *Interdisciplinary and Indigenous approaches to ecological sustainability*. Routledge.

Berkes, F. (2008). *Sacred ecology: Traditional ecological knowledge and resource management* (2nd ed.). Routledge. https://www.ncbi.nlm.nih.gov/pmc/articles/PMC6371761/

Bliege Bird, R. M., & Coppinger, R. P. (1995). Cooperative hunting and meat-sharing among Tanzania's Hadzabe. *Human Nature*, *6*(2), 113–134.

Brosius, J. P., Tsing, A. L., & Zerner, C. (Eds.). (2018). *Communities and conservation: The history and politics of community-based natural resource management*. Rowman Altamira.

Cajete, G. (2000). *Native science: Natural laws of interdependence*. Clear Light Publishers.

Cameron, E. (2012). Securing indigenous politics: A critique of the vulnerability and adaptation approach to the human dimensions of climate change in the Canadian Arctic. *Global Environmental Change*, *22*(1), 103–114.

Cassels, M., & Farr, C. (2019). Mobile applications for indigenous language learning: Literature review and app survey. *Working Papers of the Linguistics Circle*, *29*(1), 1–24.

Coulthard, G. S. (2012). An indigenous critique of neoliberalism: A roundtable discussion. *Critical Indigenous Studies*, *8*(1), 1–16.

Council of Tribal Colleges and Universities. (n.d.). Tribal colleges distance learning and technology consortium on American Indian Higher Education Consortium. https://www.aihec.org/

Cruikshank, J. (2005). *Do glaciers listen? Local knowledge, colonial encounters, and social imagination*. UBC Press.

Da Silva, S. (2018). *Indigenous peoples, human rights and development in Africa*. Routledge.

Daigle, M. (2010). *Shifting environmental governance and the impacts on indigenous peoples*. Routledge.

Diaz-Reviriego, I., Fernández-Llamazares, Á., García-del-Amo, D., & Cabeza, M. (2016). Traditional ecological knowledge and community resilience to environmental extremes: A case study in Doñana, SW Spain. *Ecology and Society*, *21*(3), 22.

Grenoble, L. A., & Whaley, L. J. (2021). Toward a new conceptualisation of language revitalisation. *Journal of Multilingual and Multicultural Development*, *42*(10), 911–926.

Homewood, K. M. (2004). Ecology of pastoralists in transitional environments: Social transformation and resilience in northern Kenya. *Human Ecology*, *32*(1), 74–102.

Jacobsen, J. A. (1970, January 1). Ciuliamta akluit/things of our ancestors: Yup'ik elders explore the Jacobsen collection at the Ethnologisches Museum berlin : Free

Download, borrow, and streaming. *Internet Archive*. https://archive.org/details/ciuliamtaakluitt0000unse

Jensen, O. P., MacKinnon, A., & Armstrong, J. (2018). Indigenous protected areas and conservation science: A literature review. *Biological Conservation, 221*, 287–295. https://doi.org/10.1016/j.biocon.2018.02.020

Kawagley, A. O. (1995). *A Yupiaq worldview: A pathway to ecology and spirit*. Waveland Press.

Langford, W. (2019). Cultural considerations for mobile tourism applications. *Tourism Management Perspectives, 31*, 100554.

MacArthur, A. R. (2017, October 12). World's largest collection of Yup'ik and cup'ik videos now available online. *Alaska Public Media*. https://alaskapublic.org/2017/10/11/worlds-largest-collection-of-yupik-and-cupik-videos-now-available-online/

McCarty, T. L. (2002). *A place to be Navajo: Rough rock and the struggle for self-determination in indigenous schooling*. Routledge.

Nakata, M. (2000). *Indigenous knowledge and the cultural interface: Underlying issues at the intersection of knowledge and information systems*. UTS ePress.

Natcher, D. C., & Davis, S. (2013). Rethinking devolution: Challenges for Aboriginal resource management in the Yukon Territory. *Society & Natural Resources, 26*(1), 65–81.

Pascoe, B. (2018). *Dark Emu: Aboriginal Australia and the birth of agriculture*. Magabala Books.

Richards, P. (2019). *Voices of a rainforest: Culture, myth, and history in the Amazonian frontier*. University of Texas Press.

Sanders, W. (2003). Indigenous communities and the Australian state: Contesting the recognition of cultural values. *Human Ecology, 31*(4), 517–536.

Simpson, L. (2014). *Land as pedagogy: Nishnaabeg intelligence and rebellious transformation*. ARP Books.

Smith, L. T. (2008). *Decolonizing methodologies: Research and indigenous peoples*. Zed Books.

Smith, L. T. (2019). *Indigenous research methodologies* (2nd ed.). Zed Books.

Smith, E. A., & Lightfoot, K. G. (2014). Evolutionary viewpoints on human prosocial behavior. *Proceedings of the National Academy of Sciences, 111*(38), 13604–13611.

Tauli-Corpuz, V. (2019). *Cornered by protected areas: Replacing 'fortress' conservation with rights-based approaches helps bring justice for Indigenous peoples and local communities, reduces conflict, and enables cost-effective conservation and climate action*. Rights and Resources Initiative.

Todd, Z. (2016). An indigenous feminist's take on the ontological turn: 'Ontology' is just another word for colonialism. *Journal of Historical Sociology, 29*(1), 4–22.

Turner, N. J., Ignace, M. B., & Ignace, R. (2008). Traditional ecological knowledge and wisdom of aboriginal peoples in British Columbia. *Ecological Applications, 10*(5), 1275–1287.

Wilcox, M. V. (2008). *The Pueblo revolt and the mythology of conquest: An indigenous archaeology of contact*. University of California Press.

Web Resources

Council of Tribal Colleges and Universities. (n.d.). Tribal colleges distance learning and technology consortium on American Indian Higher Education Consortium. https://www.aihec.org/

National Geographic. (2018). Endangered languages and scripts. https://www.nationalgeographic.com/adventure/article/130628-endangered-languages-scripts-bangladesh-indigenous-cultures-world

Native Governance Center. (n.d.). Indigenous wealth guide. https://nativegov.org/resources/indigenous-wealth-guide/

University of Hawaii. (n.d.). News and events. https://manoa.hawaii.edu/news/article.php?aId=10123

Chapter 19

The Effect of Family Financial Support on Facilitating Students' Academic Success: A Mixed Methods Analysis During University Closure and Reopening Toward Indigenous Empowerment

Enkeleda Lulaj[a], *Luis Cardoso*[b], *Muthmainnah*[c], *Gül Erkol Bayram*[d] *and Andi Asrifan*[e]

[a]University Haxhi Zeka, Kosovo
[b]Polytechnic Institute of Portalegre, Portugal
[c]Universitas Al Asyariah Mandar, Indonesia
[d]Sinop University, Turkey
[e]Universitas Negeri Makassar, Indonesia

Abstract

This study presents an empirical analysis of the impact of family financial support (FAM) on students' academic success across three countries: Kosovo (KOS), Portugal (PT), and Indonesia (ID). The primary aim was to assess differences in the extent of family financial support and self-financing (SF) of education among these countries, as well as to evaluate the significance of independent variables on student performance (FAM, SF) during and after the COVID-19 pandemic. Data were collected from 216 students enrolled in PhD, Master's, Bachelor's, and professional programs, and analyzed using SPSS (WV23). The analysis utilized descriptive statistics, one-way ANOVA, and the General Linear Model (GLM). Results revealed significant variations in family financial support, with differing levels of assistance affecting academic outcomes. Independent variables were found to have a notable influence on student performance, demonstrated through tests such as ANOVA ($p < 0.005$), multiple comparisons ($p < 0.005$),

Indigenous Empowerment through Human-Machine Interactions, 329–344

Published under exclusive licence by Emerald Publishing Limited
doi:10.1108/978-1-83608-068-820251019

Levene's test ($p > 0.005$), and between-subject's effects ($p < 0.005$). The findings underscore the need for families in all three countries to adapt to the changing educational landscape, particularly in the postpandemic era, by aligning financial support with the evolving academic demands. Limitations of the study include the scope of variables, student sample size, and geographical focus, highlighting the potential for broader research in future studies.

Keywords: Family financial support; students; personal finance; Kosovo; Portugal; Indonesia

1. Introduction

Because of the large-scale global outbreak caused by the appearance of the 2019 novel coronavirus (COVID-19), it is now a major public health concern (Lai et al., 2020). Based on the variables of financial support from the family (VARFAM1, 2, 3, 4, and 5) for students during online learning, it is emphasized that the COVID-19 pandemic affected the whole world, including the three countries of this research: Kosovo (KOS), Portugal (PT), and Indonesia (ID); many people lost their jobs due to the closure of businesses, and financial support was more difficult for their students. Education systems around the world were once again challenged to come up with new measures to immediately sustain educational efforts (UNESCO, 2020a, b, c), but this does not consider giving solutions to the problems of every individual student. Education is a fundamental human right. When it comes to dealing with a crisis, the education system appears to be ill-prepared (Bozkurt & Sharma, 2020). The purpose and main objective of this research to see if there is a difference in the financial support for students from their families, or self-financing of their studies, between the three countries: KOS, PT, and ID, as well as the influence of independent variables on the dependent variable between families, and self-financing of studies during university closure (COVID-19) and university reopening (post-COVID-19). Based on the results obtained from the econometric analysis, recommendations will be made for the future of these countries and all other countries, including families, universities/faculties, and businesses that create jobs for students.

2. Review of Literature and Hypothesis Development

The impact of financial support from families and the state on the economy and education is critical, especially for those who have lost their jobs and still need to meet their financial obligations (VARFAM1, 2, 3, 4, and 5). Sainato (2020) noted that the pandemic negatively affected businesses and jobs, impacting family well-being. For instance, Scandinavian Airlines (SAS) announced temporary layoffs for 10,000 employees (Jacobsen & Ahlander, 2020), and Wayzata Home Products closed, resulting in 141 job losses (Gilyard, 2020). Moody's analytics reported that over half of Americans faced job threats due to COVID-19 (Isidore, 2020). Van Aardt et al. (2009) and Mogaji (2020) highlighted that failing to meet financial obligations can lead to feelings of vulnerability, as resources and life challenges contribute to this situation (Perrig-Chiello et al., 2016). Unemployment and financial hardships increase the risk for families, which may force changes in their financial habits (Van Aardt et al., 2009). The pandemic could further exacerbate their precarious financial conditions (Davis et al., 2015). Financial inclusion is essential for improving living standards (Sarma & Pais, 2011). Delgado et al. (2015) emphasized that states should enhance employment opportunities for students and families to boost financial support. Lulaj (2024) observed that families often struggle to provide students with desired clothing and goods. To assist, businesses should offer reasonable prices and develop worker skills (Lulaj et al., 2024a, b; Lulaj, 2023). Families must also adapt to changes in the job market to improve their well-being through better company performance (Lulaj et al., 2024a, b). Evidence suggests that universities maintaining innovation during post-COVID recovery will gain a competitive advantage (Flammer & Ioannou, 2015). According to UNESCO (2020b, c) and the European Commission (2022), the COVID-19 pandemic threatens the educational goals of the 2030 Sustainable Development Agenda (SDG4). Even as we transition to a post-COVID-19 phase, challenges remain, including the digital divide, inadequate educational support, insufficient investment in infrastructure and connectivity, and the need for enhanced teacher training (Bradley & Green, 2020). These factors jeopardize learning outcomes (Azevedo et al., 2021). Lulaj and Dragusha (2022) highlight the importance of families maintaining tax contributions to support students, especially as public spending has risen due to COVID-19 (Lulaj, 2022a, b), and must be allocated according to citizen needs (Lulaj et al., 2022). Key challenges include concerns about debts, housing costs, and job losses (Wahab & Othman, 2021). Financial difficulties for students are significant, alongside psychological and academic impacts, as noted by Busaad and Alnaim (2021) and Bhat et al. (2020) regarding Kashmir. Muthmainnah et al. (2022) emphasize critical thinking and self-control in education. Rodríguez-Planas (2020) found that 52% of students' risk dropping out of college, while Manze et al. (2021) reported that 54.1% of students or their household members lost income due to COVID-19, with nearly half indicating increased household expenses. Other studies corroborate these findings (Akin-Odanye et al., 2021; Ali & Talha, 2021; Gallant et al., 2020; Hung et al., 2021; Lestari et al., 2022). Lulaj (2020) underscores the need for financial education to help families support students, who often lack savings for emergencies (Lulaj, 2021). Salmi (2020) discussed measures for student well-being and technology access for online learning. Students with computers at home generally come from higher socioeconomic backgrounds (Di Pietro et al., 2020). The pandemic's impact has prompted various responses (Aristovnik et al., 2020). A global approach to post-pandemic financial support is necessary (European Commission, 2022). Lulaj (2021) notes that fair public spending and transparency are crucial for supporting families and students (Lulaj, 2019). In summary, understanding family financial support for students is vital, and this study will address these aspects in subsequent findings.

H1. There is a difference between families in three countries regarding student financial support in facilitating academic success during online learning (COVID-19)
H2. The independent variables have a significant impact on the variable of financial support from the family for increasing student performance during online learning (COVID-19)
H3. There is a difference between families in three countries for student financial support in facilitating academic success after the start of learning on campus (post COVID-19)
H4. There is a difference between students for self-financing of their studies after the start of learning on campus (post-COVID-19)
H5. Independent variables have a significant impact on the variable of financial support from the family, as well as self-financing of studies to increase student performance after the start of learning on campus (post-COVID-19)

3. Methods

This research was carried out with students of the doctorate (PhD), Master's (MSc/MA), Bachelor's (BSc/BA), and professional studies (P), as well as with their families in three countries: KOS, PT, and ID through the completion of the survey during and after pandemic (2019–2022). A total of 216 questionnaires were completed (116 BSc/BA, 41 MSc/MA, and 1 PhD, P58). The methods for analyzing family financial support in facilitating students' academic success and self-financing across KOS, PT, and ID were conducted using descriptive analysis, one-way ANOVA (OWA), and the General Linear Model (GLM) via SPSS (WV23). As highlighted by Loeb et al. (2017), descriptive statistics helped examine participant demographics such as gender, education, income, and family size across the three countries. To assess differences in family financial support during online (COVID-19) and on-campus (post-COVID-19) learning, one-way ANOVA was applied, in line with Mahapoonyanonta et al. (2010), who stress its relevance for model power and transformation. The GLM, developed by Nelder and Wedderburn (Banks & Fienberg, 2003), was employed to test student performance and academic success facilitation, allowing for the evaluation, verification, and prediction of financial support issues during studies.

Fig. 19.1 shows the results for the session of financial support of students from the family and the relation of the dependent variable (FAM financial support) of the three states (KOS, PT, and ID) with the independent variables (VARFAM1, VARFAM2, VARFAM3, VARFAM4, and VARFAM5).

Fig. 19.2 shows the results for the session of financial support of students from the family and self-financing of studies, as well as the relation of the dependent variable for the three states (KOS, PT, and ID) with the independent variables (VARFSFPC1 and VARSFSPC3).

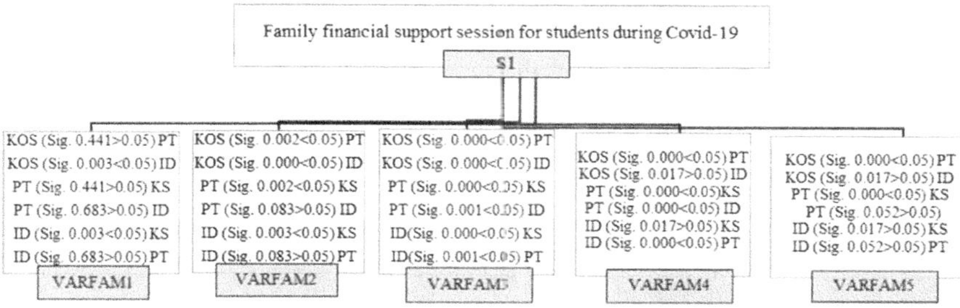

Fig. 19.1. Family Financial Support Session for Students During COVID-19. *Source:* Authors' original contribution.

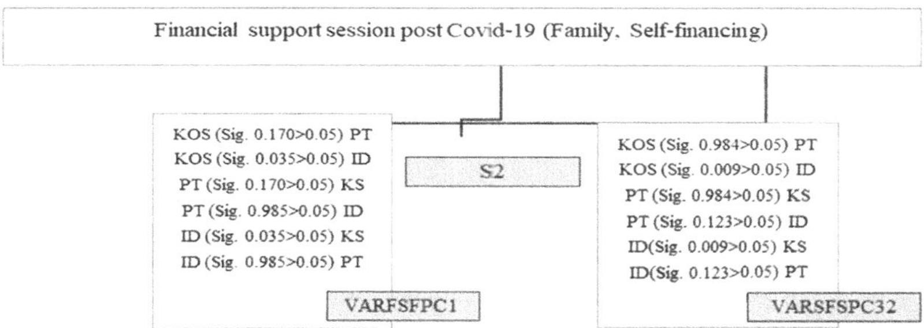

Fig. 19.2. Financial Support Session Post-COVID-19 (Family, Self-Financing). *Source:* Authors' original contribution.

4. Findings and Discussions

The economic well-being of families (VARFAM1) is crucial for purchasing equipment for online learning. Hong et al. (2021), Gu (2022), and Lulaj et al. (2021) emphasize that families need financial stability and emergency reserves. For calmness during online learning, Sainato (2020) and Lulaj (2022a, b) highlight its impact on performance. Regarding special learning spaces, Yustina et al. (2022) argue that dedicated environments enhance effectiveness. On student motivation, Chiu et al. (2021) and others emphasize its role in satisfaction. For financial support from families, Lulaj et al. (2022) and Lulaj (2021) highlight the link between family and state well-being. Regarding self-financing of studies, Aucejo et al. (2020) and Hanushek and Woessmann (2020) note that COVID-19 has led to delays and job losses.

Table 19.1 presents the descriptive analysis of variables (Co, Edu, *A*, *G*, Inc, and FAM) related to financial support from family and self-financing among 216 students in KOS, PT, and ID during and after COVID-19. Key findings reveal

Table 19.1. Descriptive Analysis for (VAR) Variables (Co, Edu, A, G, Inc, and FAM).

	Statistics																	
	State/Country VAR(Co)			**Education VAR(Edu)**			**Age VAR(*A*)**			**Gender VAR(*G*)**			**Income (Euro/The rupiah), VAR(Inc)**			**Family Members VAR(FAM)**		
N V	216			216			216			216			216			216		
M	0			0			0			0			0			0		
N Valid		*F*	*P*		*F*	*P*		*F*	*P*		*F*	*P*		*F*	*P*		*F*	*P*
	KOS	101	46.8	Basic studies	116	53.7	17–25 years	188	87.0	*F*	163	75.5	100–300 E/R	93	43.1	1–3	52	24.1
	PT	22	10.2	Master	41	19.0	26–30 years	5	2.3	*M*	51	23.6	301–500 E/R	59	27.3	3–5	88	40.7
	ID	93	43.1	PhD	1	0.5	31–40 years	12	5.6	NB	1	0.5	501–1,000 E/R	44	20.4	5–8	64	29.6
Total		216	100.0	*P*	58	26.9	Over 40 years old	11	5.1	NA	1	0.5	Over 1,000 E/R	20	9.3	+8	12	5.6

Source: Authors' original contribution.

Note: n = 216 students.

that KOS students had the highest response for VAR (Co) at 46.8%, while PT students had the lowest at 10.2%. In terms of education level, Bachelor's students responded most (53.7%), whereas PhD students had the least (0.55%). For age, students aged 17–25 provided the highest response (87%), compared to just 0.5% from the 26–30 age group. Female students had the largest response for VAR (*G*) at 75.5%, while nonbinary students recorded the lowest at 0.5%. Regarding income, those earning 100–300 Euros reported the highest response (43.1%), while students with incomes over 1,000 Euros responded the least (9.3%). Lastly, students from families with three to five members had the highest response for VAR (FAM) at 40.7%, while those from families with more than eight members had the lowest response at 5.6%.

4.1 ANOVA and GLM Analysis of Family Financial Support for Students During COVID-19

This section examines whether families in KOS, PT, and ID provided similar financial support for student success during university closures and the impact of independent variables on this support to enhance student performance. Independent variables:

VARFAM1: Did your family get the opportunity to buy equipment for online learning (computer, laptop, camera, microphone, learning applications, etc.)?

VARFAM2: Did your family create the conditions to have silence while learning online?

VARFMA3: Did your family create special learning spaces during the pandemic?

VARFAM4: Did your family help you complete the teaching assignments given by the professors during the pandemic, through additional online tuition fees?

VARFAM5: Did your family provide motivational financial support when you scored well on tests, exams, and other school activities during the pandemic?

Dependent variable: Family financial support.

Table 19.2 shows the difference between the families of the three countries (KOS, PT, and ID) for the financial support of students during online learning (COVID-19), where it is emphasized that the *p*-value for all variables is less than *0.005* ($p < 0.005$) or (Sig: VARFAM1: 0.004 < 0.005; VARFAM2: 0.000 < 0.005; VARFAM3: 0.000 < 0.005; VARFAM4: 0.000 < 0.005; VARFM 0.000 < 0.005). Therefore, hypothesis H1 is accepted, because there is a difference between the families of the three countries for the financial support of students in facilitating academic success during online learning or university closure (COVID-19).

Table 19.3 presents the results of multiple comparisons using Tukey HSD (THt) and Bonferroni (BOt) tests at the 0.005 significance level ($p < 0.005$), highlighting the impact of family financial support on the academic success of

Table 19.2. ANOVA.

Family Financial Support Session for Students During COVID-19						
		SSq.	**df**	**MSq.**	***F***	**Sig.**
VARFAM1	Between groups	10,526	2	5,263	5,732	0.004
	Within groups	195,581	213	0.918		
	Total	206,106	215			
VARFAM2	Between groups	47,096	2	23,548	29,657	0.000
	Within groups	169,122	213	0.794		
	Total	216,218	215			
VARFAM3	Between groups	44,608	2	22,304	27,123	0.000
	Within groups	175,152	213	0.822		
	Total	219,759	215			
VARFAM4	Between groups	33,655	2	16,828	17,695	0.000
	Within groups	202,562	213	0.951		
	Total	236,218	215			
VARFAM5	Between groups	34,028	2	17,014	18,683	0.000
	Within groups	193,972	213	0.911		
	Total	228,000	215			

Source: Authors' original contribution.

Note: Sum of Squares (SSq.), Mean Square (MSq.).

Table 19.3. Multiple Comparisons (KOS, PT, and ID).

Family Financial Support Session for Students during COVID-19						
			Mean Difference (*I–J*)		**Sig.**	
Independent Variable	**(*I*) Country**	**(*J*) Country**	**Tukey HSD**	**Bonferroni**	**Tukey HSD**	**Bonferroni**
VARFAM1	Kosovo KOS	PT	0.27588	0.27588	0.441	0.667
		ID	0.46503*	0.46503*	0.003	0.003
	Portugal PT	KOS	−0.27588	−0.27588	0.441	0.667
		ID	0.18915	0.18915	0.683	1,000
	Indonesia ID	KOS	−0.46503*	−0.46503*	0.003	0.003
		PT	−0.18915	−0.18915	0.683	1,000
VARFAM2	Kosovo KOS	PT	0.53195*	0.53195*	0.002	0.003
		ID	0.98552*	0.98552*	0.000	0.000

Table 19.3. *(Continued)*

Family Financial Support Session for Students during COVID-19						
			Mean Difference (*I–J*)		Sig.	
Independent Variable	**(*I*) Country**	**(*J*) Country**	**Tukey HSD**	**Bonferroni**	**Tukey HSD**	**Bonferroni**
	Portugal	KOS	−0.53195*	−0.53195*	0.002	0.003
	PT	ID	0.45357	0.45357	0.083	0.099
	Indonesia	KOS	−0.98552*	−0.98552*	0.000	0.000
	ID	PT	−0.45357	−0.45357	0.083	0.099
VARFAM3	Kosovo	PT	1.30603*	1.30603*	0.000	0.000
	KOS	ID	0.74396*	0.74,396*	0.000	0.000
	Portugal	KOS	−1.30603*	−1.30603*	0.000	0.000
	PT	ID	−0.56207*	-0.56207*	0.001	0.029
	Indonesia	KOS	−0.74396*	−0.74396*	0.000	0.000
	ID	PT	0.56207*	0.56207*	0.001	0.029
VARFAM4	Kosovo	PT	1.34113*	1.34113*	0.000	0.000
	KOS	ID	0.38805*	0.38805*	0.017	0.018
	Portugal	KOS	−1.34113*	−1.34113*	0.000	0.000
	PT	ID	−0.95308*	−0.95308*	0.000	0.000
	Indonesia	KOS	−0.38805*	−0.38805*	0.017	0.018
	ID	PT	0.95308*	0.95308*	0.000	0.000
VARFAM5	Kosovo	PT	1.16382*	1.16382*	0.000	0.000
	KOS	ID	0.63302*	0.63302*	0.000	0.000
	Portugal	KOS	−1.16382*	−1.16382*	0.000	0.000
	PT	ID	−0.53079	−0.53079	0.052	0.060
	Indonesia	KOS	−0.63302*	−0.63302*	0.000	0.000
	ID	PT	0.53079	0.53079	0.052	0.060

Source: Authors' original contribution.

*$p < 0.005$.

students from KOS, PT, and ID during online learning (COVID-19). Significant differences were observed across the variables. For VARFAM1, which addressed support for purchasing equipment, families in KOS provided more assistance than those in PT (0.27588) and ID (0.46503*). In terms of creating quiet study conditions (VARFAM2), KOS families also significantly outperformed PT (0.53195*) and ID (0.98552*). Regarding the provision of special learning spaces (VARFAM3), KOS families again showed greater support compared to PT (1.30603*) and ID (0.74396*). For VARFAM4, related to financial contributions

Table 19.4. Levene's Test.

Dependent Variable: State/Country			
F	**df1**	**df2**	**Sig.**
12,686	106	109	0.713
a. Design: Intercept + VARFAM1 + VARFAM2 + VARFAM3 + VARFAM4+ VARFAM5			

Source: Authors' original contribution.

for academic tasks, KOS families made significantly higher payments than PT (1.34113*) and ID (0.38805*). Lastly, concerning motivational support based on academic achievements (VARFAM5), KOS families provided more encouragement compared to PT (1.16382*) and ID (0.63302*). These findings indicate substantial variations in family support across the three countries during online learning.

Table 19.4 shows the assumption of homogeneity of variances is ensured because the p-value is greater than *0.005 (Sig. 0.713)*. Therefore, hypothesis H2 is accepted.

4.2 ANOVA and GLM Analysis of Family Financial Support for Students During Post-COVID-19

Did families in KOS, PT, and ID provide similar financial support for students during the post-COVID-19 university reopening? Was self-financing of studies (SF) consistent across these countries? Did independent variables significantly impact financial support and student performance? Independent variables:

VARFSFPC1: Financial support from the family after COVID-19.
VARSFSPC2: Self-financing of studies after COVID-19.

Dependent variable: Financial support from family, self-financing (post-COVID-19).

Table 19.5 shows the difference between families (FAM) and self-financing (SF) for the financial support of students after the start of classes on campus (post-COVID-19), in the three countries (KOS, PT, and ID), it is emphasized that the p-value for all variables is less than *0.005* ($p < 0.005$) or (Sig.VARFSFPC1: 0.005 < 0.005; VARSFSPC2: 0.004 < 0.005). Therefore, hypotheses H3, and H4 are accepted, because there is a difference between the FAM, and SF of the three countries for the financial support of students after the start of classes on campus (post-COVID-19).

Table 19.6 presents the results of multiple comparisons using Tukey HSD and Bonferroni tests at a significance level of 0.005 ($p < 0.005$) for financial support from families (FAM) and self-financing of studies (SF) among students in KOS,

Table 19.5. ANOVA.

Financial Support Session Post-Covid-19 (Family, Self-financing)						
		SSq.	**df**	**MSq.**	***F***	**Sig.**
VARFSFPC1	Between groups	6,652	2	3,326	3,752	0.005
	Within groups	188.83	213	0.887		
	Total	195,481	215			
VARSFSPC2	Between groups	8,110	2	4,055	5,000	0.004
	Within groups	172.76	213	0.811		
	Total	180.87	215			

Source: Authors' original contribution.

PT, and ID after returning to campus post-COVID-19. The analysis revealed that families in KOS provided significantly more financial support than those in PT (0.40009) and ID (0.33802*), while no significant difference was found between PT and ID. In terms of self-financing, KOS students significantly differed from ID

Table 19.6. Multiple Comparisons (KOS, PT, and ID).

Financial Support Session Post-Covid-19 (Family, Self-financing)						
			Mean Difference (*I*–*J*)		**Sig.**	
Independent Variable	**(*I*) Country**	**(*J*) Country**	**Tukey HSD**	**Bonferroni**	**Tukey HSD**	**Bonferroni**
VARFSFPC1	Kosovo KOS	PT	0.40009	0.40009	0.000	0.001
		ID	0.33802*	0.33802*	0.005	0.004
	Portugal PT	KOS	−0.40009	−0.40009	0.000	0.001
		ID	−0.06207	−0.06207	0.003	0.000
	Indonesia ID	KOS	−0.33802*	−0.33802*	0.005	0.004
		PT	0.06207	0.06207	0.003	0.000
VARSFSPC2	Kosovo KOS	PT	−0.03600	−0.03600	0.984	1,000
		ID	0.38433*	0.38433*	0.009	0.010
	Portugal PT	KOS	0.03600	0.03600	0.984	1,000
		ID	0.42033	0.42033	0.123	0.151
	Indonesia ID	KOS	−0.38433*	−0.38433*	0.009	0.010
		PT	−0.42033	−0.42033	0.123	0.151

Source: Authors' original contribution.

*$p < 0.005$.

Table 19.7. Levene's Test.

Dependent Variable: State/Country			
F	**df1**	**df2**	**Sig.**
2,621	51	164	0.786
a. Design: Intercept + VARFSFPC1 + VARSFSPC3			

Source: Authors' original contribution.

students (0.38433*), but not from PT (−0.03600). These results indicate notable differences in financial support and self-funding among the three countries following the return to on-campus learning.

Table 19.7 shows that according to the results of Levene's Test, it is emphasized the value of p is greater than *0.005 (Sig. 0.786 > 0.005)*. Therefore, hypothesis H5 is accepted, because for students to increase their performance, the independent variables significantly affect (financial support from the family, self-financing) after the start of classes on campus (post-COVID-19).

5. Conclusion and Future Studies

The pandemic severely disrupted education in KOS, PT, and ID. This study aimed to identify how family financial support influenced students' academic success and self-financing of studies (SF) across these countries. It assessed independent variables during university closures (VARFAM(1–5)) and reopening's (VARFSFPC1, VARSFSPC2) to evaluate their impact on dependent variables (FAM, SF). Results indicated that KOS families provided greater support for students during online learning compared to PT and ID, particularly in purchasing equipment, ensuring conducive study environments, and motivating academic achievement. After the transition back to campus, financial support differences were noted across countries, with KOS students engaging more in self-financing than their peers in PT and ID.

Recommendations include: (a) families in all three countries should adapt to the new educational landscape by creating quiet study spaces and acquiring necessary technology; (b) families need to enhance opportunities for students based on their specific needs; and (c) the overall impact of family financial support on academic success was limited, affecting performance across all countries.

References

Akin-Odanye, E., Kaninjing, E., Ndip, R., Warren, C., Asuzu, C., Lopez, I., Muiruri, C., & Vilme, H. (2021). Psychosocial impact of covid-19 on students at institutions of higher learning. *European Journal of Education Studies*, *8*(6), 112–128. https://doi.org/10.46827/ejes.v8i6.3770

Ali, S., & Talha, N. (2021, November 7). During Covid-19, impact of subjective and objective financial knowledge and economic insecurity on financial management behavior: Mediating role of financial wellbeing. *Journal of Public Affairs*, e2789. https://doi.org/10.1002/pa.2789

Aristovnik, A., Keržič, D., Ravšelj, D., Tomaževič, N., & Umek, L. (2020). Impacts of the COVID-19 pandemic on life of higher education students: A global perspective. *Sustainability*, *12*(8438), 1–34.

Aucejo, E., French, J., Araya, M., & Zafar, B. (2020). The impact of COVID-19 on student experiences and expectations: Evidence from a survey. *Journal of Public Economics*, *191*, 104271. https://doi.org/10.1016%2Fj.jpubeco.2020.104271

Azevedo, J. P., Hasan, A., Goldemberg, D., Geven, K., & Iqbal, S. A. (2021, March 17). Simulating the potential impacts of COVID-19 school closures on schooling and learning outcomes: A set of global estimates. *The World Bank Research Observer*, lkab003. PMCID: PMC8108634. https://doi.org/10.1093/wbro/lkab003

Banks, L., & Fienberg, S. (2003). Statistics, multivariate. In *Encyclopedia of physical science and technology* (3rd ed., pp. 851–889). https://doi.org/10.1016/B0-12-227410-5/00731-6

Bhat, B. A., Khan, S., Manzoor, S., Niyaz, A., Tak, H. J., Anees, S., & Ahmad, I. (2020). A study on impact of COVID-19 lockdown on psychological health, economy and social life of people in Kashmir. *International Journal of Science and Healthcare Research*, *5*(2), 36–46.

Bozkurt, A., & Sharma, R. C. (2020). Emergency remote teaching in a time of global crisis due to the corona virus pandemic. *Asian Journal of Distance Education*, *15*(1), 1–6. https://doi.org/10.5281/zenodo.3778083

Bradley, S., & Green, C. (2020). *The economics of education (A comprehensive overview)*. Elsevier Ltd. Paperback ISBN: 9780128153918. https://doi.org/10.1016/C2017-0-02304-2

Busaad, Y., & Alnaim, M. (2021). Parents' perceptions regarding the effects of COVID-19 on their children with and without disabilities. *International Journal of Instruction*, *14*(4), 997–1012. https://doi.org/10.29333/iji.2021.14457a

Chiu, T. K. F., Lin, T. J., & Lonka, K. (2021). Motivating online learning: The challenges of COVID-19 and beyond. *Asia-Pacific Education Researcher*, *30*, 187–190. https://doi.org/10.1007/s40299-021-00566-w

Davis, P. R., Trevor, C. O., & Feng, J. (2015). Creating a more quit-friendly national workforce? Individual layoff history and voluntary turnover. *Journal of Applied Psychology*, *100*(5), 1434–1455.

Delgado, M., Porter, M. E., & Stern, S. (2015). Clusters and the great recession. *Paper presented at the DRUID15*, Rome, June 15–17.

Di Pietro, G., Biagi, F., Dinis Mota Da Costa, P., Karpinski, Z., & Mazza, J. (2020). *The likely impact of COVID-19 on education: Reflections based on the existing literature and recent international datasets*, EUR 30275. EN, Publications Office of the European Union. ISBN 978-92-76-19937-3, JRC121071. https://doi.org/10.2760/126686

European Commission. (2022). Directorate-General for Education, Youth, Sport and Culture. In A. Finlayson, M. Moso-Diez, & V. Orlovic Lovren (Eds.), *EENEE Analytical Report N°02/2021 – Impact of COVID-19 on education for sustainable development (ESD) in the context of twin transition: Executive summary, 2022*. https://data.europa.eu/doi/10.2766/89568

Flammer, C., & Ioannou, I. (2015). *The dog that didn't bark: Long-term strategies in times of recession*. University of Western Ontario. Working Paper.

Gallant, J., Kroft, K., Lange, F., & Notowidigdo, M. J. (2020). *Temporary unemployment and labor market dynamics during the COVID-19 recession*. National Bureau of Economic Research. Working Paper 27924.

Gilyard, B. (2020). Wayzata home products ceases operations. http://tcbmag.com/news/articles/2020/march/wayzata-home-productsceases-operation. Accessed on March 20, 2020.

Gu, J. (2022). Impact of living conditions on online education: Evidence from China. *Sustainability*, *14*, 3231. https://doi.org/10.3390/su14063231

Hanushek, E., & Woessmann, L. (2020). The economic impacts of learning losses. *OECD*, 1–24. https://www.oecd.org/education/The-economic-impacts-of-coronavirus-covid-19-learning-losses.pdf

Hong, G., Ou, J., Zhang, Z. H., Ni, M., Zhou, X., & Liao, L. (2021). The relationship between family support and e-learning engagement in college students: The mediating role of e-learning normative consciousness and behaviors and self-efficacy. *Educational Psychology*, 1–9. https://doi.org/10.3389/fpsyg.2021.573779

Hung, C. C., Kao, H. S., Liu, H. C., Liang, H. F., Chu, T. P., & Lee, B. O. (2021, February). Effects of simulation-based learning on nursing students' perceived competence, self-efficacy, and learning satisfaction: A repeat measurement method. *Nurse Education Today*, *97*, 104725. Epub December 16, 2020. PMID: 33373813. https://doi.org/10.1016/j.nedt.2020.104725

Isidore, C. (2020). More than half of American jobs are at risk because of the coronavirus. https://edition.cnn.com/2020/03/16/economy/joblosses-coronavirus. Accessed on March 20, 2020.

Jacobsen, S., & Ahlander, J. (2020). SAS to halt most flights, temporarily lay off 90% of staff. *Reuters*. https://www.reuters.com/article/us-health-coronavirus-sas/sas-to-halt-most-flights-temporarily-lay-off-90-of-staff-idUSKBN2120YZ/. Accessed on March 20, 2020.

Lai, C. C., Shih, T. P., Ko, W. C., Tang, H. J., & Hsueh, P. R. (2020). Severe acute respiratory syndrome coronavirus 2 (SARS-CoV-2) and coronavirus disease-2019 (COVID-19): The epidemic and the challenges. *International Journal of Antimicrobial Agents*, *55*(3), 105924.

Lestari, W., Yazid, N. H., Azhar, Z. N., Ismail, A., & Sukotjo, C. (2022, February 23). Impact of COVID-19 on Malaysian dental students' physical, mental, financial and academic concerns. *BMC Oral Health*, *22*(1), 46. https://doi.org/10.1186/s12903-022-02081-w

Loeb, S., Dynarski, S., Mc Farland, D., Morris, P., Reardon, S., & Reber, S. (2017). Descriptive analysis in education: A guide for researchers. NCEE 2017-4023. In *National center for education evaluation and regional assistance*. http://ies.ed.gov/ncee/

Lulaj, E. (2019). Transparency and accountability in the public budget, empirical study (data analysis) in local governments- municipalities. *International Journal of Education and Research*, *7*(4), 69–86.

Lulaj, E. (2020). Budget education and management as a necessity for well-being and financial stability: Cluster & MDS analysis. *International Journal of Financial Research*, *11*(6). https://doi.org/10.5430/ijfr.v11n6p348

Lulaj, E. (2021). Accounting, reforms and budget responsibilities in the financial statements. *Accounting and Finance*, *91*(1), 61–69. http://doi.org/10.33146/2307-9878-2021-1(91)-61-69

Lulaj, E. (2022a). An unstoppable and navigating journey towards development reform in complex financial-economic systems: An interval analysis of government expenses (past, present, future). *Business, Management and Economics Engineering*, *20*(2), 329–357. https://doi.org/10.3846/bmee.2022.17389

Lulaj, E. (2022b). The correlation of financial stress and educational-teaching factors on students during online learning in the Covid-19 (Cov19) pandemic. *International Journal of Instruction*, *15*(2), 435–454. https://doi.org/10.29333/iji.2022.15224a

Lulaj, E. (2023). A sustainable business profit through customers and its impacts on three key business domains: Technology, innovation, and service (TIS). *Business, Management and Economics Engineering*, *21*(1), 19–47. https://doi.org/10.3846/bmee.2023.18618

Lulaj, E. (2024). Money talks: A holistic and longitudinal view of the budget basket in the face of climate change and sustainable finance matters. *Ekonomika*, *103*(1), 91–107. https://doi.org/10.15388/Ekon.2024.103.1.6

Lulaj, E., & Dragusha, B. (2022). Incomes, gaps and well-being: An exploration of direct tax income statements before and during Covid-19 through the comparability interval. *International Journal of Professional Business Review*, *7*(6), e0623. https://doi.org/10.26668/businessreview/2022.v7i6.623

Lulaj, E., Dragusha, B., Hysa, E., & Voica, M. C. (2024a). Synergizing sustainability and financial prosperity: Unraveling the structure of business profit growth through consumer-centric strategies—The cases of Kosovo and Albania. *International Journal of Financial Studies*, *12*(35). https://doi.org/10.3390/ijfs12020035

Lulaj, E., Dragusha, B., Lulaj, D., Rustaj, V., & Gashi, A. (2021). Households savings and financial behavior in relation to the ability to handle financial emergencies: Case study of Kosovo. *Acta Scientiarum Polonorum. Oeconomia*, *20*(1), 35–48. https://doi.org/10.22630/ASPE.2021.20.1.4

Lulaj, E., Gopalakrishnan, A., & Kehinde Lamidi, K. (2024b). Financing and investing in women-led businesses: Understanding strategic profits and entrepreneurial expectations by analysing the factors that determine their company success. *Periodica Polytechnica Social and Management Sciences*. https://doi.org/10.3311/PPso.22532

Lulaj, E., Zarin, Ismat., Rahman, Sh. (2022). A novel approach to improving E-government performance from budget challenges in complex financial systems. *Complexity*, 2507490. https://doi.org/10.1155/2022/2507490

Mahapoonyanonta, N., Mahapoonyanont, B., Pengkaew, N., & Kamhangkit, R. (2010). Power of the test of one-way Anova after transforming with large sample size data. *Procedia - Social and Behavioral Sciences*, *9*(2010), 933–937. https://doi.org/10.1016/j.sbspro.2010.12.262

Manze, M. G., Rauh, L., Smith-Faust, P., & Watnick, D. (2021). Experiences of college students with children during the COVID-19 pandemic. *Emerging Adulthood*, *9*(5), 631–638. https://doi.org/10.1177/21676968211020225

Mogaji, E. (2020). *Financial vulnerability during a pandemic: Insights for Coronavirus disease (COVID-19)*. SSRN. https://doi.org/10.2139/ssrn.3564702

Muthmainnah, M., Al Yakin, A., Massyat, M., Lulaj, E., & Bayram, G. E. (2022). *Developing students' life skills through edupreneurship in the digital era* (Vol. 109A, pp. 169–190). Emerald Publishing Limited. https://doi.org/10.1108/S1569-375920220000109A011

Pedro, A. P., Hasan, A., Goldemberg, D., Iqbal, S. A., & Geven, K. (2020, June). Simulating the potential impacts of Covid-19 school closures on schooling and learning outcomes. In *Policy Research Working Paper 9284*. World Bank Group.

Perrig-Chiello, P., Hutchison, S., & Knöpfli, B. (2016). Vulnerability following a critical life event: Temporary crisis or chronic distress? A psychological controversy, methodological considerations, and empirical evidence. In M. Oris, C. Roberts, D. Joye, & M. E. Stähli (Eds.), *Surveying human vulnerabilities across the life course* (pp. 87–111). Springer International Publishing. https://doi.org/10.1007/978-3-319-24157-9_4

Rodríguez-Planas, N. (2020). *Hitting where it hurts most: COVID-19 and low-income urban college students*. IZA – Institute of Labor Economics. ISSN: 2365-9793. DP No. 13644.

Sainato, M. (2020). Suddenly I have no paycheck': Layoffs and cuts for workers rocked by a coronavirus. https://www.theguardian.com/world/2020/mar/19/coronavirus-workersemployees-staff-wages. Accessed on March 20, 2020.

Salmi, J. (2020). *COVID's lessons for global higher education*. Lumina Foundation. https://www.luminafoundation.org/resource/covids-lessons-for-global-higher-education/

Sarma, M., & Pais, J. (2011). Financial inclusion and development. *Journal of International Development*, *23*(5), 613–628.

UNESCO. (2020a, May 29). The socio-cultural implications of COVID-19. https://en.unesco.org/news/socio-cultural-implications-covid-19

UNESCO. (2020b). Fiscal responses to education and training in the context of COVID-19. https://unesdoc.unesco.org/ark:/48223/pf0000374685.locale=en

UNESCO. (2020c). COVID-19 educational disruption and response. https://en.unesco.org/Covid19/educationresponse

Van Aardt, C. J., Moshoeu, A., Risenga, A., Pohl, M., & Coetzee, M. C. (2009). *A consumer financial vulnerability (CFV) index for South Africa*. University of South Africa (Unisa), Bureau of Market Research (BMR).

Wahab, M. Z. H., & Othman, K. (2021). Impact of COVID-19 on student's emotional and financial aspects in the higher learning institutions. *SEISENSE Journal of Management*, *4*(4), 1–15. https://doi.org/10.33215/sjom.v4i4.629

Yustina, Mahadi, I., Ariska, D., Arnentis, & Darmadi (2022). The effect of e-learning based on the problem-based learning model on students' creative thinking skills during the Covid-19 pandemic. *International Journal of Instruction*, *15*(2), 329–348. https://doi.org/10.29333/iji.2022.15219a

Index

Printed and bound by CPI Group (UK) Ltd, Croydon, CR0 4YY
01/06/2026
14890758-0004